Canyon Country

ARCHES & BRIDGES

and other natural rock openings

by F. A. Barnes

An illustrated guide to
the study and appreciation of
the natural arches, bridges,
windows, and tunnels of the
Colorado Plateau geologic province
in the heart of the
Four Corners States

Another Canyon Country guide book.
1987

This book is the FIFTEENTH in a series of practical guides to travel in the scenic Canyon Country region of the Four Corners States

All written material, maps and photographs in this book are by F. A. Barnes unless otherwise credited

Sketches by Terry Lan

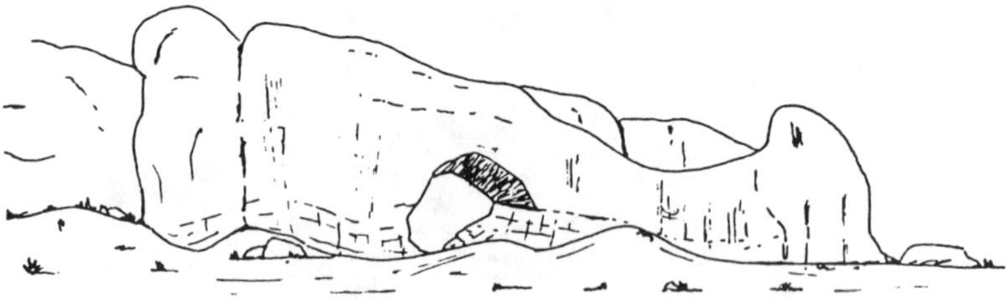

P.O. Box 963
Moab, UT 84532

ISBN 0-9615486-1-5
Library of Congress Catalog Card Number 86-50825

Canyon Country
ARCHES AND BRIDGES
and other natural rock openings
by F. A. Barnes

CONTENTS

OUTSTANDING NATURAL SPANS, PART 1 .... INSIDE-FRONT COVER

CONTENTS 3

PREFACE 6

ACKNOWLEDGMENTS 8

INTRODUCTION 11

REGIONAL GEOLOGY 15
- THE COLORADO PLATEAU 15
- ITS ORIGINS 15
- ITS STRUCTURE 16
- TECTONICS 16
- CLIMATES 17
- PRESENT CYCLE 18
- THE LAND OF ARCHES 19
- THE PRESENT SITUATION 19

HISTORY OF DISCOVERY 21
- COLORADO PLATEAU HUMAN HISTORY 21
- DISCOVERY 22
- EARLY REPORTS 23
- FIRST MAJOR DISCOVERY 27
- RAINBOW BRIDGE 33
- THE TENS AND TEENS 36
- THE TWENTIES 41
- THE THIRTIES AND FORTIES 48
- AERIAL DISCOVERY 50
- THE ESCALANTE SPANS 53
- THE FIFTIES AND SIXTIES 59
- THE SEVENTIES 61
- THE EIGHTIES 62
- THE RAINBOW BRIDGE CONTROVERSY 64
- ANATOMY OF A MODERN DISCOVERY 78

DEFINITIONS 81
- THE NEED FOR DEFINITIONS 81
- THE PROBLEM COMPOUNDED 84
- SOME SERIOUS ATTEMPTS 87
- ANALYSIS OF THE PROBLEM 95
- A DEFINITION OF "SIGNIFICANT" 99
- CONCLUSIONS 100
- SERENDIPITY 102

FORMATION MODES 105
- INTRODUCTION 105
- SOME EARLY DESCRIPTIONS 105
- THE MYTH OF WIND FORMATION 112

CONTINUED NEXT TWO PAGES

GENERAL FACTORS 119
SPECIFIC FACTORS 124
OPENING ENLARGEMENT PROCESSES 137
SPAN DEVELOPMENT RATES 144
SUMMARY AND EXAMPLES 146
DELLENBAUGH'S TUNNEL 153

AGE AND LIFE EXPECTANCY 155
HOW OLD ARE THEY? 155
WHAT OTHER WRITERS HAD TO SAY 157
FACTORS IN JUDGING AGE AND LIFE EXPECTANCY 161
SOME EXAMPLES 165
SOME SPANS THAT COULD COLLAPSE ANY TIME 166
SOME SPANS THAT HAVE ALREADY FALLEN 167
YOU CAN'T GET THERE FROM HERE 171

NAMING 173
QUESTIONS 173
ARE ALL NATURAL SPANS NAMED? 173
WHO HAS THE RIGHT TO NAME NATURAL SPANS? 174
IS THERE A SYSTEM FOR NAMING NATURAL SPANS? 176
WHICH NAMES ARE OFFICIAL? 178
WHAT DO ARCH HISTORIANS HAVE TO SAY? 179
PRESENT NAMES — SOME EXAMPLES 180
SUMMARY 188
THE CASE OF THE COVERT ARCH 190

MEASUREMENTS 193
QUESTIONS AND PROBLEMS 193
MEASUREMENT TECHNIQUES USED 194
SOME PUBLISHED MEASUREMENTS 195
SUMMARY AND CONCLUSIONS 204
MEASURING THE GEMINI BRIDGES 206

PROTECTION 209
PROTECTED NATURAL SPANS 209
LIMITS TO PROTECTION 209
HISTORY OF PROTECTION 213
RAINBOW BRIDGE ENDANGERED 216
SOME UNPROTECTED DESERVING SPANS AND AREAS 220
SUMMARY 224
THE CLAIMING OF CORONA 226

AMERIND ATTITUDES 229
THE QUESTIONS 229
OTHER SPANS 234
SUMMARY 236
LION COUNTRY 238

AESTHETICS 241
AESTHETICS 241
SOME HISTORIC IMPRESSIONS 243
SOME OUTSTANDING SPANS 253
MULTIPLE SPANS 253
AESTHETICS VERSUS PROGRESS 256
THE ENGULFED CATHEDRAL 258

CATALOGING KNOWN SPANS 263
HOW MANY ARE THERE? 263
SOME WHO HAVE TRIED 263
ARE THERE ANY UNDISCOVERED SPANS? 268
HOW TO HUNT FOR UNREPORTED SPANS 272
SUMMARY 273
FINDING FRED'S ARCH 274

ACCESS TO KNOWN SPANS 277
AFTER DISCOVERY 277
ACCESS THEN AND NOW 277
ACCESS TO KNOWN SPANS 288
ARCH COLLECTING 292
HELPFUL LITERATURE 293
FOUR ARCHES CANYON 294

PHOTOGRAPHY 297
HISTORICAL PHOTOGRAPHY 297
MODERN HAZARDS TO PHOTOGRAPHY 303
SOME SUGGESTIONS 303
PHOTOGRAPHING NATURAL SPANS 307
A LACY ARCH 310

SPANS IN SPECIAL AREAS 313
ARCHES NATIONAL PARK 314
BRYCE CANYON NATIONAL PARK 324
CANYONLANDS NATIONAL PARK 330
CAPITOL REEF NATIONAL PARK 342
DESOLATION CANYON 347
ESCALANTE CANYONS 348
GLEN CANYON NATIONAL RECREATION AREA 358
MOAB AREA 360
MONUMENT VALLEY-WHITE MESA AREA 374
NATURAL BRIDGES NATIONAL MONUMENT 382
PARIA RIVER AREA 385
PETRIFIED FOREST NATIONAL PARK 388
RAINBOW BRIDGE NATIONAL MONUMENT 390
RATTLESNAKE CANYON AREA 394
SAN RAFAEL SWELL 398
ZION NATIONAL PARK 402
FAREWELL, FAREWELL 404

BIBLIOGRAPHY 406

INDEX OF SPANS IN THIS BOOK 411

OTHER CANYON COUNTRY PUBLICATIONS 416

PERSONAL ANECDOTES .... pages 78, 102, 153, 171, 190, 206, 226, 238, 258, 274, 294, 310, 404

OUTSTANDING NATURAL SPANS, PART 2 .... INSIDE-BACK COVER

Front cover: Owachomo Natural Bridge, Natural Bridges N. M.
Back cover: Angel Arch, Canyonlands National Park

PREFACE

Author F. A. (Fran) Barnes first encountered parts of the Colorado Plateau in the late 1950s on visits to popular Bryce Canyon, Zion and Grand Canyon national parks. He was fascinated by these spectacular areas, but did not come to realize until the mid-1960s the full extent and diversity of the geologic and aesthetic marvels within the Colorado Plateau, the heartland of the Four Corners States. At that time he visited Arches and Canyonlands national parks, newly created Glen Canyon National Recreation Area and other special places within southeastern Utah.

These visits "addicted" Fran to the Colorado Plateau canyon country. He soon settled in the little southeastern Utah town of Moab, abandoning his former career for a new one -- that of a professional writer-photographer specializing in Colorado Plateau subjects. Since then he has generated hundreds of illustrated articles about the region for various periodicals, and has written eleven books on Colorado Plateau natural and human history and methods for exploring its wild backcountry. He has also prepared four special maps detailing access routes into the canyon country heartland that surrounds Moab.

This latest book in his "Canyon Country" series describes one more aspect of the land Fran Barnes has chosen to call home. Some well-meaning conservationists question Fran's goal of spreading knowledge about this unique region. To these he replies:

"Those who would exploit this wondrous land already know about it and are busily seeking to locate and develop its economic resources. My goal is to inform those who do not know about the myriad charms this land holds, those who, knowing, might help protect and cherish its treasury of knowledge and natural beauty for both present and future generations to enjoy.

"I am trying to make such people aware of what is here, of the magnificence that is all too often being lost before the American public even knows it exists. The Colorado Plateau's uncounted natural arches and bridges are but one aspect of an entire region that deserves more attention from that part of humanity that is capable of placing a high value on new knowledge and exceptional beauty."

Fran Barnes is particularly well qualified to write about arches and natural bridges. His earlier book, Canyon Country Geology, summarized the geologic history of the Colorado Plateau as it is understood to date, and provided a wealth of fascinating details about its "canyon country" heartland. He has also written numerous articles for a variety of publications about this region's natural spans, illustrating the articles from his extensive photographic files. A partial list of these articles appears in this book's bibliography.

Fran's years of exploring the Colorado Plateau hinterlands enables him to write from a wealth of personal experience and observation. The value of that observation has been enhanced by his broad education in science and natural history, and by the practical discipline of a long prior career in scientific research and aerospace engineering.

His experience with the aesthetic side of nature, with seeing and documenting its beauty, has been developed and polished through almost half a century of creative nature photography, in both monochrome and color. His scholastic abilities have been amply demonstrated by the literature research that went into this and earlier books, as illustrated by this book's carefully selected quotes from historic and scientific publications.

The ultimate product of all this is a definitive work, a book that has no parallel in print -- the only book of its kind. The reader is invited to enter the world of natural holes-in-rock, presented in such depth, clarity and detail that these unique phenomena will quickly charm and fascinate the modern reader, just as they did early Colorado Plateau explorers. For those skeptics who say "what can be said about holes in rock?" we say read on, explore vicariously, then personally sample these and other highlights of the incomparable Colorado Plateau -- and find, as so many others have, that the region is far more charming than even Fran Barnes can describe!

THE PUBLISHER

ACKNOWLEDGMENTS

A careful search of existing literature indicates that this is the first book to attempt a comprehensive general overview of the subject of Colorado Plateau natural rock openings -- its myriad arches, bridges, windows and tunnels. While this book is largely the product of personal field experience and considerable original thought, many other authors provided the foundation of knowledge so necessary to the broad synthesis attempted.

I thus want to acknowledge the indispensible value of the writings produced by those numerous authors, now dead and still living, both scientists and laymen, who have written on any aspect of Colorado Plateau natural rock openings. In particular I want to thank those whose writings I have quoted. Special thanks are due to Mr. Wilbur E. Garrett, editor of National Geographic Magazine, for granting permission to quote from a number of issues published during the adventurous first half of this century. This nation is fortunate to have had, since the late 1800s, a geographic magazine that is popular, informative and of such quality that its back issues are valuable source material for scholarly research.

The extensive quotes in this book from National Geographic Magazine and from H. L. A. Culmer's diary of his 1905 trip to the White Canyon natural bridges were made with the permission of their publishers, the National Geographic Society and the Utah Division of State History, respectively.

I also owe thanks to several fine photographers who contributed pictures of natural spans that, for practical reasons, I could not visit myself. The personnel of several state and national parks have also been very cooperative and helpful. Some even contributed photographs of natural spans under their jurisdiction, thus making the book more truly representative of the wide diversity of spans that occur in the region.

Among others to whom I am grateful are: Ed McCarrick, for his support, contributions and review comments; the Grand County Library, for its help in finding obscure references; the BLM Moab District Office for its clarification of special land withdrawal matters; tour guide Bill Crawley of Kayenta, for his cheerful cooperation and photograph contributions; Jerry Cly, for being a skillful and knowledgeable guide to spans in the Monument Valley area; my Moab friends and inveterate explorers Jack Bickers, Fred Radcliffe and Ber Knight, for contributing photographs and for reporting their new finds to me for inclusion in this book; my dear friends Doc and George Bell, the "Missouri Mules," who have long been arch hunters of great persistence; Dick Wilson, for setting an example of diligence in cataloging reported spans and in searching for others; and one potential book publisher, who inspired me with his skeptical question, "how much can you write about holes in rock?"

I am especially indebted to a young scholar of Utah history and geography, Rodney Greeno, whom I am pleased to count among my closest friends. His help in searching for, and finding, reference material related to Colorado Plateau natural rock openings in several university libraries was indispensable. Without that aid, this book would have been far less complete. Rodney's special knowledge of Four Corners States natural and human history was a valuable asset which was offered freely during the fourteen months it took to research and write this book. His review of several chapters in the draft stage was extremely helpful. In many ways, Rodney Greeno was an invaluable researcher, editor and co-author of this book, as well as a contributing photographer. He not only photographed several known spans for the book, but also discovered three previusly unreported arches, one of which may never have been seen before by anyone. My special heartfelt thanks to scholar Rodney Greeno, who beyond doubt loves the Colorado Plateau as deeply as I do.

Next, I would be remiss if I did not acknowledge the quiet, steady support of my dear wife of some three decades, Terby, who was baptized Mary Margaret. For lo! these many years she has served as a sounding board for my ideas, my syntheses, my organized words. She regularly listens patiently, offers encouragement and constructive suggestions, reviews every word I write, holds my spelling and grammar to the straight and narrow and has even typed all previous writings in final form.

Terby was relieved of the onerous task of typing with this book when, for the first time, I used a microcomputer and word-processing program in my writing, from first rough draft to final copy. It was printed out at various stages, at the rate of 120 characters per second, on a marvelous little computer peripheral called a dot-matrix printer. In truth, both my wife and I owe a full measure of thanks to the many people who have helped to develop modern microcomputer technology, which has made creative writing a pleasure instead of a chore.

For those who may be interested, this book was written on a Sanyo MBC-2000 microcomputer, using the WordStar word-processing program. The type for the book was set directly from the computer using several of the type fonts available in an Epson FX-85 dot-matrix printer, then reduced to the appropriate column-width for page layout.

To all who have thus contributed to this book, directly and indirectly, my sincere thanks. To the book's readers -- I hope you enjoy reading it at least as much as I enjoyed writing it.

Fran Barnes

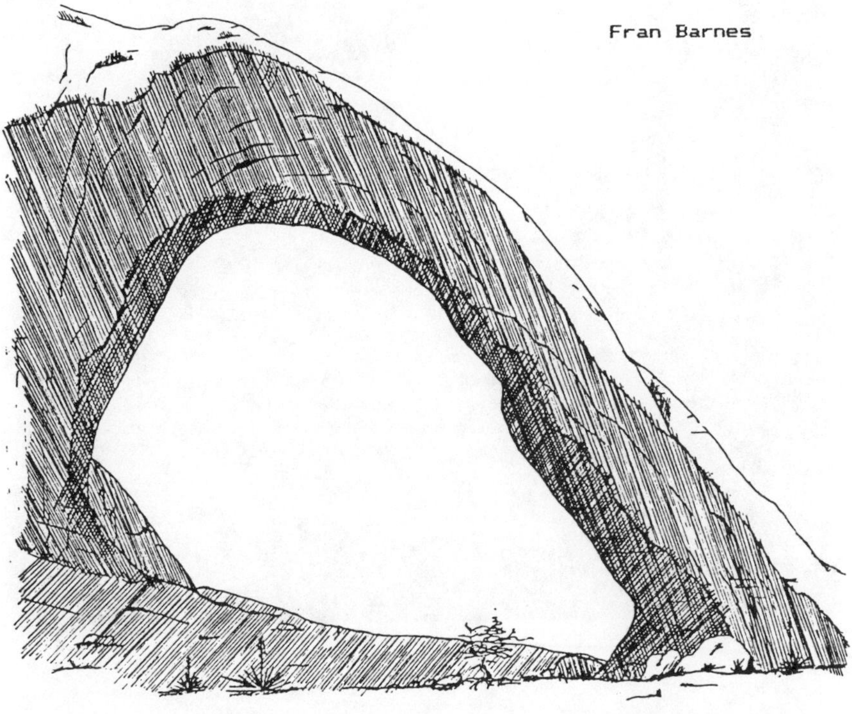

Angel Arch

INTRODUCTION

HUMAN NATURE

Humanity has been endowed by nature with an exceptional amount of intelligence, the ability to notice and think about anything unusual in the environment as a means for enhancing survival. In modern mankind, this alertness to novelty, this curiosity about everything, has led not only to the development of science and technology but to the nurturing of the sense of aesthetics, the ability to see and appreciate things of beauty, things of unusual color, form, texture and movement.

Since antiquity, or at least as long as there has been any record of human thinking, mankind has thus been fascinated by the novelties of nature, whether living, geographic, climatic, astronomic or a combination of these. Mankind has long marveled at great trees, strange life forms, vast mountain ranges, auroras and eclipses. Man has also taken special note of beautiful flowers, graceful animals, lovely lakes, arcing rainbows and colorful sunsets -- and those rarities of nature, its beautiful, graceful, lovely, arcing, colorful holes through solid rock, its arches, windows and natural bridges.

Humanity has always been fascinated by these unusual natural phenomena, and has often been either fearful or enchanted, depending on cultural background. Doubtless, individuals within many of the myriad human cultures throughout human history and prehistory have felt fear or superstitious awe at the sight of a large hole through solid rock, put there by some unknown, perhaps menacing force.

It is equally certain that individuals with other cultural backgrounds have seen only beauty and grace in natural spans. But it is a rare human being, indeed, who can gaze upward at a big natural arch or bridge of stone without feeling a sense of being in the presence of something very special.

RARE PHENOMENA

In most of the world, natural stone openings are rare indeed. On much of the world's above-water surface -- its land -- exposed bare rock is not at all common. Soils and vegetation cover the land wherever moisture is plentiful. Sands or gravels mask or bury most of the native rock in regions where water is scarce. Perpetual snow and ice cover vast regions of land, even entire continents. And all too often when extensive rock is exposed, as in mountain ranges and lesser outcrops that resist total obscurity by trees and other forms of plantlife, the rock is not the kind that readily weathers into arches, bridges and windows.

Even so, sizable natural openings in rock have been found many places throughout the world. Some kinds of openings are formed by rockfalls that have plugged narrow but deep gorges, with other rock and soil then building up durable "bridges" across chasms. Or they are natural caves or tunnels whose roofs have partially collapsed, leaving ceiling segments bridging open caverns. In a few places, persistent streams or rivers have taken advantage of weaknesses in rock to make short tunnels within their streambeds, thus creating natural bridges of sorts. Pounding waves have also punched holes through rock peninsulas, making sea arches.

But large natural rock spans that attract human attention and stimulate a unique sense of aesthetics are very, very rare throughout most of the world. A few large natural bridges in Europe have been widely documented, studied and appreciated for millennia. There are probably also spans in Africa, Australia and South America, but few have been publicized.

Before the advent of the twentieth century, a scholar who decided to list and study all of the documented natural rock openings of any significant size, would have had an easy task. His list for the eastern hemisphere would have been limited to a dozen at most. The western hemisphere would not have added many spans to such a list. A few were noted in South America, and several natural bridges were found in eastern North America. A very few others were recorded in the midwest, and a few more were noted in the far west, such as several sea arches along the coast. The volcanic islands of the Pacific contributed a number of bridges formed by collapsed lava tubes.

Few of the natural spans known in North America by the mid-1800s were worthy of more than passing note. They were local curiosities, perhaps studied briefly by geology students, but not truly inspiring in either size or shape. Exceptions to this were Natural Bridge in Powell County, Kentucky, and Virginia Natural Bridge in the state of that name.

THE COLORADO PLATEAU

Then, in the second half of the nineteenth century, Americans began systematically to explore a vast western region acquired by treaty from Mexico, a region about which little was known, a region shown as blank and "unexplored" on all honest maps. This Arizona-sized stretch of unknown geography was essentially the drainage basin of the upper Colorado River and its many tributaries, but besides being an elevated "desert wasteland," it was different in other ways. The first geologists to study the region soon realized its unique characteristics and named it the "Colorado Plateau Geologic Province," a term that has endured into modern times.

The Colorado Plateau is unusual in several major ways. It is an immense elevated region surrounded on all sides by still higher mountain ranges, making it a high basin. Its climate is mostly arid or semi-arid. It has an exceptionally high ratio of exposed bare rock, and much of that rock is of a nature that readily forms natural arches, bridges, windows and other openings.

Just how readily is still not widely known, because even now, more than a century since Americans started exploring the Colorado Plateau, it is largely terra incognita. Many Americans have heard of or visited the region's several well-known national parks, but relatively few are familiar with its many equally spectacular national monuments and state parks, or its vast de facto wilderness areas. Many have heard of its major river, the Colorado, but few know about the numerous spectacular tributaries of the Colorado -- the Green, Dolores, San Juan, Fremont, Gunnison, Dirty Devil, San Rafael, Escalante and others.

A handful of stockmen and prospectors, past and present, have learned a few things about the immense hinterlands of the Colorado Plateau, but their interests are narrow, and the knowledge they have gained is jealously guarded, not broadly shared with other Americans, especially not with the more objective segments of the scientific community, or those who place other values above economic gain. Although there are parks and monuments in the region, these represent but a tiny fraction of the total area that should be protected for its absolutely unique natural beauty, for the priceless knowledge that is there in archaeology, geology, paleontology, paleobotany, climatology, ecology, wildlife and other fields of study.

NATURAL SPANS

One aspect of study that has largely been neglected to date is the region's natural rock openings, so rare everywhere else but so plentiful within the Colorado Plateau. Since exploration of the region began in the mid-1800s, no one -- whether scientist or amateur -- has made any systematic attempt even to document the natural spans known to exist within the Colorado Plateau, let alone explore its more inaccessible areas for spans not yet reported.

A few individuals have started to log known spans in the region, but have barely made good starts before dropping the huge task. The scientific community has made little effort to assume the job, although a few individual scholars have made limited surveys of small special areas such as Arches National Park.

Looking Glass Rock

THIS BOOK

The purpose of this book is to introduce a wide readership to the fascinating subject of natural rock spans in general, and those of the unique Colorado Plateau in particular, in the hope that this will somehow inspire a more comprehensive examination of these phenomena, from both a scientific and aesthetic viewpoint. The book presents a summary of present knowledge about the natural spans in the Colorado Plateau Geologic Province -- locally known as "canyon country" because of its incredibly complex maze of canyon systems. It offers a representative sampling of the hundreds, perhaps thousands of spans within its several distinct geologic areas. Separate chapters summarize what is now known about the various aspects of natural span study -- their history, formation, ages, naming, measurement, protection, aesthetics, cataloging and photography, and how the region's prehistoric and historic natives viewed certain major spans.

No attempt has been made in this book to offer readers more than a generous sampling of the known natural spans within the Colorado Plateau. There are simply too many, and more are being found and reported all the time. For example, the author and some of his friends and professional acquaintances found a dozen or so significant spans never before reported just while this book was in preparation. Some of these are depicted in later pages.

INTRODUCTION

It was also impractical to make a comprehensive study of all the historic and scientific references related to North American natural spans, although a fair sampling of those related to Colorado Plateau spans is presented, with direct quotes where appropriate. It should be noted that some of the historic reports quoted are far from the dry, lifeless documentation normally associated with the formal study of history. Several of the accounts of early explorations made into the unknown lands of the Colorado River were colorful, fascinating and personal, and make better reading than any fictional "western."

One chapter of this book deals with the challenging problems associated with reaching known spans not near modern highways. It also offers some guidance to those who may wish to do their own exploring for unreported spans. And, for those who lack the time, strength or interest in original exploring -- in finding their own "undiscovered" arches and bridges -- the book offers several selected lists of known spans, with directions for reaching some of them. One list notes most of the natural spans that can be reached by ordinary highway vehicles, while another gives the author's favorites, a difficult choice to make. Still another chart lists multiple spans -- those that are double or triple. There is also a short list of fallen arches, and a still shorter one of Colorado Plateau natural bridges across which vehicles can drive.

The book's large special photo sections offer even armchair travelers a good look at the wide diversity of sizes and shapes that the natural rock openings of the Colorado Plateau can assume. Last but not least, an extensive bibliography lists most of the literature sources used for researching this study, many of which can provide further reading for those who find that this book has stimulated a deeper interest in the fascinating subject of arches and bridges and other natural rock openings.

Enjoy!

Fran Barnes

REGIONAL GEOLOGY

THE COLORADO PLATEAU

The Colorado Plateau Geologic Province is a large oval of land that is centered on a point about fifty miles west of the Four Corners, where the states of Arizona, New Mexico, Colorado and Utah come together. It has about the same area as New Mexico, but is defined by geologic rather than political boundaries, as shown on the map to the right.

The Plateau itself is mostly high, semi-arid desertland. Within it are several large uplifts and small mountain ranges. It is surrounded on all sides by still higher mountain ranges or highlands. Thus, the Colorado Plateau forms an enormous, elevated basin, the drainage basin of the upper Colorado River and its multitude of tributaries.

Over the eons, this river complex has carved vast canyon systems within the Plateau, such as the Grand Canyon, the Canyonlands of southeastern Utah, and others not within national parks. Because of this proliferation of canyons, the entire Colorado Plateau region has come to be called "canyon country." In this book, "Colorado Plateau" and "canyon country" are used interchangeably.

ITS ORIGINS

Geologists agree that the Colorado Plateau is geologically unique, but they are not certain what caused that difference. They largely agree on what it is, but not how it came to be that way.

They do, however, agree that the Colorado Plateau is somehow the product of the continental drift process that throughout the history of the planet has been moving continent-sized chunks of planetary crust around like enormous pieces of a jigsaw puzzle, fitting them together now this way, now that, over spans of time measured in hundreds of millions of years.

One explanation for how the Colorado Plateau became unique is that somehow during the dim geologic past it acquired its special integrity, thickness and strength by surviving a series of encounters with other chunks of planetary crust, and that these encounters left it more resistant to the geologic forces that later affected the newer land surrounding it.

REGIONAL GEOLOGY

ITS STRUCTURE

The oldest rock strata presently exposed within the Colorado Plateau are very old indeed, with the most ancient being those on the inner gorge of the Grand Canyon. There, the rock layers were deposited as sediments long before the various drifting land masses coalesced into the single supercontinent geologists call "Pangea." During that period, the huge land mass that was later to become North America was a separate continent and had been for a long time before Pangea began to form. It did not resemble the present continent in any way.

Some of the rock in the upper layers of the Grand Canyon, and the older strata elsewhere in canyon country, were deposited during the millions of years that Pangea remained essentially in one piece. The uppermost strata in the Grand Canyon, and most of those exposed elsewhere in canyon country, were deposited during Pangea's existence and after it began to break up into the present land masses.

While natural spans can and do form within the more ancient, deeper layers of the Grand Canyon, virtually all of the significant spans within the Colorado Plateau are within rock formations that were deposited just before the formation of Pangea, during its existence or after it broke up. During this period of time, the Colorado Plateau region was essentially on the west coast of the continental land mass to which it was attached. The rock deposits laid down on the Plateau during that time were at first either shallow marine sediments, or marine sediments alternating with fresh-water deposits. During most of Pangea's existence, and after its breakup, Colorado Plateau sediments were almost entirely land deposits, with occasional layers of beach sand and sandbar deposits left behind by seas that intruded briefly, then retreated.

The oldest canyon country rock layers that presently contain significant natural spans are among the last shallow marine sediments that were laid down before the region became largely dry land. Thus, the oldest significant arch-forming strata within the Colorado Plateau were deposited while most of the region was still broad, low coastal plains, with very little relief by hills or other higher features.

Later, violent geologic activity, triggered by the breakup of Pangea and the westward drift of North America, distorted the land and changed the nature of the deposits that continued to accumulate in the Colorado Plateau region until about ten million years ago.

TECTONICS

"Tectonics" is a term applied by geologists to powerful geologic forces that change and distort the land. The Colorado Plateau region suffered from such distortion even while Pangea existed, as that enormous supercontinent shifted and changed its shape in response to the powerful planetary forces that move the continents around. Great uplifts arose within the Plateau, and the earliest stages of the Sierra Nevada Mountains formed to the west, drastically affecting the Plateau climate.

But the greatest distortion began later, as Pangea separated into its present land masses, and North America headed west, impacting as it went with various sections of oceanic crust and smaller land masses that lay in its path. The various tectonic forces triggered by this continental drift produced enormous changes in what had been the open coastal plains of the Plateau. The land was distorted by gigantic uplifts, faulting and overthrusting.

The regions that surrounded the Colorado Plateau were even more affected by this geologic violence. The huge ranges of mountains that now surround the Plateau formed then, concurrent with heavy volcanic activity all around the Plateau. The forces that caused this volcanic activity were not powerful enough, however, to thrust molten lava all the way up through the thick Colorado Plateau rock strata, but lava did intrude almost to the surface several places, forming stupendous bulges. Later, when the region entered it present erosion cycle, and thousands of feet of rock and sediments had been eroded away, the cores of these gigantic bulges were exposed and became the present La Sal, Abajo, Henry, Carrizo, La Plata, Rico, Ute and Navajo mountain ranges.

Other very obvious remnants of this period of violent geologic activity are also visible within the Colorado Plateau. One is the Monument Uplift. Signs of this immense upheaval are visible in Monument Valley, the Needles District of Canyonlands National Park, along Comb Ridge, and elsewhere in the Four Corners region. Another is the San Rafael Swell in central Utah. The Uncompahgre Uplift distorted western Colorado, and other uplifts and great fault zones shaped northern Arizona and northwestern New Mexico.

It is interesting to note that the groundwork for many of the major natural spans within the Colorado Plateau was laid by the tectonic forces that have so distorted the region. Stresses far underground deformed rock layers, leaving faults and joints and stress patterns that later affected erosion patterns and played important parts in span formation, once erosion exposed the affected rock.

CLIMATES

During all these megayears of deposition and distortion, the Colorado Plateau was also undergoing continual changes in climate. Part of this was due to the gradual drift of Pangea into northern latitudes. This and other factors caused sweeping shifts in global climate patterns. But the greatest change to Colorado Plateau climate was brought on by the formation of high mountain ranges between it and the Pacific Ocean to the west.

The northward drift of the Colorado Plateau is shown by the fact that some 470 million years ago, long before Pangea came together, the equator ran through what is now the state of Utah. About 250 million years ago the equator crossed what is now Texas. Now, the center of the Colorado Plateau is about 37 degrees north of the equator.

The formation of the Sierras first affected the climate of the Colorado Plateau by placing it within a "rain shadow." Later, the upthrust of California's Coast Ranges and the Wasatch Range in central Utah increased this effect, making the Plateau region even more arid. The formation of the present Rocky Mountains did little to change this until, again due to some little-understood continental drift process, the entire western region of North America began to uplift, leaving the Plateau and the whole western half of the continent about a mile higher than it had been.

This enormous uplift started the present erosion cycle, the one that formed all of the great canyon systems within the Colorado Plateau Geologic Province.

REGIONAL GEOLOGY

The Maze, Canyonlands National Park

PRESENT CYCLE

The big climatic change that created the present scene within canyon country began about ten million years ago, when the Colorado Plateau and the mountain ranges that surround it were thrust to higher elevations. This caused increased precipitation in the Rockies and in various higher areas within the Plateau. That in turn caused continual runoff to penetrate the lower areas of the arid Plateau for the first time, forming a system of rivers and tributary streams. Geologists can find no trace of the Colorado River earlier than about ten million years ago, indicating that all of canyon country has been shaped by erosion since then.

The great western uplift was, of course, quite slow by human standards. Thus, the first rivers and streams that flowed into the region from its highlands and the surrounding mountains were quite gentle. They meandered across the fairly flat terrain in great loops and curves. As the runoff continued to increase, these meandering streams cut their original channels deeper and deeper, entrenching themselves and confining the flowing water to patterns established long before the present canyon systems existed.

During the ten million years in which canyon country was being shaped by water, the amount of water flowing in its rivers and streams varied. Some of these variations were due to changes in overall global climate. Local climatic factors caused others. The main effects of these climatic changes were cooler temperatures, varying seasonal precipitation and the glacial periods these brought. The glaciers from the last great North American continental ice cap are still retreating in Alaska and parts of Canada, and in the higher elevations of some mountain ranges farther south. Several small mini-glacial periods have also affected the Plateau region within the last few thousand years.

The glaciers within and adjacent to the Colorado Plateau affected it in two major ways. The melting glaciers added runoff to the region's rivers and streams, and the colder, wetter climate that formed the glaciers increased the general erosion caused by direct precipitation. The glaciers themselves did not, however, play a direct part in forming the canyon systems in the region, as they did in shaping the valleys and gorges of many mountain ranges, including some of those within the Plateau.

The end result of all this changeable, cyclic erosion within the Colorado Plateau was not only the production of the vast, complex canyons so characteristic of the region, but also the creation of natural rock spans in unprecedented profusion. The Plateau became a land of natural arches and bridges that has no parallel anywhere.

THE LAND OF ARCHES

Thus, it is a totally unique combination of factors that makes the Colorado Plateau a "land of arches." These factors are its exceptionally sturdy foundation, its multiplicity of marine and land rock strata, its violent distortion and uplift by continental drift forces, and its high-desert climate punctuated by recent glacial periods -- plus a ten million year period of erosion during which these factors could work their intricate but natural magic.

Of course, none of the present natural spans has existed for nearly that long, but spans have been forming in the region ever since water first began cutting into its older, harder sandstone layers. Within the last several million years, flowing and falling water has slowly created, then just as surely destroyed, generation after generation of Colorado Plateau natural spans. At times, during long periods of aridity, this formation and destruction of spans was very slow. During wetter times, the process was more rapid.

Thus, the rate of formation and aging of natural spans within the Colorado Plateau has varied drastically, in step with regional climatic changes. Because of this climatic variable in span formation, it is quite probable that virtually all presently existing spans were formed during the last two mini-glacial periods, that is, within the last few thousand years. This subject is discussed in more detail in later chapters on span formation and age.

THE PRESENT SITUATION

Where does the Colorado Plateau stand today, and where is it going in the future? The massive uplift that started the present erosional cycle has essentially reached its limit, at least in this region, but are other factors due to come into play within the foreseeable future?

Today, the variable climate that has shaped the Plateau during its present erosional cycle is in one of its drying phases. This means that the natural erosion that creates, ages, then destroys natural spans is slowing. Tree ring studies show that over the last several thousand years the region's climate has roughly followed 400 year cycles, varying from wetter to drier and back again, and that it is now well into a more arid phase that will probably continue for another century or more.

Such cycles are not smooth, however. There are always minor fluctuations from year to year and from decade to decade, but always with a distinct general trend.

This natural climatic trend, however, is not the only factor involved in determining how fast Colorado Plateau natural spans form, age and fall. Humanity has now entered the scene. Some geologists claim that humanity itself is the most violent geologic factor on the planet, barring the few really big volcanic eruptions that occur now and then. There is a lot of evidence in support of that concept. To explore it fully would take an entire book, but just within the Colorado Plateau it is easy to see that anything man does that affects precipitation or water runoff rates there also affects erosion rates and hence the formation, aging and collapse of natural spans.

Utah and other western states have been seeding clouds for decades. If this has not affected precipitation rates in and west of the Rockies, then the public servants in these states have certainly wasted a lot of public funds. Although it is almost impossible to prove, scientific studies have concluded that these weather modification programs have increased precipitation in parts of the region by as much as 20%, and may even have contributed to some of the destructive flooding that has plagued Utah and Colorado.

Other human activities have seriously affected precipitation runoff rates, whether the precipitation is natural or manmade. In arid areas, anything that disturbs the natural vegetation accelerates runoff. Among the human activities that contribute to this are the clearing of land for agriculture, the grazing of domestic livestock, road and trail building, logging, pinyon-juniper "chaining," mineral search and development activities, construction projects, and the slow but steady growth of the towns and cities in the region.

Two human factors that produce less obvious effects are air pollution, which can adversely affect soil chemistry and the plant life that binds the soil, and dam construction, which can cause land subsidence, slippage along fault zones and the massive collapse of such steep land forms as cliffs and mountain slopes. The removal of mineral deposits from underground and the lowering of water tables can also cause land settling. The shock waves associated with the blasting that accompanies many kinds of human activities can cause rockfalls, even at considerable distance. The sonic booms of supersonic aircraft can do the same. There are many other human activities that affect erosion rates within the Colorado Plateau, and hence the life expectancies of the many natural spans in the region.

To summarize, for the foreseeable future human activities will play a greater part than natural forces in the rate of aging and destruction of Colorado Plateau natural spans. Man aside, the region is entering a period of very slow aging for such spans. But human activities are bringing very rapid destruction, on both the geologic and human time scale. Some spans have already been lost because of the human factor, and many others are doomed to premature collapse.

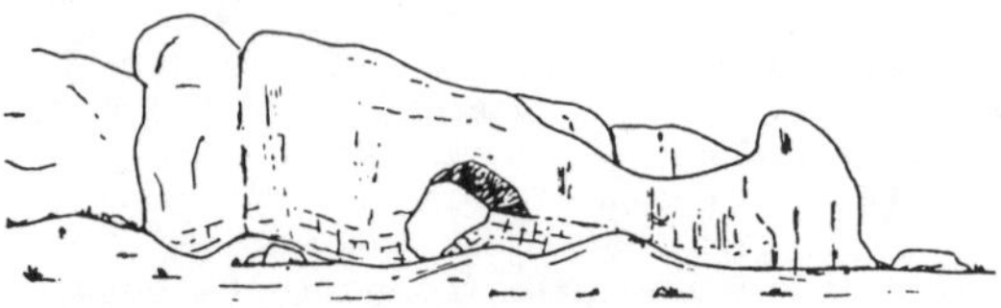

HISTORY OF DISCOVERY

COLORADO PLATEAU HUMAN HISTORY

Humans first penetrated the Colorado Plateau region thousands of years ago, when nomadic, lithic-age Amerinds wandered into it seeking game and forage. The prehistoric human occupancy and use of the general Four Corners region makes fascinating reading. It is quite probable, but impossible to substantiate, that the first impact of the region's larger natural spans upon humanity began in dim prehistory. Some of the primitive Amerind predecessors of the Anazasi culture, which later occupied much of the region, doubtless knew about many of its arches and bridges, and quite possibly incorporated some of them into their religious beliefs.

Or, perhaps the first interface between man and span was with the nomadic forebears of the prehistoric Fremont culture that later occupied much of what is now Utah north of the Colorado River. There are many spectacular and very impressive natural spans both north and south of the Colorado River, which halves the Colorado Plateau. The subject of possible interfaces between the prehistoric and historic Indians of the Colorado Plateau and its many natural spans is discussed further in a later chapter.

White men first entered what are now the Four Corners States in the mid-1500s, as Spanish explorers came seeking the fabled "Seven Cities of Cibola," whose streets were supposedly paved with gold. In 1540, the Spanish explorer Coronado crossed the southeastern corner of what is now Arizona and penetrated as far east as central New Mexico, but did not travel far enough north to enter the Colorado Plateau itself. The first well-documented penetration of the Plateau occurred much later, when an attempt was made to establish a route between the thriving town of Santa Fe, New Mexico, and other Spanish settlements in California, far to the west.

In 1776, the Dominguez-Escalante expedition headed north from Santa Fe, dipping in and out of the flanks of the Plateau as it went north along the western slopes of the Rockies, then west through the Uinta Basin. From the area of Provo, Utah, by then in serious trouble, the expedition headed south then east, to cross the Colorado River upriver of the present Glen Canyon Dam. It then returned to Santa Fe via the southern borderlands of the Plateau.

Accounts of this ill-fated expedition are fascinating, but apparently its members either did not see any natural spans, or considered them to be unworthy of note. This is not surprising. The expedition's route did not go near any significant spans, and its goals did not include listing scenic attractions. Further, its members were too beset with serious troubles to care about any curious holes-in-rock encountered along the way.

Later, other explorers did manage to establish a route from Santa Fe to California, crossing the Colorado River in Moab Valley and the Green River near the present town of Green River, Utah. Although the Spaniards did not use this "Old Spanish Trail" route to any great extent, others did, at least until better routes to southern California were established across Arizona. The users of the original route, however, encountered so many hardships and dangers that they did little exploring along the way, and left scant record of the region's natural spans.

In the mid-1800s, the Plateau was penetrated by a handful of trappers, who floated down its rivers seeking fur-bearing game. One such trapper left dated "D. Julien" inscriptions on the walls of the Colorado and Green river gorges all the way down to Glen Canyon in southern Utah, but left no written record of the numerous natural spans he must have seen while drifting down the Green River from Wyoming.

Official American exploring parties began to penetrate the Colorado Plateau region after it was acquired from Mexico following the war of 1848. The first expeditions crossed the southern borderlands of the Plateau, seeking practical routes west to California. Then, in 1859, a U.S. Army mapping expedition penetrated the Plateau almost to the Colorado-Green confluence, but made no note of natural spans, even though it passed within sight of at least three of significant size, and close to three others.

D. Julien inscription, Hellroaring Canyon

DISCOVERY

The concept of "discovery" needs to be examined here, because it soon became a matter of bitter controversy, and to this day arouses irrational thought and action among some who are personally involved in finding and cataloging the region's natural spans.

Most historians simply note, for example, that "Rainbow Bridge was discovered on August 14, 1909, by Professor Byron Cummings and party," then go on to other matters without examining the implications of their use of the word "discovered." The reader is left with the impression that Cummings was the first human in history to gaze upon that mighty span, a highly questionable assumption. With the region occupied for thousands of years by American Indians before white men even appeared on the scene, it is obvious that Cummings could not have been the first human to see the span. It simply had to have been known to, and seen by, countless Indians and possibly a few white men too, long before Cummings' official "discovery."

The implication of the 1909 "discovery" date is that even though Rainbow Bridge was widely known to exist before Cummings got to it, all those who had known about or seen it before did not count, that official discoveries can be made only by professionals, that discoveries made by others are not recognizable. This kind of scientific-academic-bureaucratic chauvinism persists even today, a pernicious form of discrimination that says, in effect, that discoveries made by common folk do not count, that only proper authorities who document them properly can make discoveries.

This attitude has a logical but limited basis, of course, in that if a "discovery" is not thoroughly documented, on a practical level it doesn't exist as an historic record available to the public. Unfortunately, no Indians and few private citizen explorers prepare such documentation. Professional explorers do, and thus garner the lion's share of the credit for "discoveries." This has led to no little humor, and occasional outrage, among individuals who have known about certain natural spans within the Colorado Plateau for decades, when "discovery" claims are made by individuals who are in a position to publicize them.

The matter can be further complicated if discovery is not recognized, even if accompanied by a published report and photographs, unless a signed and dated register is left under the span. Of course this rather limited definition of discovery allows a zealous arch collector the personal satisfaction of "discovering" spans long since documented, but it has not been widely accepted.

The subject of "discovery" is discussed further in the chapter on definitions, in order to clarify the tricky subject for readers, but no attempt is made to resolve the many conflicting claims about natural span discoveries within the Colorado Plateau. A few examples are given, however, and the biggest controversy over discovery, that surrounding Rainbow Bridge, is summarized later in this chapter.

The bottom line, however, is that those who take an active interest in trying to pin down the "truth" about who actually discovered any particular natural span should be prepared for an exasperating experience, whether the span is a major one and well documented, or remote and little-known. The subject is a "can of worms" that is best left to professional historians, or ignored entirely, and even serious historians sometimes become confused.

EARLY REPORTS

The first note of a natural span within the Colorado Plateau appeared in the report of Lt. Joseph C. Ives describing his 1857-1858 expedition up the lower Colorado River in the makeshift paddlewheel steamboat, "Explorer," then across northern Arizona by land. His report contained a sketch of a "natural bridge" near their camp 100, which was about forty miles west of Fort Defiance near the Arizona-New Mexico border.

The second Colorado Plateau natural spans to be documented were noted by members of the famous Powell expeditions down the Green and Colorado rivers in 1869 and 1871. Major Powell and two men in his party, Frederick Dellenbaugh and Jack Hillers, reported seeing some of the many natural spans in the Desolation-Gray canyon system in the northern reaches of the Plateau. In his account of the 1869 venture Powell noted that:

> "In several places these lateral canyons are separated from one another only by narrow walls, often hundreds

of feet high, so narrow in places that where softer rocks are found below they have crumbled away and left holes in the wall, forming passages from one canyon to another. These we often call natural bridges."

Dellenbaugh's 1902 report of the expedition made two observations about natural spans within Desolation Canyon:

"Through some of the upper narrow promontories of sandstone there were large holes, or arches, some of them probably a hundred or more feet in diameter."

And, a bit farther down the canyon:

" -- we saw, a couple of thousand feet above on the right, a gigantic sample of the natural arches."

Dellenbaugh's book also had photographs of a "Hole in the Wall" window "near Fort Defiance, Arizona," and noted that the kind of sandstone shown weathered to produce "large arches, alcoves, etc." There was also another photograph of an arch in the book. The pictures must, however, have been taken some time after the first Powell expedition. In a 1926 book about the 1871 Powell expedition, Dellenbaugh described Chandler Arch, one of the largest spans in Desolation Canyon:

"A mile below (Chandler Creek) -- we could see a remarkable topographic feature -- a gigantic aperture, or natural arch, in the cliff. It had a span of at least 300 feet with a height of about half as much. It was 1500 or 1800 feet above the river."

Jack Hillers' diary of the 1871 Powell expedition also noted Chandler Arch in his entry dated August 20, 1871:

"(One boat) stayed behind for Beaman to get a photograph of a natural bridge which spanned a gulch some 200 feet wide and 300 feet deep, about 1500 feet from the river. This bridge has been formed by erosion."

Ives Arch

It should be noted here that these and many other early reports of Colorado Plateau natural spans simply mentioned their existence, and perhaps made a few comments about how they might have formed. These chroniclers apparently did not consider their finds to be anything special, at least not enough to claim their discovery. The idea of "discovering" a particular span came later, as public awareness grew that this last region to be explored of a greatly expanded United States of America was a treasurehouse of phenomena very rare elsewhere -- natural arches and bridges.

Nevills Arch, Owl Creek Canyon. Harvey Halpern photo.

In the mid-1800s Mormon pioneers moved into what is now Utah and established settlements several places within the Colorado Plateau. Prospectors soon followed, seeking their fortunes in the virtually unexplored land. Certainly, as the Mormon stockmen and the prospectors became familiar with the region's wild backcountry, they must have seen many of its natural spans, just as the Indians must have earlier. But for the most part such rugged men were interested in other things than odd holes in the stupendous masses of rock that dominated the landscape. The same could be said of the outlaws who hid from justice in the region's maze of canyons. All of these early occupants of the Plateau considered it to be desolate wasteland with few redeeming features. They simply did not see its incomparable beauty.

In 1877, a large natural span was reported in Arizona. Although it is not within the Colorado Plateau proper, it needs to be described in order to gain perspective on the problem of definitions discussed in a later chapter.

The span's "discoverer" obviously knew it was known to the area's Indians, because he found it while trying to escape from some Apaches who were seeking to kill him for encroaching on their territory. In seeking to elude his pursuers, Dave Gowan, a sailor-turned-prospector, sought refuge in a strange tunnel out of which a creek flowed. This tunnel was some 60 miles south of the present city of Flagstaff and turned out to have another end, making it a natural bridge of sorts. Inside, the tunnel was the maze of alcoves, grottos and ledges that allowed Gowan to hide from the Indians. Later, he homesteaded the land around what became known as Tonto Natural Bridge, which is still privately owned. This curious span is discussed further in later chapters.

HISTORY OF DISCOVERY

Perhaps the next report of a natural span within the Colorado Plateau was made by Frederick Gardiner in an 1885 article in Science. The article reported a large natural bridge to which he was guided by a local Indian. The span was located about twenty miles north of where the "Atlantic and Pacific railroad" crossed the Arizona-New Mexico border. Obviously, the span had been known by others already, at least the Indian guide, but Gardiner was the first to report it. He did not claim "discovery." In a 1917 report, U.S. Geological Survey geologist Gregory mentioned a "Black Rock Bridge, 2 1/2 miles south of Fort Defiance," noted that it was probably the one reported by Gardiner, and described how it had formed.

Arch historian Vreeland (#11) notes that the span, which he called Black Rock Natural Bridge, "-- was probably first seen in 1885 by Frederick Gardiner." In fact, Gardiner's report was dated 1885, but it also named a railroad that only existed under that name in that location from 1881 to 1883, which meant that he probably visited the span during that period. His report also had a photograph of the span, discussed it in detail, and even described how it probably formed, so it is certain that he was there. This example serves to illustrate how difficult it can be to pin down exact discovery dates.

Landscape Arch, Arches National Park

Although most historians are now convinced that the White Canyon natural bridges in southeastern Utah were first seen by white men in the 1880s, the next published report of a large span within the Colorado Plateau appeared in the April 22, 1898 issue of Science, authored by Arthur Winslow. Apparently this was also the first report describing any of the many arches within what later became Arches National Park. The report showed a picture of the immense span now called Landscape Arch, with these comments:

> "The remarkable natural bridge illustrated in this article has, so far as I know, never been called attention to before, and is, therefore, entitled to rank as a new discovery among the curiosities of nature. It is an object rivaling the celebrated natural bridge of Virginia in magnitude -- "

Although the article's author used the word "discovery," he did not claim the honor himself, but noted that he had obtained the photograph "-- from the original photographer in Moab," thus leaving the actual discoverer of Landscape Arch a mystery. The article also had a picture of another large arch in the vicinity, with a man on horseback beneath it.

Sipapu Natural Bridge, Natural Bridges National Monument

FIRST MAJOR DISCOVERY

As already noted, there is also some question about who first "discovered" the three spectacular natural bridges that are now within Natural Bridges National Monument. General Land Office surveyor William B. Douglass noted in his 1908 survey notes that a local Paiute reported knowing about the bridges "-- for half a century, being born near them. His father knew them before his day." In addition, there were oral reports of men who claimed to have seen the bridges in the 1880s. There is no doubt about who authored the first published reports of these magnificent spans, but their "discovery" by white men depends upon whether hearsay oral accounts are given as much weight as personal written accounts.

Professor Byron Cummings of the University of Utah noted in his 1910 report on "The Great Natural Bridges of Utah" that:

> "The White Canyon bridges were first visited by white men in September, 1883, when Mr. Cass Hite, accompanied by Indian Joe, Edward Randolf, and Scotty Ross, made a

trip into that region. Mr. Hite gave these bridges the names of the President, the Senator, and the Congressman."

A 1941 WPA guide to Utah qualified this somewhat:

"The first white visitors (to the White Canyon bridges) were anonymous. A few trappers may have traversed the country, but they left no record. In the 1880s, Cass Hite, a desert misanthrope whose name still graces a one-man town in the Colorado River, worked his way through this country, and according to historical gossip of San Juan (County) old-timers, brought back accounts of many wonders including three 'whoppin'' natural bridges."

A 1957 book about the region by the San Juan County Daughters of Utah Pioneers took the same position:

"They were discovered by Cass Hite in 1883. So far as known, he was the first white man to recognize their grandeur. It is not improbable that prospectors and trappers roamed the bridge country earlier but failed to realize its importance as a world wonder."

There is some evidence that two men visited the bridges in the winter of 1882 and 1883, but again no written report was made. Most historians agree, however, that a local rancher was the next to "discover" the bridges, in 1895. They do not, however, agree on who the rancher was. Some say it was Al Scorup. Others claim that Scorup learned about the spans from another local cattleman named Emery Knowles. But, as noted in a 1947 article in Desert Magazine, "Being more interested in grass than rock formations, neither of these men publicized their finds." Most historians agree, however, that from 1895 on, other local cattlemen saw the big bridges fairly often.

This local knowledge first came to public attention through Mr. Horace M. Long, a mining engineer. According to W. W. Dyer, the writer who later documented the first "official" visit to the great spans, Long heard about the wonderful "arches" in 1902 from one of the Scorup brothers but was unimpressed. At that time, the term "arch" was often applied to the arch-topped alcoves that were fairly common in the sandstone cliffs that dominated the region, so Long naturally supposed that Scorup's "arches" were of this sort.

However, as Scorup continued to describe his "arches" and enthuse about their size and beauty, Long came to realize that he really meant "natural bridges," and bridges far larger than any known to exist elsewhere. Dyer's report noted that:

"From Scorup's statements it seemed probable that they had never been seen by any of the white race, save perhaps a half-dozen cattlemen and cow-boys and possibly an occasional fugitive from justice. So far as Scorup knew, they were first discovered by Emery Knowles, in 1895. Scorup himself saw them in the fall of the same year, in company with two cow-boys, Tom Hall and Jim Jones. Scorup greatly desired to revisit this remarkable region, and wished especially to obtain photographs of the bridges. He believed himself to be the only person then living in Utah who knew their

> exact location. He offered to conduct Long thither, but annexed to his offer the condition that one of the arches be named the 'Caroline' in honor of his mother. On March 13, 1903, Scorup and Long set out from Dandy Crossing, on the Colorado River."

The exact identity of "Scorup" in this and other reports is not certain. Some indicate that it was Al Scorup. Others say it was Al's brother, Jim, who guided Long to the bridges. It might be noted here that the "Dandy Crossing" of the Colorado River was established and operated by Cass Hite. For many years, this was the only practical way to cross the river in that region of Utah. The site is now beneath the waters of Lake Powell, several miles downlake from the present development that bears Hite's name.

The trip to the big natural bridges was easier from this direction than via the approach from the settlements of Blanding and Bluff later used by others. The route taken by the explorers closely parallels the present scenic highway, Utah 95, as it travels between the Hite area on Lake Powell and Natural Bridges National Monument. As Scorup and Long traveled, they got water from lower White Canyon and camped in cliff-dweller ruins in some of the canyon's many tributaries. Long made notes describing the geology, archaeology and natural beauty he saw along the way.

Once they reached the spans, they photographed and measured them. Long found it so easy to get to the top of one span that "(we) rode our horses over," thus becoming the first to do this with the span now called "Owachomo." Long wrote that "I am the first white man who has ever ridden over this bridge," but he was far from the last. During the ensuing years, many photographs appeared in public print showing as many as eleven horses with riders atop the span.

Scorup and Long also renamed the three spans "Caroline," after Scorup's mother, "Augusta" after Long's wife, and "Little Bridge," due to its smaller size in proportion to the others. They either did not know, or did not care about, the names Cass Hite had given the bridges earlier. While at the bridges, Horace Long took the first photographs of them, while Mrs. Long later made the first painting of the span named after her.

The story of Long's expedition to the White Canyon natural bridges, together with some of his photographs, was published in the August 1904 edition of Century Magazine. The account was authored by W. W. Dyer. A sketch was printed with the article showing the largest span, "Augusta," together with the Capitol building in Washington D.C. and the Great Pyramid of Egypt. The bridge soared above the Capitol dome and was higher than the pyramid. Dyer's account was extracted in the September 1904 issues of the American Geologist and National Geographic magazines. Cummings, in his 1910 report, "The Great Natural Bridges of Utah," noted that while accounts of the expedition:

> "-- did a great deal to awaken interest in these natural bridges and direct attention to the scenery of Utah -- no careful measurements were made, their size was overestimated, and the comparisons with the United States Capitol and other buildings (were) erroneous and misleading."

The 1904 accounts of the "discovery" of these three immense natural bridges did, indeed, attract a lot of public attention. Among those who became keenly interested in the spans was Colonel Edwin F. Holmes of Salt Lake City, "a member of the Salt Lake Commercial Club, and booster of Utah's scenic beauties" according to C.R. Steen's 1972 preamble to the diary of the man who led the 1905 expedition to the spans that Holmes organized and outfitted. Holmes asked H.L.A. Culmer, a fellow member of the Commercial Club and artist whose work Holmes admired, to lead an expedition to the great spans to measure, photograph and depict them in paintings.

Two other men joined Culmer on the expedition, S.T. Whitaker, an architect, and C.W. Holmes, the son of Colonel Holmes. Neither of these men kept journals of the trek, but Culmer's diary later served as a basis for several published accounts, including a summary by Colonel Holmes in the March 1907 edition of National Geographic, and another by Culmer himself in the September 1908 edition of Technical World Magazine. The diary also served as a basis for a series of radio broadcasts in the winter of 1936 and 1937. Culmer's diary was first published as a supplement to the June 1937 edition of the Southwest Monuments Monthly Report, and appeared again in the Winter 1972 edition of the Utah Historical Quarterly. It makes fascinating reading. It drew an intimate and detailed picture of the rugged land and people of southeastern Utah at that time. It is quoted several places elsewhere in this book.

The Culmer party approached the bridges from Bluff, rather than via the easier route from Dandy Crossing taken earlier by Scorup and Long. It took a week to get from Salt Lake City to Bluff. At Bluff, four more men joined the expedition -- Al Scorup as guide, two men named Perkins and Nielson as packers, and a fellow named Adams as cook. From Bluff, it took the group two more adventurous days to reach the first bridge.

Once there, they spent five days measuring, photographing, sketching and admiring the bridges and exploring the many archaeological sites nearby. Some of Culmer's observations are quoted later in this book, in appropriate chapters. The expedition's members accepted two of the names given the spans earlier by Scorup and Long, but changed "Little Bridge" to "Edwin," after the expedition's organizer, Colonel Edwin F. Holmes.

The group left the bridges by a route that took them into Grand Gulch, where they explored a bit but evidently saw none of the natural spans that are there. It took the main party ten days to get back to Salt Lake City, with numerous adventures along the way.

It is interesting to compare the coverage given this expedition by the two Salt Lake City newspapers. The April 1, 1905 edition of the Deseret News noted the expedition, discussed the expected archaeological findings and went into great detail about how these might relate to Mormon beliefs, but virtually ignored the natural bridges which were the prime subject of the expedition.

The April 22 edition of the Deseret News described some "snapshots" taken by the party that were the last "-- taken of them before they penetrated the wilderness where they now are amid the bleak and crested wilds." One photograph depicted the party "-- crossing a turbulent and dangerous confluent of the San Juan (river) on their way out into the dark and unknown sections of the San Juan County."

The May 1 edition of the Deseret News offered a summary of the expedition's findings under the headline, "No Such Thing As Cliff Dwellers," a conclusion supposedly reached by Culmer after his examination of the many prehistoric ruins encountered on the trip. The short report detailed many of the expedition's archaeological findings, but only mentioned the natural spans in the last paragraph:

> "The party spent some time at the natural bridges, about which so much has recently been written, and carefully measured each bridge."

The May 1, 1905 edition of the Salt Lake Tribune contained an article by Culmer which briefly summarized his expedition's findings about the natural bridges and described a few of the prehistoric artifacts found. The article used the term "cliff dwellers" several times, but made no mention of the "no such thing as cliff dwellers" headlined by the Deseret News.

The May 14 edition of the Salt Lake Tribune gave a lengthy and highly descriptive, even poetic, account of the expedition, again written by Culmer. The article had artwork and photographs depicting the natural spans and other nearby scenes, plus a map of the region. It discussed several archaeological sites and artifacts in detail, called their creators "cliff dwellers," and mentioned several obvious uses of the ruined structures. It also mentioned that there was no proof that the structures had been used as dwellings, but did not emphasize this point as the Deseret News had.

One might conclude from these widely different coverages of the same expedition by two contemporary newspapers that selective, interpretive news reporting is not a recent invention.

The next significent trip to these spectacular natural bridges was made in 1907 by a group headed by Professor Byron Cummings of the University of Utah, and under the direction of the Archaeological Institute of America. It consisted of Cummings and four other men by the names of Scranton, Driggs, Brown and Judd. Together, these men surveyed the land around the natural bridges and prepared detailed maps which located the spans and the major archaeological sites near them. These maps were later forwarded to Washington D.C., where they played an important part in the establishment of Natural Bridges National Monument by President Theodore Roosevelt on April 16, 1908.

Cummings later wrote reports about this expedition. One version appeared in the February 1910 issue of the National Geographic Magazine. Another became part of a University of Utah Archaeological Bulletin dated November 1910. As noted in this bulletin:

> "An additional survey was made in 1908 by Mr. W.B. Douglass, Examiner of Surveys for the United States government. In September, 1909, President Taft issued a second proclamation on the Natural Bridges National Monument by which the former boundaries were changed somewhat, two large cave springs a few miles away included, and the names changed to the Owachomo, the Kachina, and the Sipapu."

The report went on to say that it would be "-- very difficult to secure a general recognition of the Hopi names now applied by the government." Now, three-quarters of a century later, few recall or even know about the earlier names given these three magnificent natural bridges. The two cave springs included in the monument in 1909 were deleted in 1962, but still appear as annexes to the monument on older U.S.Geological Survey topographic maps of the region.

A man named Zeke Johnson became the first "custodian" of Natural Bridges National Monument. The full tale was told in an article in the 1947 issue of Desert Magazine. It seems that in 1908 Zeke, a new settler near Blanding, wanted to find a shorter route from there to the Utah town of Tropic, near Bryce Canyon, where his mother lived. Local cattlemen and Indians discouraged him, but he decided to try anyway. He hoped to find a route that would go near the big White Canyon bridges he had heard about.

He headed west from Blanding, skirting along the southern flanks of the Abajo Mountains. On the third day, he came across one of Al Scorup's cattle trails and followed it to one of the bridges. Upon his return to Blanding, Johnson met some men who were interested in locating a feasible route for a railroad between New Mexico and California. They asked him to guide them through the country he had just scouted. The three men spent the next sixty days exploring the region, visiting all three natural bridges along the way.

Zeke found that he enjoyed guiding people to the Monument more than he did herding cattle, and was soon doing this much of the time. The Desert Magazine article tells what happened next:

> "When Park Service officials began looking for a custodian for Natural Bridges National Monument, Zeke Johnson was their natural choice. He served twelve years at the munificent salary of one dollar a month, plus the privilege of renting horses to what he always called 'sightseers.' He spent two years laying out 12 miles of trail (from Blanding to the Monument). In 1920 Zeke was given a regular salary and remained as custodian until 1942, when he was retired and pensioned after serving two years beyond the prescribed age limit. But his retirement was not due to physical infirmities, even though he was then 73 years old."

The next documentation of previously unreported natural spans within the Colorado Plateau appeared in the Volume XIV, 1907 edition of The Mineral Collector. The article, entitled "Some Unknown Natural Bridges," stated that there were "dozens of natural bridges, varying in size from a few feet to hundreds of feet across" on the southwestern slopes of the Blue (Abajo) Mountains in San Juan County, Utah, with the largest being the three White Canyon bridges.

The article's author, a man named Parsons, also noted that there was a "curious natural bridge in New Mexico" with a span of 60 feet that he believed to be "remarkable because it is carved not by water, but by sand-laden winds." This misconception is discussed in the chapter on span formation modes. Parsons gave no further information about the "dozens" of spans on the southwestern slopes of the Abajos, but he could have been referring to those in Arch, Grand Gulch, Owl, and Fish Creek canyons, all of which are in that area. The New Mexico span could have been La Ventana Arch, which is visible from New Mexico 117, not far south of Interstate 40.

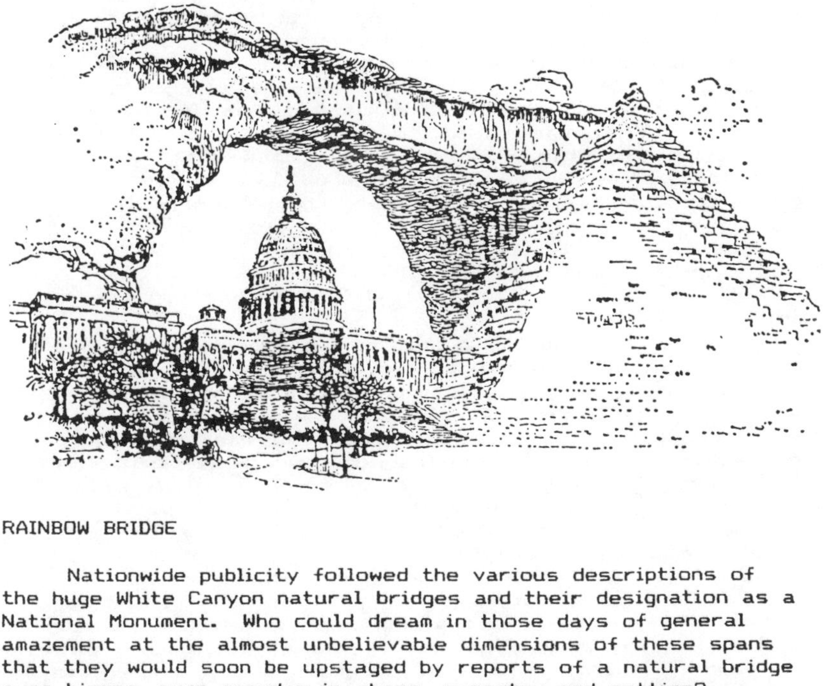

RAINBOW BRIDGE

Nationwide publicity followed the various descriptions of the huge White Canyon natural bridges and their designation as a National Monument. Who could dream in those days of general amazement at the almost unbelievable dimensions of these spans that they would soon be upstaged by reports of a natural bridge even bigger, even grander in shape, symmetry and setting?

The story of the "discovery" of Rainbow Bridge is surrounded by controversy. The generally accepted version of the story is presented in the following paragraphs, while various aspects of the controversy are examined in some detail in a section later in this chapter. In the 1960s, a different kind of conflict developed over Rainbow Bridge, one which continues today and which may have a tragic ending. This second major battle over Rainbow Bridge is summarized in the chapter on protection.

The American public was first introduced to Rainbow Bridge by an article in the February 1910 edition of the National Geographic:

> "We used to be much interested in the descriptions and illustrations in our readers and geographics of the Natural Bridge of Virginia. People travel from all

> parts of the world to behold this strange natural phenomenon and enjoy the picturesque scenery of the Appalachian Mountains, and feel well paid for their effort. But Utah is the home of really great natural bridges. She astonishes us with not merely one but half a dozen, any one of which surpasses the Virginia structure in grandeur and beauty. The three remarkable bridges -- the Edwin, the Caroline, and the Augusta -- which were discovered in Utah several years ago and described in this magazine, are now eclipsed by another marvelous structure in the same state."

The Rainbow Bridge discovery story that is generally undisputed goes like this:

Some time in the very early 1900s, either John Wetherill or his wife learned about a large natural arch from one or more of the Indians who visited their trading post in Kayenta, Arizona. One of the Wetherills shared this information with Professor Byron Cummings of the University of Utah, who was doing an archaeological survey in the vicinity. Someone later gave the same information to Mr. William Douglass, surveyor for the U.S. General Land Office, who was in the area making a further survey in connection with Natural Bridges National Monument.

Both Cummings and Douglass expressed keen interest in locating the alleged span. According to Douglass, he made a first attempt to get to the reported span "-- in December 1908, but owing to the snow, it was found impossible to do so." Both Cummings and Douglass made plans for 1909 expeditions. One arranged for a Navajo Indian guide, while the other chose a local Paiute to lead the way.

The details are disputed, but the two expeditions started within a few days of each other, and soon joined forces. It quickly became evident, however, that Douglass' Navajo guide had been more than a little overconfident. The party became lost and discouraged, even after the Paiute guide, who was late joining the expedition, caught up. The fearful Indian guides and helpers agreed to continue with the trip only after being ridiculed by John Wetherill, who was along as the expedition's leader.

The Paiute guide then led the party over a perilous and difficult route to the east, then north of Navajo Mountain to Rainbow Bridge. As Douglass noted in his official report: "The bridge was reached on Saturday morning August 14, 1909, at 11 o'clock." The expedition members didn't spend much time at the span because they were almost out of food, having cached much of it earlier to lighten the loads of the pack animals and for use on the return trip.

While there, several members of the party, led by Wetherill, managed to reach the top of the immense sandstone arch. This was necessary so that the Douglass team could measure the span. As Douglass told it:

> "Mr. Wetherill had succeeded in scaling the south walls of the cliff and returned with the information that the top of the bridge could be reached by means of a long rope. I dispatched my two chainmen and flagman with the necessary ropes. They lowered themselves over a 50-foot ledge to a point on the south arch of the bridge where sufficient footing was afforded to enable them to clamber to the top. Two steel tapes having a combined length of 333 feet were lowered over the edge to the bed of the stream below. Later I was lowered to

the bridge and made measurements as to width. We were the only ones to reach the top of the bridge --"

Cummings told a somewhat different story in one of his later descriptions of this 1909 expedition:

"-- we found a way of climbing over the bare rocks to a point above the bridge, then letting ourselves down a steep bare slope by rope to the level of the top of the arch so we could reach the crest. The necessary crawling and sliding was rather hard on pants and shirts, but gave us great satisfaction."

After making the first measurements of Rainbow Bridge, the expedition prepared to return along the same route. As Douglass told it:

"It required more time to make the segregation survey than was anticipated. Our supply of food was exhausted. At the close of the last day of laborious cliff climbing, we ate our last food which consisted of one biscuit and a spoonful of boiled beans per man. For the next days we had nothing whatever, but we succeeded in finishing the survey and reaching the supplies we had left on the way."

Upon his return, Douglass made an official report, but returned to the bridge two months later, in October, to complete his survey. This time he laid out a 160 acre area surrounding the span for possible withdrawal as a national monument. On May 30, 1910, President Taft proclaimed this tract to be Rainbow Bridge National Monument. Its boundaries have remained unchanged since that time, but later became the subject of a second dispute.

Cummings made a somewhat sketchy first public report about the 1909 expedition in the February 1910 issue of National Geographic Magazine. He also wrote a more complete description, which was published in a November 1910 Bulletin of the University of Utah, plus other accounts in later years. At the request of the National Park Service, a member of the Cummings party, a young archaeological student and "dig foreman" named Neil M. Judd, wrote a clear and detailed report of the 1909 joint expedition from his own recollections. This was published in 1919 and offered many details that did not appear in other accounts.

As Professor Stephen Jett of the University of California wrote in 1980, "The story of the 1909 'discovery' of (Rainbow Bridge) is -- a tale of rivalry, personality clashes, and distortion of the true facts." This tale is summarized later in this chapter. In contrast to the conflicting claims of the two official "discoverers" of Rainbow Bridge, one man who might have found it much earlier flatly admitted that he had not. In his 1910 report of the expedition, Cummings noted that "Mr. Cass Hite of Hite, Utah, who has traversed this region more extensively, probably, than any other living white man --" denied having seen the span. He quoted Hite as saying:

"The bridge found near Navajo mountain is located in about the only spot in that region that I did not explore or prospect. No, I did not see the bridge you sent me the picture of, and I don't think any white man ever saw it until your party did."

Quite a few people had other opinions about this, as explained later.

Rainbow Bridge

THE TENS AND TEENS

The publicity generated by the several reports published in 1904 and 1910 about the four great natural bridges in Utah stimulated not only public interest but scientific investigation of such spectacular phenomena. One such investigation led a member of the Geological Society of America, Herdman F. Cleland, to present a paper on the subject in December 1909. The paper was published the following year by the Society, under the title "North American Natural Bridges, With a Discussion of their Origin."

This study attempted to present a comprehensive picture of all such phenomena then known. It named and described the few in Europe, mentioned a sampling in South America, and devoted considerable space to an analysis of how natural bridges form. It then presented a supposedly complete list of such spans then known to exist in North America. The list noted only seven spans within the Four Corners States, Arizona, Colorado, New Mexico and Utah -- the four newly reported bridges in Utah (Kachina, Owachomo, Sipapu and Rainbow) and three in Arizona (Tonto, Black Rock and a petrified log bridge that is now within Petrified Forest National Park). None were listed for Colorado or New

Mexico, and the large spans reported by Major Powell and Dellenbaugh in Desolation Canyon were not even mentioned. This was an odd omission, since a Dellenbaugh article in the May 20, 1898 issue of Science had clearly stated that within Utah there were "-- a large number of these natural arches. In the canyon of Desolation, on Green River, they are particularly common, and from the surface of the river some of them seemed of huge proportions."

It is interesting to note that had Cleland's report been made a mere ten years earlier, he would have listed only one natural span within the Colorado Plateau -- Black Rock Natural Bridge in Arizona, reported by Gardiner in 1885. Almost none of the many hundreds of other large natural spans now known to exist in this geologic fantasyland had been reported in print by 1900, except for those in Desolation Canyon, and Landscape Arch north of Moab, although doubtless others were known to the stockmen, prospectors and outlaws who had by then penetrated many parts of the region.

Professor Byron Cummings wrote an article titled "The Great Natural Bridges of Utah," which appeared in the February 1910 issue of National Geographic Magazine. In this he summarized the recent discoveries of the White Canyon bridges and Rainbow Bridge in Utah, then went on to describe "-- two smaller arches that would each be an attraction by itself were they not overshadowed by the grander Nonnezoshi (Rainbow Bridge)." This article also depicted a "Pritchett Valley Bridge" and described an arch "-- below the western slope of Elk Ridge (with a) span of more than 100 feet --".

A Cummings report, also titled "The Great Natural Bridges of Utah," was published by the University of Utah in November 1910. This report also described the three White Canyon bridges and Rainbow Bridge and summarized their discovery, but perhaps its greatest value was its graphic description of the region in which these magnificent spans stood. This report was probably the first "travel promotion" writeup touting southern Utah's spectacular scenery, including its giant natural spans.

The report also told about the trip Cummings had made in the vicinity of Moab following his August 1909 expedition to Rainbow Bridge. He again noted, but didn't name, the arch in the foothills of the Abajo Mountains, then described in some detail his trip into the wild backcountry to the southwest of Moab:

> "In November, 1909, under the guidance of Dr. John Williams of Moab, we visited a bridge in the edge of Grand County that deserves to be classed with those of San Juan County and counted among the great natural wonders of our continent. This is a graceful arch with a total elevation of 62 feet and a span of 122 feet long and 49 feet high. It stands beside the cliff on the western edge of Pritchett valley -- "

This was probably the first written report of this span, although it was obviously known to a few Moabites earlier. The span is not one of the largest in the region, but is undoubtedly one of the most graceful. Cummings mentioned that "-- we have christened (the arch) Pikyabo (Pee-kya-bo), the Ute name for water tank," a different name than he gave in his National Geographic article. Neither name endured. The span is now called Pritchett Arch.

Pritchett Arch

Window Arch

Wigwam Arch

Cummings went on to describe several other natural openings in the vicinity. One was a large tunnel-like span near "Pikyabo" which they named "Wigwam Cave." Another was a window farther down Pritchett Canyon. A photograph in the report named the span "Block Bridge," another name which did not last. It is now called Window Arch.

Another photograph in the report, together with a mention that "Mr. Frank Hall very kindly took the time to show us a peculiarly balanced rock -- ", indicates the approximate discovery date of a large arch named after Mr. Hall. The photograph shows upper Pritchett Canyon. In the distance, the huge sandstone fin that contains a very large and beautiful arch can be seen edgewise, so that its opening is not apparent. But since Frank Hall made no mention of this span to Cummings, it is quite probable he had not found it yet. A picture of the span appeared in the November 17, 1911, edition of the Moab newspaper, with Hall's name on it, indicating that he discovered it sometime in 1910 or 1911. Surely had the big span been known either to Hall or Williams in 1909, they would have shown it to Cummings, especially since they passed within a few hundred yards of it. After its report in 1911, "Halls Bridge" was apparently "lost" again to local residents because of its obscure location. It was rediscovered again fifty-seven years later, as described in a later section of this chapter.

Joseph E. Pogue, of the United States National Museum, wrote an article about Rainbow Bridge which appeared in the November 1911 issue of National Geographic Magazine. The article noted that:

> "Recently a United States Geological Survey party, consisting of H.E. Gregory, in charge, John Wetherill, K.C. Heald, and the writer, stood upon the summit of Navaho Mountain and looked over a country of wildness and grandeur."

He also noted that while Rainbow Bridge was:

> "-- only four miles distant -- (it) was only reached after two days time and a journey of 35 miles over a very indirect route."

The report went on to summarize the official story of Rainbow Bridge's discovery and describe the U.S.G.S. group's journey to the span. It also provided several photographic views of the bridge, discussed how it formed, and gave the dimensions taken by the party's surveyors. These were somewhat different from those recorded by Douglass in 1909, but Pogue concluded that:

> "A mere recitation of figures must fail to convey an adequate idea of the imposing nature of the bridge. It is not the size alone, though this far exceeds the greatest masonry arches constructed by engineering skill, nor is it solely the graceful lines or curvature of maximum stability, but rather all of these, that combine to make this the most remarkable single arch now known."

Not to be left out of all the excitement about Rainbow Bridge, western author Zane Grey made a trip to the span in 1913, guided by John Wetherill. In his book, "Tales of Lonely Trails," published in 1922, Grey described his trip to Rainbow Bridge in almost poetic terms, and noted that:

> "-- my party was the second one, not scientific, to make the trip. Later this same year Wetherill took in the Roosevelt party and after that the Kolb brothers. There is no guide except Wetherill, who knows how to get there."

The 1913 expedition organized by Theodore Roosevelt explored around in northern Arizona before heading into southern Utah to visit famous Rainbow Bridge. Roosevelt described the trip in the October 11, 1913, issue of The Outlook. His party came from the west, via Lee's Ferry, Tuba and --

> "-- Kayentay, where John Wethrill, the guide and Indian trader, lives. On August 10, (1913) under Mr. Wetherill's guidance, we started for the natural bridge -- "

Roosevelt also mentioned that "Wetherill has been with every party that has visited the Bridge from the time of its discovery four years ago," and that his party was being guided by the same "Ute" that led the Douglass-Cummings expedition in 1909. The Roosevelt article made interesting reading, not only for its description of his trip to Rainbow Bridge, but for his commentary on the prehistoric ruins and living Indians he saw along the way, and his perceptive comments about the impact the encroaching American culture was having on the region's Indian cultures. Roosevelt's report is further quoted later in this book, in appropriate chapters.

Herbert E. Gregory, the leader of the U.S. Geological Survey party to Rainbow Bridge that Pogue reported about in 1911, authored a technical paper in 1917 describing some of his field research into the "Geology of the Navajo Country." In this report, Gregory described several natural spans within the general Four Corners region, including "Black Rock Bridge, 2 1/2 miles south of Fort Defiance," an arch he called "Defiance Window," Rainbow Bridge and the White Canyon bridges. He then went on to analyze how each probably formed. The report also had photographs of two spans captioned "Owl Bridge, Nasja Canyon, Utah" and "Hope Window, Carson Mesa, Arizona."

A 1922 report on "Natural Bridging in High Plateaus of Utah" by Professor Frederick J. Pack of the University of Utah provided a lengthy description of the geology of southern Utah as it related to natural span formation in general and to the several major spans in particular. It also noted that two large spans near Cedar City had not been found until 1917, "-- within 10 miles of a flourishing settlement." One of these was in what is now Cedar Breaks National Monument. The other was in a tributary of Ashdown Gorge, the canyon that drains the colorful Cedar Breaks amphitheater. The report showed photographs of both spans. The span in Ashdown Gorge was reported as being "-- 100 feet long (with the opening) slightly less." The length of the other arch was given as about 50 feet. Although the title of this report suggested broad regional coverage, it actually described only the three White Canyon bridges, Rainbow Bridge and the two near Cedar City. It made no mention of the many other spans then known to exist in southeastern Utah.

THE TWENTIES

The early part of the 1920s saw considerable active exploration of the more remote regions of the Colorado Plateau, especially in the immense maze of Colorado River tributary canyons in southern Utah and northern Arizona. Perhaps one of the most ardent explorers during that decade was Charles L. Bernheimer, who explored the Navajo Mountain region and various routes to Rainbow Bridge in the years 1920, 1921 and 1922.

Bernheimer's account of his first trip to Rainbow Bridge was published in the November-December 1920 issue of Natural History. Earl H. Morris, a member of the second expedition, described it in the November-December 1922 issue of the same publication, and Bernheimer summarized all three trips into the region in an article that appeared in the February 1923 issue of National Geographic Magazine.

One of Bernheimer's goals was to find an alternate route to Rainbow Bridge, other than the one pioneered by Wetherill and subsequently used by others. As Bernheimer noted in his 1923 article, his first attempt failed:

> "In order to reach Rainbow Natural Bridge in 1920, we climbed Navajo Mountain, but finally had to skirt it by a circuitous trail to the east. In 1921 we organized our expedition to reach the Rainbow Natural Bridge by a route to the west of Navajo Mountain, utilizing Forbidding Canyon. We failed -- "

Morris's account of the 1921 expedition gave more details. Bernheimer drove from Flagstaff, Arizona, via Tuba City to the Wetherill trading post at Kayenta where, he observed:

> "-- Colonel Roosevelt and other men of distinction have sat and discussed with the courteous hosts the craft of the frontiersman and the lore of the aborigines, and departed enriched by the experience."

The party of six, including Wetherill and a Navajo helper, first explored the cliff dwellings of Navajo National Monument, then headed northwest into the maze of canyons to the west of Navajo Mountain, seeking a way to Rainbow Bridge from that direction. The account made several observations about how changing climate had affected the land:

> "A generation ago the bottom of the Segi (Canyon) was for miles a chainlike series of pools, hence the Spanish name, Laguna; but the only remaining traces of them are hardened bands of black muck in the arroyo banks and occasional patches of moribund reeds, the dwindling remnant of the greenery with which they were fringed."

Later, in describing a verdant canyon found on the western slopes of Navajo Mountain, Morris noted that:

> "These conditions were a sample of the past which had survived into the present."

This second expedition also failed to find an alternate route. As Morris told it:

HISTORY OF DISCOVERY

> "We had failed in our attempt to find a second route to the Rainbow Bridge by following West Canyon around the end of Navajo Mountain. Inasmuch as we were going to the bridge, there was no alternative but to strike eastward until we intersected the known trail. (We got within) six or eight miles in an air line --"

They finally reached their goal, but only after "-- we had ridden fully fifty miles, and in so doing had made almost the complete circuit of Navajo Mountain." Bernheimer's third attempt was more successful, as he noted in his 1923 National Geographic article:

> "For many years it has been my ambition to find a passage west of Navajo Mountain to the Rainbow Natural Bridge, thus penetrating one of the most appalling rock jumbles on this continent. This feat was consummated by our expedition of 1922. To John Wetherill, of Kayenta, Arizona, who discovered the Rainbow Natural Bridge in 1909, and who has been my guide for three successive years, is also due the discovery of a feasible route to the Rainbow Natural Bridge, the object of our expedition of 1922 --"

A few of the problems the expedition encountered in reaching the remote span by a new route are described in this book's chapter on access to known spans.

Some of the 1909 Cummings-Douglass "discovery" party had hiked down to the Colorado River from Rainbow Bridge and found indications that others had traveled the river and even camped near the mouth of the canyon that led to the bridge. This led to speculation about earlier visitors. It also led later explorers to use this approach. A U.S. Geological Survey party, which was exploring and mapping the region in 1921, may have been the first to report reaching the span from the river although, in a 1980 report, Stephen C. Jett of the University of Calfornia wrote that "A few river runners, such as Ellsworth Kolb, reached the Bridge from the Colorado as early as 1921 -."

The U.S.G.S. survey was reported in the October 1923 issue of The Geographical Review:

> "In October 1921, K.W. Trimble, of the United States Geological Survey, with his assistants, mapped Bridge and Aztec canyons and the adjacent portions of the Colorado Canyon, in connection with the mapping of the San Juan and Colorado canyons for a power and reservoir project at Lees Ferry, Arizona. Three or four government parties, including H. D. Miser and Sidney Paige, that descended (the) Colorado River in 1921 also visited the bridge by ascending Aztec and Bridge canyons from the river. Another party in 1921 ascended the Colorado from Lees Ferry in a flat-bottomed boat equipped with a motor-driven paddle wheel at the stern; the upstream trip of 68 miles with this boat required five days. Besides the parties just mentioned only a few others have visited the bridge from the Colorado, though several have made the attempt and failed on account of inability to identify the mouth of Aztec Canyon."

The Geographical Review report, which was jointly authored

by Miser, Trimble and Paige, also noted and printed photographs of two spans in addition to Rainbow Bridge. One was near the top of the 700-foot inner gorge of the Colorado, "-- six miles below the mouth of the San Juan River." Its photo caption noted that the other span was "-- in San Juan Canyon 3 miles below Piute Creek -- (it) has an estimated span of 50 feet and stands on an inaccessible rim of the canyon." The photograph was credited to Miser.

Miser's own report of this survey, issued as a U.S.G.S. technical paper in 1924, reported the San Juan span plus another one very near the San Juan-Colorado confluence, in a short canyon connecting the two river gorges. To illustrate how even authorities and historians can go astray when it comes to the "discovery" of natural spans, arch historian Vreeland notes in his Volume 7 that the span near the river confluence was "-- probably first seen by Jack Morehead, Chief Ranger for the National Park Service, in 1971," thus missing the span's actual discovery by fifty years. In writing about the San Juan arch, Vreeland stated that:

> "Although this arch was known to river runners on the San Juan for many decades it wasn't until 1971, when the waters of Lake Powell reached about 3620 feet above sea level, that visitors were aware that this peculiar arch had two more openings."

Miser, however, did not miss this feature either. His report noted that:

> "The bridge when viewed from the bottom of the canyon -- appears as a tiny arch, yet its span and height are each estimated to be 50 feet. The bridge has apparently been formed by the partial collapse of the roof of an alcove or small cave in the massive Navajo sandstone. A distant view of the alcove indicates that two other parts of the roof have fallen in."

The conclusion to be drawn from this is that early explorers, especially those with technical educations, were astute observers who produced remarkably detailed reports under difficult conditions.

An October 1983 article in the Deseret News magazine about the remeasuring of Kolob Arch in Zion National Park commented on early knowledge of this immense span:

> "-- old-timers maintain that sheepmen and cattlemen had long known of the arch's existence, even though in southern Utah it was seldom a topic of conversation. This could well be true, since such people were more given to talking about the condition of grass and water than to any of the area's abundant scenic aspects. Otto Fife, for many years the sheriff of Iron County, recounted that his brother -- visited the arch many times, including one hike to the spot with Dr. Frederick J. Pack, a geologist at the University of Utah, in the early 1920s."

MAIN TEXT CONTINUED ON PAGE 46

HISTORY OF DISCOVERY

CLARA S. BERNHEIMER NATURAL BRIDGE

Text and photographs by
Gary Topping, Curator of Manuscripts
Utah State Historical Society

Among the many achievements of the Charles L. Bernheimer expeditions between 1919 and 1930, Bernheimer himself seems to have been most proud of two, the building of a trail to Rainbow Bridge around the south and west of Navajo Mountain through Redbud Pass, and the discovery of a large natural span in Monument Valley which he named Clara S. Bernheimer Natural Bridge after his wife. The span, which is located in an unnamed box canyon to the south of Hunt's Mesa, is a natural rock opening of impressive scale and symmetry.

Charles L. Bernheimer (1860–1944) was a wealthy Manhattan businessman and leader in the field of commercial arbitration who became interested in the Southwest through the novels of Zane Grey and the geological reports of Herbert E. Gregory. Guided by John Wetherill and Zeke Johnson, and often accompanied by the archaeologist Earl Morris, Bernheimer made a number of significant additions to the geographical and archaeological knowledge of Utah and Arizona, as recorded in elaborate field notes and reported in a book, "Rainbow Bridge" (1924), and several popular articles.

For a fee of two hundred dollars, a Paiute guide called Old Mike led Bernheimer to the Monument Valley span on June 8, 1927. Even though Old Mike had misrepresented it as being larger than Rainbow Bridge, Bernheimer was well pleased with the discovery. In addition to naming the natural bridge after his wife, Bernheimer sometimes referred to it as "The Veiled Lady" because of its difficulty of access, a problem that has scarcely diminished today. Bernheimer admired the nearly perfect symmetry of the bridge and estimated both its height and span to be 150 feet. Careful recent measurements indicate it has a maximum opening width of 135 feet and a total height of 135 feet. Difficult as the span is to find, it is even more difficult to reach and photograph. Most visitors have had to be content, as was Bernheimer himself, with a view from some distance away.

A register placed under the bridge in 1936 indicates that since then only a few dozen people have negotiated the perilous climb to stand beneath it, and fewer still have emulated the intrepid John Wetherill in standing atop the lofty arc of rock. In 1986 the register was removed from its original location to prevent further water damage, and is now protected with other historic Bernheimer documents.

A spring near the span, enlarged and lined with stones by Bernheimer's men to water their huge pack train of more than forty animals, can still be seen, but an olive bottle bearing a record of the party's visit and placed atop the span by John Wetherill has long since vanished. A large bronze plaque, designed by Bernheimer and installed by Wetherill nearby in 1937, is still there, close to an inscription made by the 1927 discovery party.

Clara S. Bernheimer
Natural Bridge

Photo by
Gary Topping

Climbing to Clara S. Bernheimer Natural Bridge
Photos by Gary Topping

HISTORY OF DISCOVERY

Kolob Arch, Zion National Park

In 1923, Neil Judd, who had been a member of the 1909 expedition to Rainbow Bridge, led a National Geographic expedition into a region largely neglected by earlier government surveyors, the remote triangle of barren land wedged between the San Juan and Colorado rivers. The expedition traveled first to Kayenta, where they met John Wetherill who was to serve as guide. They had trouble, however, getting local Indians to accompany them, once the Indians heard where they were going. From Kayenta, the expedition traveled through Monument Valley to the settlement of Mexican Hat on the San Juan River. It then explored westward through the unknown territory until it again reached the San Juan. With some difficulty, the party forded the treacherous river and continued on west to Rainbow Bridge.

The expedition was reported in the March 1924 issue of National Geographic Magazine, but the only new spans mentioned were three large arches in Monument Valley, now called Ear-of-the-Wind, Eye-of-the-Sun and Moccasin Arch. This may have been the first documented report of these spans, but there is little doubt that these and others in Monument Valley had been known to both Indians and white men for a long time.

In November 1923, Mr. F. A. Wadleigh of the Denver and Rio Grande Western Railroad System wrote a letter to Stephen T. Mather, Director of the National Park Service, noting the "discovery" of an area in southeastern Utah, dubbed "Devils Garden" by its discoverer, an area that Wadleigh thought should be set aside as a national monument:

> "This place was accidentally discovered by Alexander Ringhoffer, his two sons and son-in-law on the 24th of December last while they were prospecting the rocky ridge mentioned. They returned the following day, Christmas, with their families and spent several hours in wandering around among these rock formations. They made two subsequent trips into the region, -- once with an engineer from Los Angeles who was doing some

surveying in the vicinity. No one else, other than Beam and myself, is known to have been in the so-called 'garden.'"

Evidently neither Wadleigh nor Ringhoffer was aware of Winslow's 1898 article in Science, with its photographs of Landscape and another arch near the area described in Wadleigh's letter. Although Ringhoffer's "Devils Garden" is now called Klondike Bluffs, this is within sight of the area that contains Landscape Arch, which was later mistakenly named Devils Garden.

In a letter to Steven Mather dated March 13, 1926, a Dr. Frank R. Oastler of New York City reported the results of a survey in southeastern Utah he had performed at Mather's request. The letter described, and included photographs depicting, three areas which Oastler recommended be granted national monument status. Two of these areas were protected within Arches National Monument in 1929. The third area was Arch Canyon, which lies between Blanding and Natural Bridges National Monument.

In 1927, a book was published in which its author, Clyde M. Kluckhohn, described in great detail a trip he had taken to Rainbow Bridge earlier in the decade. The book made such good reading that it was reprinted in 1967 and again in 1980 by Rio Grande Press. Parts of this very readable account are quoted in appropriate chapters of this book.

With few new reports of major spans in the 1920s, and the establishment of Arches National Monument in 1929, it would be easy to assume that by then all significant natural spans in the Colorado Plateau region had been discovered, duly reported, and given special protection. This illusion could have been maintained during most of the next two decades as well. Although trickles of new reports continued to appear in public print, it seemed certain that the glamorous days of spectacular discoveries were over. However, the vast canyon maze of the Colorado Plateau still held some surprises in store.

Keystone Arch, Arch Canyon, with 1938 sketch

HISTORY OF DISCOVERY

THE THIRTIES AND FORTIES

A 1931 U.S. Geological Survey paper authored by Gregory and Moore reported that "Arches -- were noted near the head of Cottonwood Creek, at the 'narrows' on Halls Creek and on tributaries to the Paria." These are all tributaries of the Colorado River gorge within southern Utah and northern Arizona. The report also noted several natural spans within the Escalante Canyon system, and was possibly the first written report of these spans. The Escalante spans are discussed further later in this chapter.

A March 1933 University of Colorado study of "caverns, ice caves, sinkholes, and natural bridges" was largely a rewrite of Cleland's 1909 report that was discussed earlier, but it did recognize the growing proliferation of natural spans in America due to the many discoveries in the Colorado Plateau region:

> "Instead of being very rare, as most people suppose, natural bridges of one sort or another are not at all uncommon. There are many in foreign lands, and bridges of natural origin have been reported in at least 22 of our states, many of them insignificant, but others of much scenic and scientific interest."

This report, authored by Junius Henderson, Professor of Natural History at the university, provided an update to Cleland's list of 1909. For Arizona he only listed a couple of additional spans that had already been reported earlier. For Colorado, his home state, he noted that "There are a few insignificant arches or windows in the state --" and mentioned "-- Ela Natural Bridge in Colorado National Monument --," but said nothing about the many beautiful arches in and near Rattlesnake Canyon, or massive Juanita Arch in a tributary of the Dolores River gorge near Gateway.

Henderson also listed the natural bridge in New Mexico that had been reported in 1907, but had quite a bit more to say about Utah:

> "This state is preeminently the land of natural bridges, arches and windows. They are not only very numerous in southern Utah, but among them are some of the finest and most stupendous examples yet discovered anywhere."

He went on to list Utah's three national monuments devoted to protecting some of those spans, plus others in Bryce Canyon, which was also a monument by then, and Cedar Breaks, which was proclaimed a monument later that year. He also mentioned an arch "-- close to the main highway about 25 miles southeast of Moab," now called Wilson Arch, and quoted several earlier reports of spans already mentioned in this chapter.

In a 1938 U. S. Geological Survey Professional Paper, geologist Herbert Gregory described a number of arches and natural bridges in the San Juan River area, including those in Arch Canyon, the area that Oastler had recommended be protected. The report contained both a sketch and a photograph of a span Gregory called "Cathedral Arch" and a sketch of a "Keystone Arch," both in Arch Canyon, but did not depict the third span he mentioned seeing there, which he called "Angel Arch."

The 1941 WPA guidebook to Utah mentioned earlier also noted the discovery in 1940 of another natural bridge in the Escalante canyon system, and the discovery of "Broad Arch" in Capitol Reef National Monument that same year by three men named Hibbert, Chesnut and Fagagreen. This beautiful span is now called Hickman Natural Bridge. The book also mentioned Cassidy Arch, "-- the second largest natural bridge in the Monument --," but didn't name it, plus another smaller span in Capitol Gorge.

A geologic report on "Nature's Bridges" in the September 1943 issue of The Scientific Monthly, authored by Raymond E. Janssen, discussed how natural bridges form and gave as one example the "Great Arch of Zion," even though that feature is an arch-topped alcove, not a free-standing arch.

An article, "Motoring into Escalante Land," appeared in the September 1949 issue of National Geographic Magazine. Its author, Jack Breed, heralded the discovery of "-- an arch - a new arch, uncharted and unnamed -- actually it is a double arch" not far south of "Kodachrome Flat," the present location of a Utah state park development. The expedition members named the span "Grosvenor Arch," after Dr. Gilbert Grosvenor who was then president of the National Geographic Society. The name has lasted, probably because of the publicity given the span under that name by the article, but there is considerable doubt that the span was "new." As with many other such Colorado Plateau "discoveries," local stockmen had probably known about the span for decades, but had simply not reported it.

The expedition continued driving south into "-- one of the least known wilderness areas in the United States --" seeking, among other things, to locate and reach some of the big natural spans rumored to exist within the labyrinthine Escalante canyon system. They failed in this goal, as described later in this chapter.

During the 1940s, the Colorado Plateau region was invaded by hordes of prospectors seeking first vanadium, then uranium. The vanadium was needed for making high-strength steel alloys for use in the manufacture of weapons and other products in support of America's war effort. The uranium was needed first for the manufacturing of nuclear warheads, then also for the developing nuclear power industry. Uranium prospecting and mining boom-and-bust cycles continue to this day.

All of this search for the rare minerals that are to be found within the Plateau led to the penetration of its wild backcountry by every means and manner. Four wheel drive trails and dirt airstrips proliferated in regions previously touched only by the feet of men, horses and domestic livestock. This inevitably led to the discovery of many more natural spans -- but not necessarily to their being reported. Men seeking the eerie glow and sudden wealth of uranium oxide simply saw no value to odd holes-in-rock. Even the detailed reports generated by university geologists in support of uranium prospecting failed to mention these geologic oddities in areas where they are plentiful. From the perspective of several decades it is interesting to note that while the true value of uranium is diminishing, and its use for any purpose being questioned, the aesthetic value of the region's natural spans steadily increases, as more and more people come to know about and appreciate their unique natural beauty.

HISTORY OF DISCOVERY

Aerial view of Canyonlands National Park

AERIAL DISCOVERY

In the 1940s and 1950s, Colorado Plateau country also came under the surveillance of "bush pilots." These daring fellows could better be called "rock pilots" because they learned to land their light planes almost anywhere in this vast land of rock, and cruise down its narrow canyons and between its soaring towers with their wingtips just feet from the vertical rock. At first, most such daring flight was in support of mineral search activities, but later some pilots offered scenic cruises over their favorite regions.

All this overflight led, quite naturally, to the spotting of features not easily seen from the tortuous vehicle trails below, including many natural spans. One such pilot, Jim Hurst, would spot an arch from the air, then try to get to it on the ground for a closer look. Others no doubt did the same, but many chose to ignore such findings. For example, a pair of twin natural bridges was "discovered" near Moab, Utah, sometime in the 1960s by a local backcountry tour guide, and reported in a 1970 magazine article. When another report appeared in the Moab newspaper soon thereafter, many locals claimed to know about the spectacular bridges, but only one could prove his claim -- a bush pilot who flew directly over the spans to show that he knew their location. Arch historian Vreeland claims that the twin bridges were "-- probably first seen by Clive Stewart in 1933 or 1934," but if so, the discovery was not documented, and the spans' location did not become known to very many local people.

A second National Geographic expedition seeking to reach the reported spans in the Escalante canyon system was triggered by a letter from H.W. Bement, Director of the Utah State Aeronautics Commission, who had spotted several of the spans from the air. As noted in the September 1955 issue of National Geographic Magazine, in an article by W. Robert Moore:

> "Flying alone earlier in the year, Bement had spotted one of the arches. Later he had gone back to explore the area more carefully and had located the others."

Bement was quoted as saying that:

> "No one in Salt Lake City quite believed me when I told about the arches. Only Burnett Hendryx and a few persons in Escalante seemed to know anything about them. That's why I wrote to the National Geographic Society."

The expedition that resulted from this letter is described later in this chapter.

Pilot Royce Knight, flying in the Glen Canyon area in the early 1960s, reported a huge span in the upper reaches of Wetherill Canyon. At the time, this span was very difficult to get to by land, since Lake Powell did not exist. A photograph of this "Keyhole Arch" appeared in the July 1967 issue of National Geographic Magazine.

In 1969, Moab resident and "arch collector" Dick Wilson wrote an article for the local paper in which he mentioned the big spans reported by Powell and Dellenbaugh in Desolation Canyon:

> "After reading these statements about natural arches being found in Desolation Canyon by the Powell expeditions, this writer attempted to find persons in Moab who had seen them, but to no avail. Certainly an arch as large as described by Dellenbaugh would not remain unnoticed for 100 years but that seemed to be the case. A flight with pilot Dick Smith last week, however, confirmed the existence of at least three of them -- although they appeared to be from the airplane, smaller than what Dellenbaugh described."

Wilson's aerial explorations also played a part in the rediscovery of Hall Bridge. As mentioned earlier, this big span was first reported in 1911, then apparently faded into obscurity because of its well-hidden location. In 1971, arch historian Robert Vreeland was flying over the maze-like region to the southwest of Moab, Utah, looking for spans new to him when he spotted one in upper Pritchett Canyon. The pilot identified the span as Pritchett Arch, but this proved to be wrong when Vreeland and several others got within sight of the span on the ground later that same day. A second ground trip to the span was made a few weeks later. A photograph taken then was published in the Moab newspaper, together with a request for any available local knowledge of the span.

Dick Wilson replied to this request in the form of a letter to the editor, noting that the span had been depicted in a 1911 edition of the paper and saying:

> "I first learned of the discussed arch in 1968 while

> doing historical research in the files of the Times-Independent. In that same year I -- flew over it and recognized it by its picture. I don't want to deflate the enthusiasm of those who come from afar or to minimize authentic discoveries. On the other hand we must give full credit to the old timers who explored every nook and cranny here."

This claim to having flown over the span three years before Vreeland's "rediscovery" was verified by Wilson's original arch log, which notes "Hall Bridge" in a March 1968 entry, followed by mention of the 1911 photograph and the comment "I saw it from the air." Arch historian Vreeland chose to ignore Wilson's claim of prior rediscovery by noting in his own log of Moab area spans that "The arch was rediscovered on June 9, 1971, from the air in the morning and on the ground that afternoon," demonstrating once again that arch historians are both human and fallible.

Halls Bridge, Pritchett Canyon

HISTORY OF DISCOVERY

THE ESCALANTE SPANS

As mentioned earlier, the first report of the several large natural spans in the Escalante River's complex canyon system was a U.S. Geological Survey paper dated 1931 that summarized a survey of the Kaiparowits region made by H.E. Gregory and R.C. Moore. Although it is certain that a few of the settlers in the vicinity of the town of Escalante knew about some of the spans much earlier, their knowledge did not get into public print. There were also indications that Gregory Natural Bridge might have been visited as early as 1921 by another Geological Survey party, but this visit apparently did not get into any formal report. A still earlier visit is mentioned in the September 1955 National Geographic Magazine article that documented its second expedition:

> "Though cattleman Allen Cameron -- had camped under the arch in 1918, its discovery was not announced until 1940, when Norman D. Nevills, of Mexican Hat, named it and placed a register at one end of the arch."

Apparently, neither Bement, who spotted most of the Escalante spans from the air in the mid-1950s, nor the leaders of the two National Geographic expeditions that tried to get to the spans, were aware of the two earlier reports. Because of these reports, however, the second and successful National Geographic expedition was one of rediscovery, rather than discovery. The "first report" credit must go to Gregory and Moore, whose report said, in addition to the earlier quote:

> "The largest (arch) observed is in Escalante Canyon above the mouth of Sand Creek. This Escalante arch, which stands about 30 feet out from the wall of Navajo sandstone, has a span of about 130 feet, a height of 70 feet above the floor of an ancient alcove, and a thickness of 13 feet. The -- top is 12 feet wide. On Fortymile Creek a beautiful natural bridge has resulted from the undercutting of a meander spur and has reproduced the history of the well-known Rainbow Bridge in the Navajo country. As estimated -- the Fortymile Bridge has a span of 200 feet and a height from stream bed -- of 75 feet. Above the mouth of Willow Creek on the Escalante River another bridge is in process of formation. A narrow wall that separates the compressed limits of a stream curve 2 miles long is already perforated by a window."

Nevills' discovery was summarized in the 1941 Utah guide book:

> "In midsummer, 1940, Nevills announced discovery of Gregory Natural Bridge, a huge span in a side canyon of the Escalante River, which enters the Colorado River from the west, at a point about 15 miles north of Rainbow Bridge. Nevills reported that the bridge has a height of 305 feet, only three feet less than that of Rainbow Bridge, and a span of 293 feet, nearly 20 feet longer than that of Rainbow. Nevills was at the time on a canyon expedition from Green River, Wyoming, to Boulder Dam. The newly discovered bridge was named for Herbert E. Gregory, Government geologist and authority

on the Navaho country."

The span Nevills named after Gregory was one of those reported by Gregory and Moore in 1931, which are now called Escalante Natural Bridge, Gregory Natural Bridge and Stevens Arch, in the order they were described by Gregory and Moore. Apparently, their report was in error, in that Gregory Natural Bridge spans Fiftymile Creek, not Fortymile, and the third span is in the Escalante Canyon above the mouth of Coyote Creek, not Willow Creek. There is no big river meander near the Willow Creek confluence, but there is where Coyote Creek meets the Escalante. The quite different span heights of 75 and 305 feet given for Gregory Natural Bridge are probably because Gregory gave the opening height, and Nevills measured the total span height.

Grosvenor Arch

HISTORY OF DISCOVERY

The stories of the two National Geographic expeditions make interesting reading. The first expedition made a good attempt in 1949, even though it failed. Although it did accomplish its other goals, it wanted the glory of "discovering" the legendary natural bridges of the Escalante:

> "We had heard tall tales about many natural bridges and arches that supposedly were hidden in the numerous tributary canyons of the Escalante. One of our major objectives was to check the veracity of these reports, especially those that claimed the existence of bridges greater than the famous Rainbow Bridge."

After "discovering," measuring and photographing Grosvenor Arch, the expedition worked its way south along The Cockscomb, without realizing that this long, colorful monoclinal ridge hides a number of arches just as lovely as Grosvenor Arch. They then headed for the historic "Crossing of the Fathers," where Dominguez and Escalante crossed the Colorado River gorge in 1776. The National Geographic expedition hoped to reach the crossing with their land vehicles, something never done before.

> "Expeditions that reached this point before us had all been by boat or pack train. Previously, no one had dared to risk a vehicle in this wild country."

After considerable difficulty, a few of the group reached the Crossing of the Fathers with one of their vehicles -- a four wheel drive Jeep -- where they saw the steps cut into the rock by the ill-fated 1776 party in order to get their horses down the slickrock slopes to the river.

After this accomplishment, the National Geographic expedition returned to civilization and took established roads to the town of Escalante. From there, it again headed toward the Colorado River, approximately following the route of the Hole-in-the-Rock party of Mormon pioneers. Part of the group descended that horrendous cliff-hanging trail to the river, where they were met by riverman Art Greene with the air boat he used for hauling sightseers and others up the Colorado from Lees Ferry in Arizona. It had taken Greene three days to fight his boat up the 86 miles of river.

From the Hole-in-the-Rock crossing, the party boated still farther upriver to the mouth of the Escalante River, there to meet disappointment:

> "The mouth of the Escalante proved too shallow to admit a boat and its bottom too treacherous with quicksand to permit us to enter on foot. Our only alternative was to backtrack up the Hole and enter Escalante Canyon overland through one of the side gulches."

This was done, but without the success they were seeking. One group descended Davis Gulch, "-- only to be driven out by a wild bull that was hiding in a narrow glade." Others went down Soda Gulch, now called Fiftymile Canyon, but soon had to leave their horses and continue on foot because of trouble with quicksand and stretches of water in very narrow places. They managed to pass the quicksand and swam the deeper stretches of water with their cameras on their heads. They eventually reached Gregory Natural Bridge, but were disappointed that it was not as large as expected, noting "-- it is no Rainbow."

A third party descended Willow Gulch and explored the "center section of Escalante Canyon," but reported that only "-- small windows, arches, and one bridge were found -- nothing to bear out the claims we had heard." Unfortunately, most of the larger Escalante spans are many miles upstream from Willow Gulch and the stretch of main canyon they explored. The expedition covered its disappointment by noting that:

> "In talking with ranchers who range cattle in this desolate section, we soon realized that no one could claim discovery of any of these features. The Mormons have known about them for generations. True, few people have seen them, and few ever will --"

This last assertion proved to be somewhat pessimistic. Hundreds, perhaps thousands of people have seen the big Escalante spans in recent years, some from the ground and some from the air. Still more will visit them in the future, now that Lake Powell's waters have penetrated far up the Escalante canyon system. Some of the spans can now be seen from boats. Others are only short hikes beyond where the water ends when the reservoir is at full flood. Gregory Natural Bridge has long since vanished beneath the rising water, a casualty of "progress."

Phipps Arch, Escalante canyons, Rudi Lambrechtse photo

As mentioned already, the National Geographic's second search for the Escalante spans was triggered by a letter from a pilot who had seen some of them from the air. This second expedition was reported in the magazine's September 1955 issue. This effort started with an aerial survey, flown by pilot Bement who had rekindled interest in the search. From the air, Robert Moore, who headed the expedition and authored its report, soon saw why the spans had managed to elude explorers for so long:

"It is understandable that few persons other than the cattlemen who have pastured herds along the Escalante know anything about the river or its natural arches. The canyons are accessible only on horseback or foot -- the (Escalante River) twists and turns snakelike through the eroded Navajo sandstone -- and threads a harsh land, a wilderness of eroded slickrock and patches of sandy desert where little more than scrub brush and prickly pear thrive. It is one of the emptiest places in all Utah. During our flight over the river and its side canyons -- Coyote, Willow, Soda, and Davis Gulches -- we saw eight natural arches and bridges hewn by frost, water, and wind in the flaming Navajo sandstone."

Thus encouraged by its aerial reconnaissance, the group set out on horseback, guided by Burnett Hendryx of Panguitch. They descended into the Escalante canyon system via Coyote Gulch, which has three sizable spans in it. As each span was reached, it was measured and named. The first, a large opening through a canyon meander wall, they named Jacob Hamblin Arch, after an early Mormon scout, although they thought it unlikely that Hamblin had ever seen the arch. A short distance downstream from this arch, they noted the indications of a fallen span:

"At one of the kinks in the Coyote, a short distance downstream from Jacob Hamblin Arch, we saw what remained of the pedestal and the broken arm of a once mighty arch that had collapsed. As time is measured in geologic formations, it had fallen only recently, for the broken edges were little eroded. Numerous large rocks that once formed its span lay strewn over the talus slopes and on the canyon floor."

Farther down Coyote Gulch, they came to another big hole through a meander wall. The stream flowed through this one, so they named it Coyote Natural Bridge. The third span was high on the canyon wall. They named this one Cliff Arch. The names of all three Coyote Gulch spans have been widely accepted.

In the main Escalante gorge, the party soon encountered a third span, this one high in a thin-walled cliff formed by a meander just downstream from another major tributary, Stevens Canyon. They named this one Stevens Canyon Arch. When measured, with some difficulty, the span's opening was 225 feet wide, even though it seemed much smaller when viewed from the canyon floor some 800 feet below. This proved to be the largest of all Escalante spans, even though Gregory had referred to it only as "a window."

The party continued down the Escalante to Soda Gulch (Fiftymile) which they ascended to Gregory Natural Bridge. Moore noted that he was the fiftieth person to sign the register placed under the span by Nevills in 1940, more than a decade earlier.

It was not practical to get to the other spans seen from the air via the main gorge:

"Within both Davis Gulch and Willow Gulch, whose outlets we had passed, lie other arches. But to reach them we had to climb out of Escalante Canyon, cross the desert, and descend again into the gulches."

This they did, but not without difficulty. In Davis Gulch,

they recalled the misadventure that had turned back the earlier Geographic party:

> "Near here, a few years ago, a wild bull turned back an exploring party. Later he was killed; his hide still hangs on the corral fence."

They entered Davis Gulch between two of its arches, so started downstream seeking one of them. The canyon soon became impassable to horses, so they had to continue on foot. Before long, they came to another large hole through a meander cliff, this one high above the canyon floor. They named this span La Gorce Arch, after Dr. John La Gorce, who was then editor of National Geographic Magazine.

The next day they reached the other Davis Gulch span, a large rubble-filled hole in a flying buttress of sandstone. They named this span Bement Arch, after Harlan W. Bement, the director of the Utah State Aeronautics Commission who was the pilot who had reported seeing the Escalante spans.

The eighth big span of those seen from the air was in Willow Gulch, which the expedition descended without difficulty. It, too, was in a sandstone fin, but acquired an impersonal name:

> "Like Bement Arch, Broken Bow Arch projects conspicuously from the left wall of the canyon. The arch gains its name from a broken Indian bow found near its base; oddly enough its shape reminds one of a bow bent nearly double."

The expedition ended with Broken Bow Arch. It did not visit the spans in the upper Escalante gorge and its many tributaries, such as Phipps Arch, Bowington Arch, Maverick Bridge, Escalante Natural Bridge, Cedar Wash Arch, Lamanite Bridge and a number of smaller spans that have not been given names. The first four named are within easy hiking distance of where the present highway, Utah 12, crosses the Escalante river. There are also a number of spans in the higher country into which the Escalante canyon system is cut, but none of any great size.

Aerial view of Broken Bow Arch, Escalante canyons

THE FIFTIES AND SIXTIES

After the exploration of the Escalante canyon system, and the thorough documentation of the natural spans found there, the balance of the 1950s and most of the 1960s became another period during which few new significant spans were formally reported, but public awareness of existing spans slowly grew. Many arches and bridges simply came to be known to more and more people, without individuals trying to claim discovery rights or making public reports. It is quite probable that of all the spans known at this writing, the majority have not been formally discovered and reported, including some that are quite large and beautiful. Several examples are given in this book in which arches that were reported as "new discoveries" turned out to be known to local residents, although in all fairness it should be noted that some such claims to prior knowledge proved to be untrue.

One interesting new discovery made in the 1950s was, surprisingly, in Arches National Monument. Workmen constructing a gas pipeline through the monument in 1956 reported finding a large span in the head of the Clover Canyon drainage, one not previously known to the National Park Service. Park rangers named the big span Clover Canyon Bridge. Although Arches is not a very large park as western parks are measured, and it is dedicated to the natural spans in its strange, geologically distorted terrain, previously unreported spans continue to be found, sometimes by park rangers, sometimes by park visitors or dedicated "arch collectors."

A 1959 vacation guide to Colorado gave a tantalizing hint of the many beautiful arches to be found in part of that state's segment of the Colorado Plateau. It showed a photograph of a "Glade Park Arch," but gave no exact location and made no mention of the many others nearby. Arch historian Vreeland claims that the span was discovered in 1880 by a W.P. Ela, for whom it was later named Ela Natural Bridge, but the 1959 guide book did not use that name. There are two possible reasons why the other spans in the general Glade Park area were not also widely known and reported.

One is that the terrain in which they are located is remote, deeply cut by Colorado River tributary canyons, and almost inaccessible. The other possible explanation is probably apocryphal. A local story asserts that the Glade Park, Colorado, vicinity was occupied during early years by persons having something less than the usual respect for the law, and that they and their descendents took a dim view of outsiders entering "their" country for any reason whatsoever. If true, this would have tended to discourage exploration of the area, and discovery of the arches there by people who care about such things.

For whatever reason, the Glade Park-Rattlesnake Canyon arches were not reported very early, despite their numbers and beauty, and despite the fact that they were just a few miles from the largest city in western Colorado.

In the late 1960s, Moab resident Dick Wilson began making a survey of natural spans in southeastern Utah, starting with the Moab vicinity and spreading out from there. He obtained leads to spans by consulting local residents, especially those with knowledge of the area's backcountry. As he heard of each span, he recorded it and the source of his information. He then made every effort to confirm the span's existence and location by visiting it, or at least spotting it from the air. In several cases, he found still other spans while seeking to verify those reported to him. Wilson then reported in the local newspaper any spans that seemed to be "new," thus documenting new finds, and making public what had been private knowledge. Among those reported was his rediscovery of Hall Bridge, as described earlier.

Dick Wilson's log of southeastern Utah spans was never completed or published, but is available to serious scholars. It served as a valuable research reference for this book, and is discussed further in appropriate chapters.

East Rim Arch, Rattlesnake Canyon

THE SEVENTIES

Huge twin natural bridges not far from Moab, Utah, were first documented in 1970, although they were doubtless known to exist long before then. They were named Gemini Bridges, after the mythical Gemini Twins.

A short article about the Rattlesnake Canyon arches near Grand Junction, Colorado, appeared in the February 27, 1972 edition of Colorado West, the Sunday magazine of the Grand Junction Daily Sentinel. The article noted that:

> "It is quite probable that these great arches are less well known, and have fewer visitors, today than 50 years ago when folks were more inclined to take long hikes and horseback rides on Sunday to see such out-of-the-way places."

Between 1970 and 1974, the author of this book photographed and wrote articles about a number of natural spans within southeastern Utah, without particular attention to who might have "discovered" the spans. The articles appeared in the Moab newspaper. Other articles by this author about the region's natural spans appeared throughout the 1970s in various magazines.

Later in the decade, author David Sumner wrote about visiting the Rattlesnake Canyon arches two times. His write-up of both visits appeared in the Fall 1976 issue of Mariah Magazine, in support of granting the area wilderness status. The article also noted a "pioneering reconnaissance in 1973" by the University of Colorado's Wilderness Study Group. Sumner noted that "the photographs accompanying this article are the first to appear in a national publication."

In the winter of 1974, the author of this book explored a canyon to the northwest of Moab, Utah, that proved to contain four sizable arches quite near each other. A report describing the spans and showing photographs of them was published in August 1975. The canyon subsequently became known as "Four Arches Canyon," and was further described in a guidebook to the area published in 1978.

Beginning in 1976, arch historian Robert Vreeland began to print segments of his index of known spans. Most of these booklets were out of date before they were even printed, as more spans continued to be found and reported throughout the Colorado Plateau. The subject of cataloging the region's natural spans is covered in a later chapter.

In 1977, two major spans were reported in southeastern Utah. One was first spotted from the air in a very remote location by a Bureau of Land Management survey team. The span, subsequently named Aqueduct Arch, was first reached by land vehicle by this author. Later in the same year, another large span was reported for the first time by this author. The arch was so well hidden that it was named Covert Arch. Photographs of both Aqueduct Arch and Covert Arch were published in 1978, in backcountry guidebooks and maps produced by this author.

The May 7, 1977 edition of the Salt Lake Tribune reported the finding of a previously unreported arch in the Kaiparowits Plateau, "some eight miles" from the town of Escalante. The span was discovered by a Mr. Edson B. Alvey:

> "Mr. Alvey, explorer, educator and guide, has explored the mountains, plains and cliffs throughout the Kaiparowits-Fremont area over the past 60 years, and is

> know by local authorities to have discovered numerous arches, bridges, fossils, minerals, Indian campgrounds and cliff dwellings, and other works of nature and man."

Mr. Alvey, who named this small span "Horizon Arch," was quite probably the first to see it, because its location is very remote. A photograph of the arch appeared with the newspaper article, and another was published in January 1979 by the Bureau of Land Management, on a hiking-guide map of the Escalante River canyon system.

The May 27, 1979 edition of the Salt Lake Tribune reported the discovery of another "new arch" near Kodachrome State Park by park superintendent Tom Shakespear, "just two miles from his home." One oldtimer in that area later indicated that the span was known to local stockmen. Other local residents disclaimed earlier knowledge, once again bringing up the question -- is an arch truly "discovered" if it is not adequately reported? A local contest to name this arch resulted in its being dubbed Shakespear Arch. A marked hiking trail now leads to this hidden span.

Shakespear Arch, Kodachrome Basin

THE EIGHTIES

The discovery and reporting of Colorado Plateau natural spans continued in the 1980s. A number of enthusiastic amateur "arch collectors," some living within the region, others from elsewhere, continued their searches of the region's vast expanses of arid wilderness, probing into labyrinthine canyon systems never before seen except, perhaps, by prehistoric Indians and a very few stockmen and prospectors, men with little interest in reporting natural wonders.

Most discoveries made by such collectors were not formally documented, but some were. Quite a few spans found during the 1980s are appearing in print for the first time in this book. Examples of these are Skywalk Arch, Cliffhanger Bridge, Newberry

Window, Pocket Arch, Wow Arch, Undine Bridges, Harts Draw Bridge, Dellenbaugh Tunnel, Armijo Arch, Boulder Bridge, Tusher Tunnel, Cavern Arch, Silhouette Window, Dimmocks Jughandle and Inchworm Arch. There are others. Some of these are so remote and well hidden that they may never have been seen by earlier explorers, or even by local cattlemen or prospectors.

Three other spans, Jacks Arch, Periscope Arch and Lin-Lynne Arch, were first described by Moab resident Jack Bickers in a 1983 trail guide to the Arths Pasture Rim area to the northwest of Moab, Utah, with photographs of the larger two. A number of smaller but interesting spans were reported in Pritchett Canyon, Bull Canyon, and Sevenmile Canyon, all within a few miles of Moab, Utah, with their first published photographs appearing in this book.

In late 1983, a well hidden arch, not previously known to the Park Service, was discovered in Bryce Canyon National Park and reported in the January 17th edition of the Salt Lake Tribune. The span was within 300 yards of the paved highway through the park, Utah 12, but had escaped notice during decades of public use of the road. The span was found by a local resident, quite by accident:

> "Mr. Brinkerhoff had been watching for a five-point deer he had regularly seen in the area when he spotted the natural arch and it did not take him long to start checking with old timers in the area about its existence. Everyone knew about the buck, but no one he talked to could remember having seen the natural bridge, and some, though hesitant to say so, even may have doubted his sanity."

Mr. Brinkerhoff's sanity need not have been doubted. In a region as immense, undeveloped, geologically distorted and eroded as the Colorado Plateau, there are certainly going to be many things still hidden, still undiscovered by anyone, including arches and natural bridges. This is discussed further in the chapter on cataloging known spans.

Inchworm Arch, Lost Spring Canyon

HISTORY OF DISCOVERY

THE RAINBOW BRIDGE CONTROVERSY

Earlier in this chapter, the official story of the discovery of Rainbow Bridge was summarized. The full story of the discovery of this spectacular natural bridge in southern Utah is far more complex. Many details of the official version are controversial, and there is reason to doubt that the 1909 Douglass-Cummings expedition was actually the first to reach the span.

The controversy over the Douglass-Cummings expedition to Rainbow Bridge can be summed up as several questions:

1. Which local Indians knew about Rainbow Bridge before the Wetherills learned about it?

2. Who first learned about the span from the local Indians, John Wetherill or his wife?

3. Of the two Indian guides, the one with the Douglass party and the one Cummings hired to guide his party to the bridge, which one actually provided the critical guidance?

4. Who actually saw and reached the span first?

5. Who really deserves credit for the official discovery of Rainbow Bridge?

There are other questions too, but these were the most debated. Before going further, however, two points should be made. First, earlier in this century the English spelling of southwestern Amerind speech sounds had not been standardized. Thus, the spelling of tribal and individual Indian names tended to vary with the writer. Second, it was not uncommon among certain southwestern tribes for individuals to go by different names at different times of their lives. Thus, Douglass' guide, Mike's Boy, also known as Paiute Jim, later became Jim Mike.

Taking first the question about prior Indian knowledge of the span, here are what several writers have to say.

Professor Byron Cummings, who had already done considerable archaeological work in northeastern Arizona before the 1909 expedition, noted in his 1910 National Geographic article that:

> "Few even of the Indians are acquainted with (the region north of Navajo Mountain). It is celebrated as the place where Hoskinimi, one of the most revered leaders among the Navajo, successfully evaded Kit Carson, in 1866, when the latter taught the Navajo such a terrible lesson; but not even Hoskinimi seems to have penetrated as far as Nonnezoshi (Rainbow Bridge)."

Writer Robert Frothingham, in his 1932 book "Trails Through the Golden West," noted that the Indian who had told the Wetherills about the span had confided that it had been seen "-- by only a few Indians, who had stumbled on it inadvertently -- and got away as quickly as possible because it was a very holy spot."

Author Clyde Kluckhohn, who made a trip to Rainbow Bridge in the early 1930s, noted in his 1933 book about that adventure:

> "-- it is a misnomer to speak of discovery and discoverers

> when Navajo songs of the Bridge have existed for many centuries, and when generations of cliff-dwelling men had built their propitiatory smokes beneath it long before the Navajoes came."

A number of early reports agree that there was some kind of stone altar beneath or near the stone bridge in 1909 and for some years later. The altar could have been built by the prehistoric Anasazi cliff-dwellers who occupied the area for centuries, by the Paiutes who sparsely occupied the region to the northwest of Navajo Mountain in early historic times, or by the Navajos. The matter will never be settled, because the altar has long since disappeared, but there is little doubt that native Indians knew about Rainbow Bridge long before white men arrived in the region.

The second question -- who first learned about the bridge, John Wetherill or his wife -- will also remain in doubt forever, although a number of writers have expressed opinions on the subject.

Theodore Roosevelt, in his account of his 1912 trip to Rainbow Bridge, noted that it was Mrs. Wetherill "-- who first got from the Indians full knowledge of the Bridge."

Earl Morris, in his 1922 article about the second Bernheimer expedition, gives the credit to John Wetherill:

> "-- it was he who learned of the natural marvel from the Piutes."

Zane Grey, in his 1922 book "Tales of Lonely Trails," claimed that Mrs. Wetherill obtained the information about the bridge from the Indians:

> "Wetherill owes the credit to his wife, who, through her influence with the Indians finally after years succeeded in getting the secret of the great bridge."

Kluckhohn, in his 1927 book "To the Foot of the Rainbow," noted that:

> "As early as 1900 rumors came to Mrs. Wetherill of a wonderful multicolored arch of solid rock, and finally she met Indians who had penetrated the rock fastness of the region and actually seen the marvel."

Frothingham's 1932 book alleges that:

> "-- one day early in the spring of 1907 -- a tribesman bearing the nickname 'Sharkie' -- took John (Wetherill) for a long walk away from the village into the desert, where he unfolded to him a deep secret. It was all about a wonderful bridge of red rock, shaped like a rainbow, hidden in a deep canyon --"

All writers acknowledge that Mrs. Wetherill was very close to the local Indians and was even a highly respected honorary tribal member. She spoke their language fluently. It is thus likely, although not certain, that she was the one the Indians first told about the great stone rainbow.

The third question -- which Indian actually did the guiding that enabled the Douglass-Cummings party to reach the span -- is disputed by the two expedition leaders as well as by others.

One of Cummings' accounts noted that early in the trip

Douglass' guide "--Mike's Boy plainly did not know the country and was becoming frightened," but that Cummings' guide "-- Noscha Begay rode into camp about 10 o'clock that night so all was well."

Douglass' official account simply noted that, having met the Cummings party, "From this point we proceeded as one party all under the guidance of Jim (Mike's Boy). Later, we jointly employed an additional guide, a Paiute named Nasja Begay, supposed to have a better acquaintance with the local trails." Douglass made no mention of his guide having difficulty, nor did he mention his own efforts to scuttle the Cummings party before the joint expedition to find Rainbow Bridge. Cummings mentioned this, however, in his book, "Indians I Have Known." He noted that before the two parties joined:

> "-- W. B. Douglass was in Bluff telephoning and telegraphing to Washington to get our permit from the Department of the Interior cancelled and have us driven out of the country. Mr. Wetherill had also received an order from Mr. Douglass that he confiscate all relics we had taken from the ruins. This seemed strange, since we had regular permits from the Washington office to explore and excavate in the region where we were working."

Neil Judd, a young archaeologist who was with the Cummings party, noted in his 1919 report to the National Park Service that Mike's Boy "-- confessed during the first few days on the trail that he had never been within a day's ride of Navajo Mountain, but that he had hoped to find the bridge through directions from other Indians."

John and Louisa Wetherill Byron Cummings
Photos courtesy Utah State Historical Society

Frothingham, in his 1932 book, went into more detail:

"They had been on the trail only a few hours, however, when Wetherill's suspicions were confirmed; it was quite apparent to him that Jim the flagman (Mike's Boy) didn't know where he was going. With considerable difficulty John convinced Douglass of the fact, and the outfit halted to make camp. John then retraced his steps to Oljeto and sought out Nasjah Begay once more, with a last heroic appeal to that reluctant Paiute who, it appeared, was more gratified by seeing Jim go ahead and make a fool of himself than he would have been by making a bit of money.

"It developed that Nasjah had been in to the bridge as recently as June of that year, but had kept this to himself. His fears as to the possible resentment of his Navajo brethren had dwindled to reasonable apprehensions about horses and supplies, and about the chances of so large an expedition's getting in to the Bridge, and if it did -- of getting back again. These doubts having been cleared, Nasjah Begay returned to the desert camp with John Wetherill and the expedition resumed the trail --

"No one in that party but Wetherill was qualified to discover Paiute Jim's unfitness for the job for which Mr. Douglass had hired him, and but for his altruistic interference the expedition would have fizzled out (as Douglass') previous one had. That, however, did not deter Mr. Douglass, in his official report -- from ascribing to Jim the credit for finding the trail to the Bridge which, as a matter of fact, belonged to Nasjah Begay who, as has already been stated, was Wetherill's man. And it is owing entirely to Wetherill that Nasjah eventually received the credit which was his due."

In her version of the story, co-authored in 1934 with Frances Gillmor, Louisa Wetherill mentioned that on the discovery trip Nasjah Begay knew the terrain well enough to predict accurately to John Wetherill that "Around this bend you'll see the bridge."

In a 1973 article in the Denver Post's "Empire Magazine," Zeke Scher arranged for an interview with Paiute Jim/Mike's Boy, who was then "at least age 101." His daughter claimed he was 109 years old, and he then went by the name of Jim Mike. Jim Mike claimed that "Rainbow Bridge belongs to me." Translated, the interview went like this:

"When did you first see Rainbow Bridge?"

"We lived in Paiute Canyon. I was a boy and on this day we were looking for grass, for feed for the horses, in the canyons beyond the north side of Navajo Mountain. It was me, my father and Nasja who lived nearby. They were setting up camp and I went out looking for feed. I went into this canyon and saw this bent rock with a hole in it. I never saw anything like that. I ran back scared and told my father. He and Nasja left without going to see it."

"Nasja's son, known as Nasja Begay, wasn't there?"

"No. He was a small boy. I was older than him. He never saw it then."

"What year was that?"

"I don't know. It was many years before I told Mr. Douglass."

"Did you give it a name?"

"I called it the bent rock with a hole in it."

"Was it the most beautiful feature you've seen?"

"It's just a pile of rocks -- "

"You're aware there's a plaque honoring Nasja Begay as the first guide to Rainbow Bridge?"

"I was there first. I told his father about it. I found it. In 1909 he didn't meet us until we were almost there."

Of course, seeing the bridge first and as a boy does not preclude getting lost trying to return to it decades later. It is unfortunate that Nasja Begay was not still around to tell his side of the story. Cummings noted in his 1952 book "Indians I Have Known" that "-- he and most of his family fell victims to the ravages of the flu in the winter of 1918."

Rainbow Bridge discovery party, with Wetherill, Cummings and Douglass in the front row.

In his account of the 1909 expedition published in the August 1967 issue of Arizona Highways, Neil Judd claimed that it was neither Indian guide but Wetherill who deserved the lion's share of the credit for reaching Rainbow Bridge:

"Actually -- it was John Wetherill, with descriptive information supplied by old Nashja, who successfully led the joint expedition across the most difficult and dangerous portion of its journey. Thus, in my opinion, as Nashja and his son together deserve the honor of having previously discovered Rainbow Bridge and marked its location, it was really John Wetherill who brought success to the dual adventure. Without his leadership, I am sure, neither Cummings nor Douglass would have reached Rainbow Bridge in August 1909."

It is doubtful that anyone will ever know the truth about which Indian guide actually saw Rainbow Bridge first, or knew best how to reach it. Judd and Cummings, both members of the 1909 expedition, were firm in their positions that Douglass' guide was incompetent. Douglass' story evaded the question. His guide, as a centenarian, claimed first knowledge of the bridge, implying that Nasja Begay did not know the way to it, but did not settle the dispute about who was the best guide. The 1941 WPA guidebook to Utah noted that Wetherill cast his vote in the matter in 1927, when he "-- erected a bronze plaque at the bridge to Nashja-begay, 'who first guided the white man to Nonnezoshi.'"

Much has been written about the fourth question -- who in the Douglass-Cummings party actually saw and reached Rainbow Bridge first? Was it Douglass or Cummings? All accounts are hearsay, of course, except for the stories of those who were actually there. Five members of the 1909 expedition left their versions of the first sighting of Rainbow Bridge in print -- Byron Cummings, William Douglass, Neil Judd, Malcolm Cummings and John Wetherill.

Byron Cummings, in his book "Indians I Have Known," said that on the fourth day out Douglass tried to get him to go back to the rear of the long pack train, in order to look after Cummings' son, Malcolm:

> "It was soon evident why Mr. Douglass was anxious that I fall behind with the packs. Before long we were in Nonnozoshiboko (Big Arch canyon). Noscha Begay suggested that I ride ahead with him, saying that when we rounded a certain bend ahead, we could see the big arch, Nonnezoschie. As we reached the point and I saw the object of our long trek just a little way ahead, I turned and shouted, 'Eureka, here she is!'
>
> "We had to climb a short slope to reach a bench along which we must travel to get to the arch. Our ponies were tired, so Mr. Wetherill and I jumped off to lead our horses up the slope; but Mr. Douglass put spurs to his horse and made it lope up the hill. Mr. Wetherill was too quick for him, however; springing on his pony, he reached the goal and passed under the arch first. Thus, I was the first white man to see the Rainbow Bridge and John Wetherill was the first man to pass under this great arch. Its real discoverers were the two Pahute Indians, Noscha and Noscha Begay."

William Douglass' story, as printed in a 1955 Bureau of Land Management publication, is quite different:

> "On the morning of the last day's travel, when we were told by the Indian guides that the bridge would be reached by noon, the excitement became intense. A spirit of rivalry developed between Professor Cummings and myself as to who should first reach the bridge. The first 3 places of the

single file line were of necessity conceded to the 3 guides. For 3 hours we rode an uncertain race, taking risks of horsemanship neither would ordinarily think of doing, the lead varying as one or the other secured the advantage over the tortuous and difficult trail.

"Fortune favored me at the close, the Professor being some hundred feet in the rear when I reached the bridge. To Jim is due the credit of giving to the world the first knowledge of this remarkable monument: to the General Land Office belongs the credit for the discovery to civilization, and for its preservation as a National Monument."

Neil Judd's 1919 account written for the National Park Service contends that "-- Professor Cummings has never engaged in this petty squabble over the discovery of the Rainbow Bridge." Judd's narrative describing the day of discovery generally substantiates Cummings' version, but adds more details:

"Throughout the last day's travel, Mr. Douglass exhibited the uncontrolled enthusiasm of the amateur explorer and he was so utterly disregardful of possible danger to other members of the party as to arouse the disgust of all. He seemed to lead the party and crowded the other riders from the narrow trail as he repeatedly forced his tired horse to the front. Mr. Douglass was the only member of the expedition engaged in this wild race; time and again necessity compelled him to turn back from ledges he had unwisely followed but he always charged blindly forward again.

"The jagged rocks had torn the shoes from the horses' feet and cut their hoofs to the quick; they were extremely weary and the men also showed signs of fatigue as the party made its way over the hot sand and rocky ledges down into Tsei-nonnezoshi boko (canyon of the rainbow-turned-to-stone). The party progressed slowly and Mr. Douglass was some distance in the lead when Dean Cummings suddenly drew rein, called to those near him and pointed out the stone bridge, half hidden among the purple shadows far to the left. Douglass, who is totally deaf, had continued alone several hundred yards before he noticed that the others were already admiring the graceful curve of the bridge. Yet to those not familiar with this pioneer journey to the Rainbow Bridge, and his small part in its discovery, Douglass has always contended that undivided honors belong to him."

In his account in the August 1967 issue of Arizona Highways, Judd provided still more details about the 1909 expedition.

In 1958, Byron Cummings' son Malcolm wrote his version of the discovery day, noting that "The details of those explorations with my father are as vivid and clear as the day of their happening." His brief account:

"Shortly before noon on the fourth day, August 14th, Noscha-Begay led Dr. Cummings and John Wetherill down into Nonnezoshie Boko and to the great arch which spans the canyon. Eventually the other members of the party arrived, and at long last my pony and I straggled in and I had my first glimpse of Nonnezoshie."

John Wetherill related his version of the story in a letter quoted in Frothingham's 1932 book, "Trails Through the Golden West":

> "On the morning of August 14th we broke our last camp in the Bald Rocks Canyon and lined out for the Bridge. Dean Cummings, Douglass, and myself pushed ahead with Nasjah Begay in the lead. As we came to the bend of the canyon about three quarters of a mile above the Bridge, Nasjah Begay said we would see it shortly. Douglass was next to the guíde, his eye fixed on a branch canyon that led off to the right. Cummings was riding behind Douglass and I was behind him. I noted that he was keeping watch to the left, from which direction the arch is first seen. He saw it first and rode up beside Douglass and pointed it out to him.
>
> "Douglass didn't seem to appreciate Cummings' interest. I told the dean that I would put him in ahead of Douglass if he wished, but he said he didn't want to appear rude. I thought it was just about up to someone to 'appear rude' so I took it upon myself to ride ahead of Douglass. I was the first white man to pass under the bridge and was followed by Messrs. Douglass and Cummings in the order named. But Dean Cummings was the first white man to set eyes on it. There was plenty of glory to go around, and I have always been content with my little part. The real credit belongs to the Paiute Nasjah Begay, without whose knowledge of the trail the Bridge would probably not have been discovered for some years to come."

The 1934 account of the 1909 expedition co-authored by Wetherill's wife, Louisa, and Frances Gillmor also supported Cummings' story.

No two versions of the story agree on all critical details, but the consensus seems to have been that of the white men on the expedition, the elder Cummings saw the bridge first, and Wetherill was the first under it. From the perspective of three quarters of a century, it all hardly seems to matter. As John Wetherill concluded, there was "plenty of glory to go around."

Answering the fifth question -- who actually deserves the credit for discovering Rainbow Bridge? -- is also complicated and uncertain, and depends upon how the word "discovery" is defined. As Dr. Stephen C. Jett wrote in his introduction to a reprint of Kluckhohn's 1927 book:

> "The Cummings-Wetherill-Douglass rivalry over who would be the first White man to see or pass beneath Rainbow Bridge was not only petty, it was also futile. Unquestionably, the stone span had been seen previously by prospectors, perhaps by Mormon pioneers, and possibly even by fur-trapper Mountain Men. At least one prospector, W. F. Williams, claimed to have seen the Bridge in 1884 and '85; he also stated that he had found the names of several previous visitors scratched into the rock."

Other scholars who have delved into the controversy have also concluded that the official version of the discovery leaves much unsaid. Historian Gregory Crampton, in his 1965 book "Standing Up Country," had this to say:

> "By 1909 the gold rush in Glen Canyon had spent itself and

nearly all the miners had gone. Unquestionably, Rainbow Bridge had been visited before 1909 by prospectors who were working their way up from the Colorado, and by those who were exploring the slopes of Navajo Mountain. The reported Merrick-Mitchell mine had brought men to the Navajo Mountain-Monument Valley region in 1882 or earlier. W. F. Williams some time afterward said that he had seen the bridge in 1884 or 1885, in company with his brother, Benjamin, and their father, J. P. Williams, who organized the Williams Mining District on the San Juan River in 1895. There are other reports of the discovery of the bridge before 1909 but they count for little."

In Judd's 1967 Arizona Highways article he noted that when members of the 1909 expedition hiked from Rainbow Bridge to the Colorado River they saw plentiful evidence of earlier river usage:

> "After note-taking and photography, a walk down canyon to see the big river was suggested. The 'Silvery Colorado' proved to be a muddy flood; wreckage of a huge wooden dredge lay half submerged on the opposite shore. In a shallow cave at the north tip of Bridge Canyon, a small (prehistoric) dwelling had been demolished by placer miners to provide camping space. Broken tools and utensils were present --"

There are also reports of failed attempts to reach Rainbow Bridge before the successful expedition of 1909. In Kluckhohn's 1927 book, he claims that Mrs. Wetherill, upon learning about the span from her Indian friends:

> "-- induced her husband to set forth in quest of this solid rainbow. Two expeditions reached Navajo Mountain, only to be turned back by mountains of stone which seemed insurmountable barriers to further progress."

Jett says that these two Wetherill expeditions "-- are not otherwise known to history," but in his 1932 book, Frothingham relates the details of one secret Wetherill attempt to reach the bridge in the fall of 1907:

> "About two-thirds of the way into the Bridge canyon Nasjah Begay lost the trail, and then spent a week in a vain endeavor to recover it."

After running short of food and water for themselves and their horses:

> "There being absolutely no forage and water in that section, their situation was indeed desperate. Nasjah Begay was terror-stricken, convinced that they would never get home alive. As for Wetherill, he was dismayed by the tragic outcome of their venture, but he realized that there was nothing to do but turn back. They had been gone for nearly a month and by the time they reached home both the men and the horses were completely played out."

According to Frothingham's account, Wetherill wanted to try again in 1908, but "-- Nasjah Begay had had his fill." He did, however, refer him to another Paiute who claimed to have made the trip to the great bridge and who he thought might undertake the job. "It appears that this Paiute, 'Mike's Boy' by name, familiarly known as Jim, was working at the time as a flagman for William B. Douglass --." Mike's Boy agreed to do this the following December, but in the meantime told Douglass about the great natural bridge. This apparently aborted Wetherill's second attempt:

> "John realized that his secret was out, however, when word leaked through a few weeks later that Jim (Mike's Boy) had changed his mind and was going to guide Mr. Douglass in to the bridge."

Frothingham contended that Mike's Boy did not show up to guide Douglass that fall, but Douglass told a slightly different story:

> "An attempt was made to reach the bridge in December 1908, but owing to the snow, it was found impossible to do so."

Jett tells of another 1908 attempt that also did not succeed:

> "Louisa (Wetherill) did relate, however, that early in the spring of 1908 Wetherill's partner, Clyde Colville, undertook a search for the Bridge in the company of the Navajo, Hosteen Luka, but got no further than Cha Canyon, north of Navajo Mountain."

It is thus pretty clear that there were at least two secretive attempts to reach Rainbow Bridge before the 1909 expedition. There is also convincing evidence that the bridge had already been seen by several men in the late 1800s. In the October 1955 issue of the Sierra Club Bulletin, W. F. Heald announced that:

> "On my desk is a statement which contradicts the claim that John Wetherill and his party in 1909 were the first white men to see southern Utah's famed Rainbow Bridge. The statement was dictated by William Franklyn Williams at Winslow, Arizona, May 22, 1929, and typed by his daughter, Bernetta ('Billie') Yost. In it he describes two trips he made to Rainbow Bridge in 1884 and 1885. But he makes no claim to being its discoverer, for he states that he found evidence that other white men had been there before him and he gives the names of some of them. I first heard of William's early visit to Rainbow Bridge in the summer of 1954 when I read a manuscript entitled 'Bread Upon the Sands' by his daughter Billie."

HISTORY OF DISCOVERY

Billie Yost's book was published in 1958. It contained an interview with her father, W. F. Williams. That fascinating interview is quoted here in part:

> "Mr. Williams, your sons tell me you visited Rainbow Bridge in 1884. I'd like to hear about it."
>
> "Sure thing. Glad to oblige. But it wasn't called Rainbow Bridge then. The Navajos had named it Tse-Na'-Ahi, meaning Rock Bridge. It wasn't until 1909 that Tse-Na'-Ahi became Rainbow Bridge. That year a government survey party, headed by William S. Douglass of the General Land Office, and Professor Byron Cummings, then of the University of Utah, gave it its official name."
>
> "Why didn't you report it, Mr. Williams, if you saw it so many years before anyone else?"
>
> "Lord, Sir, it never would have entered any of our heads! First place, we weren't the discoverers of it. The Indians knew all about it, and plenty of trappers and Mormons had been there before us. There were names scratched all over the arches. Second place, while it was a wonderful sight, we didn't think it anything so awful special. There is some soul-stirring scenery in that country."
>
> "You weren't hunting for the bridge, then?"
>
> "No, indeed! We were looking for lost mines. My father was always looking for lost mines. And someone had told him about the rich Merrick and Mitchell silver diggings in Navajo Mountain. So he took my brother Ben and me with him. We were prospecting. It was just by accident we saw the bridge.
>
> "It was November(1884) when we went there. Hosteen Hoskinny, a Navajo Indian, acted as guide. He said he could show us where the (lost) mine was located. We reached the Navajo Mountain country about the twelfth of November. When we reached Paiute Canyon, we stayed a couple of days with some friendly Indians at their farm. We wanted to rest our horses and get some fresh meat to eat.
>
> "After leaving Paiute Canyon, we hit out for a place named Owl Bridge, a natural arch on the south side of Navajo Mountain. We prospected around there for a while, then went up on top of the mountain, camping where the soldiers had the telegraph station in 1864."

Williams then went on to describe in detail the route they took down from the mountain to the Crossing of the Fathers, then to Cummings Mesa and on to the highlands "south of West Canyon" where they found prehistoric ruins, two natural bridges and "a great number of tracks in sandstone." Farther on, they found some more ruins and near them three more natural bridges. Williams did not have much to say in the interview about actually reaching Rainbow Bridge, nor about his 1885 visit, but did indicate still another trip into the vicinity in 1900:

> "Was (Rainbow Bridge) hard to get to?"

"No. We had absolutely no difficulty in reaching it -- we were on an Indian trail- one they used all the time. However, when I went back there in 1900, to take the Navajo census, I was advised to go by way of Lee's Ferry and work across from that direction. Hosteen Indische told me no one was using our old trail, through Broken Leg Canyon, because the east one was so much better."

Heald got in touch with Billie Yost in 1954. His interview with Billie added some interesting details to her father's account:

"They found names cut on the free end of the arch, also some more names written in charcoal on the cliff walls farther down Bridge Canyon. The names he definitely recalled were those of Billy Ross, a man named Montgomery, Jim Black, George Emerson, Ed Randolph, and another man named Wydel. He thought the names had been carved there at different times, because some of them were weather-worn, others more fresh."

"How do you account for their disappearance?"

"I don't know for sure, but a friend of mine went to Rainbow Bridge in the 1920s and told me the names had been purposely obliterated. He said you could see where they had been taken off. It would be easy, you know. It's soft sandstone."

"Did your father ever say anything to (John Wetherill about the bridge)?"

"I don't think so. They weren't at all close. You see, my father and grandfather were not the first white men to see the bridge. Father always said it was some lone prospector or wanderer who really saw it first. When he learned of the Wetherill-Cummings-Douglass claim he only laughed and said, 'Why I saw that bridge twenty-five years ago with Dad and Ben.' Neither he nor Uncle Ben seemed to care if others claimed to be the first white men to visit Rainbow Bridge. They knew it wasn't so, and let it go at that."

In Judd's 1967 account he commented on Billie Yost's mention of the obliteration of earlier names:

"Some have even claimed that the names of earlier visitors were erased by the 1909 party. The only names erased prior to June 1, 1925, as Custodian John Wetherill wrote the National Park Service on that date, were those of visitors in 1913 and 1916; he, himself, had erased them to avoid the necessity of reporting the fact and thus making the perpetrators liable to fine or imprisonment."

Billie Yost went on to say that both of her parents had been killed in an auto accident in 1940, and that her uncle Ben had died in 1943 or 1944. Heald concluded his report by saying:

"So now we have only William Franklyn Williams' statement -- seven legal-sized pages of double-spaced typescript. Little of it deals with visits to Rainbow Bridge, but it contains a detailed account of wanderings in search of gold or silver through the maze of canyons north, northwest, and west of Navajo Mountain. The complicated topography is accurately described and there isn't a doubt that Williams was thoroughly familiar with this rough desert country -- his statement has the ring of authenticity and truth."

Perhaps some day a young doctoral candidate searching for a subject for a history or geography thesis will check out this story in detail, and settle once and for all the controversy over who really discovered Rainbow Bridge. Meantime, the official 1909 discovery date will remain in question.

William Franklyn Williams, standing, with his father and three sisters, 1890.

An interesting footnote to the controversy over the 1909 expedition to Rainbow Bridge appeared in the 1934 book co-authored by Louisa Wetherill:

> "When Herbert E. Gregory, of the United States Geological Survey, came to the Navajo country in 1909 on a geological reconnaissance, John Wetherill told him of the undiscovered bridge. But Gregory, busy with his survey, did not seek it."

Had Gregory gone along with the Cummings-Douglass party later that year, the question of who should be credited with the discovery of Rainbow Bridge might have been even more complicated.

Karl Luckert in his 1977 book, "Navajo Mountain and Rainbow Bridge Religion," took the long view about the "discovery" of Rainbow Bridge:

> "The discovery of Rainbow Bridge is an event to which, I am convinced, many people can lay claim. The ancient Desha people may have seen it as early as 7,000 to 8,000 years ago. Basketmaker peoples may have found it between A. D. 1 and 600 -- Anasazi farmers should have found it -- between the eleventh and thirteenth centuries. Some Paiute scouts may have chanced (on the bridge) during the eighteenth century or earlier. Navajos came into the specific area --

in 1863 when they fled from United States troops. Their discovery of Rainbow Bridge could have happened at any time after that date. All my Navajo informants -- gave credit for the (Navajo) discovery of Rock-arch to the Navajo, Blind Salt Clansman."

Luckert also confirmed the part the Williams family played:

"(The Navajo leader) Hashkeniinii had no difficulty finding (Rainbow Bridge) when he led the Williams brothers and their father to the place in 1884. At the time, however, the Williamses thought nothing about their discovery. They had mineral treasures on their minds --"

Neil Judd, who was with the 1909 expedition, put the whole matter of "discovery" into proper perspective quite early, when he wrote in 1927:

"Who actually discovered Nonnezoshe? Nobody knows. Some Indian in the pre-Columbian past, when man roamed widely over this continent of ours but left no written record to prove it. Some Indian was the real discoverer. But we whites have a conceit all our own which frequently tempts us to ignore the achievements of those of a different hue. A thing clothed in the traditions of a thousand years remains unknown until we, ourselves, have seen and recorded."

Rainbow Bridge

ANATOMY OF A MODERN DISCOVERY

We had been invited on an all-day 200-mile cruise on a new 48-passenger jet boat down the Green River from the town of the same name to that river's confluence with the Colorado, then up that river to Moab.

It was early June and the day was hot, but the boat's speedy passage created a continuous cooling breeze. We studied the changing shapes of the passing cliffs that soared ever higher as we entered Labyrinth Canyon, so named in 1869 by the river's first explorer, Major John Wesley Powell. In addition to enjoying the passing scenery for its own sake, I was watching, camera ready, for sizable natural openings in the sandstone cliffs.

I especially wanted to find out if a strange double natural bridge I had photographed a week earlier from high on the canyon rim could be seen from the river. We had told the boat captain of our interest and as I carefully followed our progress on a river-runners map, so did he, knowing where I wanted to look -- about halfway between two particular tributary canyons. As the boat sped by the first canyon and rounded the bend where the twin bridges were, halfway up a sloping, terraced 400-foot cliff of salmon-hued sandstone, we all spotted a large arch that we could not have seen from the high canyon rim. A bonus!

Then, I saw our twin bridges -- or at least one of them. The round, horizontal hole that defined the larger span cast a bright spot of sunlight on the rubble slope below, but the smaller opening we knew existed behind the big bridge was not visible from the river. The twin nature of the bridges could only be detected from high above, from a short stretch of the opposite canyon rim, and then only with powerful binoculars and good eyes -- or through the 250mm telephoto lens I had used to photograph the "Undine Bridges." We had named the double span after one of the early paddlewheeler steamers that had plied the Green and Colorado rivers.

As we sped on down the river, my wife and a friend continued to watch the opposite cliff, their eyes and minds now tuned to natural rock openings. Within less than a mile, they gasped, almost as one -- "there's another one!" The boat's captain, also interested and always ready to oblige, quickly throttled back and brought the boat around so everyone could study the spot where the sighting had been made. Even then, we could not be certain that we were seeing an opening in the rear of the huge alcove in the lofty cliff. As the boat eased closer, so we could perhaps get a glimpse of sky through an opening, the alcove rim cut off any further view.

Finally, we gave up and headed on down the river, to see many more fascinating spans, especially within the long stretch of rivergorge in Canyonlands National Park. We reached Moab late that evening, after a glorious day, but by then I had already made plans for another trip -- to visit that remote stretch of river gorge by land, to see if that elusive patch of light in that reclusive sandstone alcove was, indeed, cast by the hole behind a natural bridge. If it was, it was a big one -- large enough, as I told my wife, to run a train across!

Three days later I led a small safari of off-road vehicles into the remote and seldom penetrated country beyond the rivergorge rim where our elusive bridge might be. At the rim, we all scattered, searching for the span in small groups. An hour later, tired from clambering along the broken rimrock and hot from the blazing midday sun, we all gathered for lunch under an overhanging ledge of rock. All except Tom, a friend who was along to help with the search. He was too excited to stop for anything. As the rest of us ate, Tom disappeared from view, heading out to clamber along still another stretch of rim.

Sometime later he returned, took a long drink of cool water, then sat down beside my wife and quietly started eating his lunch. Shortly, she heard Tom mumble something that sounded like "you really could run a train across it." Not quite believing her ears, she asked him what he had said. Again, he mumbled something around a bite of sandwich. Excited and a bit impatient, my wife said, in a firm voice -- "Tom, did you find the bridge? Why didn't you tell us when you first got back?" Tom replied in his soft, southern accent -- "It was scairy!"

After a bit more insistent quizzing, the story came out. He had, indeed, found the bridge. But while scrambling around the lofty rim near the span looking for picture angles, he suddenly realized no one knew where he was, and that he was on very hazardous terrain. Although normally not frightened by heights, he was still a bit shaken on his return, hence his uncharacteristic reticence.

A little while later, the entire party stood gazing in wonder at a spectacular natural bridge in white Kayenta sandstone, one never before reported and even unknown to the boat captain who had been plying the Green River for almost three decades. The span was huge, but not truly visible from the river, even though we had been able to see the reflected sunlight through its opening from a very short stretch of river at a certain time of day.

Was "Cliffhanger Bridge," as the huge span was soon named, really an original discovery? There was no way to be absolutely certain, but in all probability it was. Similar sightings could have been made from boats, but without closer investigation, there would be no certainty that a span existed. Perhaps uranium prospectors who combed the region in the 1940s and '50s had seen the bridge, but if so they didn't report it.

Thus, by all objective rules of "discovery," Cliffhanger Bridge, which graces the end of a hidden alcove in a remote rim of Labyrinth Canyon, was discovered in June of 1986.

Cliffhanger Bridge, Labyrinth Canyon

Pine Tree Arch, Arches National Park

DEFINITIONS

THE NEED FOR DEFINITIONS

Several words need to be defined in any serious discussion about the natural spans within the Colorado Plateau. The term "discovery" has been subjected to limited analysis in the previous chapter, and is discussed further in the chapter on cataloging known spans.

Other words important to the subject of this book are "arch," "bridge," "window" and "tunnel," all terms that are commonly applied either to natural holes in rock or to the expanse of rock surrounding that hole. Openings that might fall within the general terms "arch" and " bridge" can be further divided into a number of types, each with its own special definition.

With most word definitions, the meanings assigned are arbitrary, standardized for the purpose of mutually understandable communications. After they are established, such meanings are carefully summarized and placed in dictionaries, encyclopedias and other books for formal teaching and for handy reference.

Unfortunately, the special words used in the cataloging, study and appreciation of natural rock openings have not to date been standardized. Most people who use the words casually do not have firm, consistent, technically-based definitions for the words, including all too many technical and non-technical writers. A few serious students of natural spans have attempted to produce a practical set of definitions, but with few exceptions their results were either narrowly applicable, or were based upon limited analysis and limited knowledge of the immense variety of natural spans that exist within the Colorado Plateau. Some of these attempts at standardizing definitions are discussed in more detail later in this chapter.

What about existing specialized dictionaries and encyclopedias, such as those devoted to the geological sciences? Surely they have covered the subject of natural rock openings. Unfortunately, they have not. Geologic scientists do not take this subject seriously.

To give a couple of examples, the "Illustrated Encyclopedia of the Earth Sciences - The Planet We Live On" defines "natural bridge," but notes that "they are of little geologic importance, however, because they are short-lived, incidental features doomed by the processes that created them." Of course in geologic terms, even a span that lasted for twenty thousand years would be "short-lived." The "Encyclopedia of Geomorphology" also defines the term "natural bridge" but goes on to say that they are "topographic features whose human interest greatly exceeds their geologic importance." It then spoils the definition given by depicting some examples, one of them Landscape Arch, in Arches National Park, which is certainly not a natural bridge as the encyclopedia defines the term.

It is doubtful if geologists will ever take much of an interest in the contentious business of defining and classifying natural spans. To them, holes in rock, no matter how shapely, are merely minor erosional phenomena. They are the inevitable results of varying combinations of rock composition and erosion, plus Murphy's Law -- anything that can happen will, over and over, given enough time. With the present major erosional cycle in the Colorado Plateau now well into its eleventh megayear, there has been ample time for almost every conceivable type and shape of hole-in-rock to form, many times over, and be destroyed by the same forces that shaped it.

It could safely be concluded then, that natural spans of all sorts are of little interest to most of the scientific community. They are, however, valued by others for the infinitely complex ways they form, and for their beauty and novelty of shape, so standardizing the terms related to those shapes will largely be the responsibility of those who appreciate natural spans for these characteristics.

There are three aspects to the problem of establishing useful, objective and consistent definitions for words like "arch" and "bridge" -- inappropriateness, semantics, and cultural inertia. These words are inappropriate when applied to most Colorado Plateau natural rock openings because of semantic confusion, but are commonly applied because of cultural inertia, the tendency to use existing words in new situations.

Although it is too late to change now, it can be argued that because of their original meanings, the words "arch" and "bridge" are inappropriate for use with natural rock openings. In the Colorado Plateau there are many stone "arches" that are not even vaguely arch-shaped, and there are numerous "natural bridges" that could never function as bridges, at least for humans. There are also arches that are bridge-shaped, and natural bridges that are arch-shaped. Unfortunately, the terms were first applied to natural rock openings long before anything was known about the incredible diversity of form exibited by the natural spans of the Colorado Plateau. They were then misapplied as this region was explored and its natural spans discovered.

Semantics plays a part in that both "arch" and "bridge" already have definitions for other uses, and these definitions lead to confusion when the words are carelessly applied to natural rock openings. "Arch" is an architectural term of long standing. It also has meanings in the science of geology that have nothing to do with holes in rock. "Bridge" commonly means a human construction across a water course of some sort. Many early western settlers and writers also applied the term "arch" to any arch-topped alcove in a cliff , regardless of whether the alcove had an opening in the rear. Many rural Utahns still use the word that way. This can easily lead an amateur arch-collector to waste a lot of time looking for splendid "arches" mentioned by local old-timers.

Cultural inertia plays a part in the confusion in that public usage of established words is not easily changed. Before the exploration of the Colorado Plateau, most American natural spans were of one type -- natural bridges that spanned flowing streams or rivers. They resembled human constructed bridges and could sometimes be used for the same purpose. This narrow but predictable use of the word "bridge" even led early writers to record the heights of such natural spans in human-bridge terms, as in "the height to the roadway" and "the road bed is 28 feet wide." Although there are a few "natural bridges" in the Colorado Plateau over which it is actually possible to drive with four-wheeled vehicles, these are the exceptions, not the rule. For a short list of such natural bridges within the Plateau, see the chapter on access to known spans.

Before the discovery of the first Colorado Plateau spans, there was little application of the word "arch" to natural spans, because there were few if any such spans in America that were not clearly "bridges." The few exceptions were "windows," or holes in cliff buttresses from which chunks of rock had fallen. None of these was arch-shaped, so the widespread use of the term "arch" had to wait until explorers began to penetrate the heartland of the Four Corners States late in the Nineteenth Century.

Musselman Arch, which is not arched, as Rainbow Bridge is not clearly a bridge as the term is generally used.

Thus, because of cultural inertia -- public habit -- the term "natural bridge" was unthinkingly applied to the first major spans reported in the Colorado Plateau, the magnificent spans of White Canyon, Utah. This inappropriate use was reinforced by the coincidental fact that these first spans were, indeed, natural bridges. The confusion began soon thereafter, however, with the "discovery" of "Rainbow Natural Bridge." Byron Cummings, who was co-leader of the official 1909 discovery party, wrote in a National Geographic article published the following February:

> "-- this remarkable freak in the earth's crust is hardly a bridge in the true sense of the term, but is more properly an enormous flying buttress that has been chiseled out by the ages -- it is a graceful arch --"

In a University of Utah archaeological bulletin published November 1910, Cummings said much the same thing:

> "In appearance (Rainbow Bridge) is not so much a real bridge as the structures in White Canyon, because the top of the span is not level. It is a graceful arch of magnificent proportions --"

Earl Morris, the chronicler of the 1921 Bernheimer Expedition that was seeking an alternate route to Rainbow Bridge, also noted this inappropriate use of the word "bridge" in his 1922 Natural History report:

> "The name (Rainbow Bridge) is confusing, for the strange product of erosion to which it is applied is in no sense a bridge, but an enormous arch --"

DEFINITIONS

In his 1922 geologic report, "Natural Bridging in High Plateaus of Utah," Frederick Pack came to the same conclusion:

> "So far as is known (Augusta/Sipapu) is the largest natural bridge in the world. The Nonnezoshie (Rainbow Bridge) is an arch."

These early objections to the misapplication of the word "bridge" did not, however, have much effect upon public usage -- cultural inertia. The situation only became more confusing as time went on and the terms "arch" and "bridge" were repeatedly misapplied, used interchangeably, and assigned almost at random to various Colorado Plateau spans. The confusion generated during the first part of this century is still with us today. Following is a sampler of how writers have compounded public confusion over the use of the words "arch" and " bridge."

THE PROBLEM COMPOUNDED

Major John Wesley Powell first noticed that the term "bridge" was not appropriate to some of the spans in the new country he was exploring. In his report describing the 1869 expedition down the Green and Colorado rivers, he noted seeing some odd natural openings:

> "In several places these lateral canyons are separated from one another only by narrow walls, often hundreds of feet high, so narrow in places that where softer rocks are found below they have crumbled away and left holes in the wall, forming passages from one canyon to another. These we often call natural bridges; but they were never intended to span streams. They would better, perhaps, be called side doors between canyon chambers."

Later in the same report Powell used the word "arch" in the more traditional manner by describing Glen Canyon as "-- a curious ensemble of wonderful features -- carved walls, royal arches, glens, alcove gulches, mounds and monuments." He also used the term "royal arch" several other places, but in each case meant arch-topped alcoves without openings.

An article in the July 17, 1924 issue of the Moab Times-Independent announcing the establishment of "The Windows" as a National Monument, showed six photographs of the nearby attraction. Only one of them was a natural span -- Turret Arch -- and that was labeled a "natural bridge."

In a 1932 book, "Canyon Country," which described some of the erosional features within the Colorado Plateau, author Julius Stone straddled the fence about Rainbow Bridge by using the terms "arch" and "bridge" interchangeably:

> "About six miles from the river stands Rainbow Bridge, probably the most remarkable arch of rock in the world. Its symmetry and graceful proportions, as well as its size and beauty of color, give to this arch a commanding position among the natural bridges of the world."

In a description of Arches National Moument in southeastern Utah, an anonymous writer in the 1941 WPA guidebook to the state recognized the problem with terminology and tried to clarify the matter, but also reinforced another public misconception:

> "Although the terms 'arch' and 'bridge' are, in keeping with western informality, used somewhat loosely by Grand County people, there is a difference between them. A bridge usually occurs in a canyon as the effect of stream erosion, and it spans something. An arch occurs in reefs, often where there is no stream course, and is in the main the product of wind weathering."

Arches are not, of course, the "product of wind weathering." This commonly believed fallacy is discussed further in the chapter on formation modes. In a 1943 article in "The Scientific Monthly," geologist Raymond Janssen related natural bridges to manmade bridges:

> "Whether fashioned of stone by Nature or designed of architectural steel by man -- a bridge always leads somewhere --"

He also described how the concept of bridges and bridging is built into human thinking and American colloquial language, thus illustrating the semantic basis for the confusion over definitions for natural spans. A number of other early students of the subject also took the position that a "natural bridge" must be flat on top like a human bridge.

In an article, "Arches and Bridges of Stone" in the July 1964 issue of "Natural History," author Willard Luce produced rather simplistic definitions for "arch" and "bridge":

> "The most obvious difference between a natural bridge and a natural arch is that a bridge spans a watercourse, although the watercourse may be dry most of the year."

Since Owachomo Natural Bridge, in Natural Bridges National Monument, no longer bridges the stream that formed it, is it still a natural bridge?

DEFINITIONS

Geologist Don Baars in his 1972 book, "Red Rock Country," gave the same definition:

> "-- bridges actually span stream channels, in contrast to 'arches' which form without the aid of stream erosion and stand alone."

These were starts, but the authors did not give enough consideration to two facts: there are very few spans in the Colorado Plateau that do not have at least some water flowing under them at one time or another, and there are few spans that flowing water in some amount has not helped form. How much flow does it take to make a span a "bridge," and how much water makes a "stream" in semi-arid desert country? Without further clarification, their definitions would leave very few "arches."

In a July 1971 article in the Moab Times-Independent entitled "Arch, Tunnel, Bridge -- or Hole in the Rock?" the author admonished public servants in general for their laxity in naming natural spans and placing signs by them that confused the public. He described the problem this way:

> "A person, either visitor, resident or expert, stands staring upward at a piece of sky surrounded by rock and asks himself, 'is the rock surrounding that chunk of sky an arch or a bridge? Or maybe the thing is a window? Or perhaps just a simple hole in rock? A tunnel, maybe?'"

The author then went on to say that if there is a little sign below the span then all is well because, after all, "-- little signs are always made by world authorities on the subject at hand, aren't they?" He then proceeded to demonstrate by example how often the little signs are made with little or no thought given the matter, with many of them dead wrong. He went on to analyze the problem of definitions in some depth, then reached three conclusions that are still fair statements of the situation:

> "1. By whatever set of definitions you care to establish, a great majority of the hole-in-rock phenomena now labeled on maps, in brochures, and on little signs, are mislabeled.
>
> "2. It would be impossible to set up a consistent set of definitions that would fit all the signs, maps and brochures now thrust upon the canyonlands visitor.
>
> "3. The whole subject is a scientific mess beyond compare."

In 1974, geographer Dale Stevens published a paper in the Great Plains-Rocky Mountain Geographical Journal entitled "A Classification of Natural Sandstone Rock Openings in Southeastern Utah." He summarized the problem with definitions, then quoted those of the American Geological Institute:

> "There has been very little agreement on nomenclature for natural rock openings, but traditionally the natural bridge is 'a natural stone arch that spans a valley of erosion,' while a natural arch is 'a similar structure which does not span an erosion valley.' The term 'window' is also frequently used to describe small arch-like natural openings."

Stevens then went on to outline a system of definitions based on "morphology -- aperture shape will be the major object of examination." His system is described and analyzed later in this chapter, but he took a questionable position by noting that "size of aperture or dimension of rock span is irrelevant, all sizes and shapes are considered." This may be true from the viewpoint of science, but the general public certainly is concerned with size. Geologist Stan Lohman, in his 1975 U.S. Geological Survey booklet "The Geologic Story of Arches National Park," took a more conservative view:

> "-- there is no universal agreement as to how large a rock opening must be to qualify as an arch."

On the practical side size does count, even though it is not easy to set a minimum standard, because unless a lower limit on size is set, the term "arch" loses any meaning it may have. There are literally tens of millions of small natural rock openings in Colorado Plateau sandstone formations. In some formations it is not uncommon to see five or six small openings within a few square feet. The subject of opening size versus significance is discussed further later in this chapter.

Lohman did not try to establish a set of definitions for natural rock openings, but outlined three classes of "arches" within Arches National Park. His classes were based upon the geologic formations in which the openings appeared, and whether the plane of the opening was horizontal or vertical. Such a system was essentially meaningless even within the park, and useless elsewhere. His book mentioned one way the park had served to obscure the matter of definitions, by recalling that an early Arches visitor center display had shown a pothole arch as an example of stream-action, rather than pothole seepage.

The massive 1977 McGraw-Hill "Encyclopedia of the Geological Sciences" contains no entries of any kind for natural rock openings.

In a 1979 report entitled "The Development of Natural Rock Arches at Arches National Park, Utah," J. David Rogers concluded that:

> "The delineation between natural arch and natural bridge seems artificial since the mechanics involved in formation of both are identical."

This recognition that the terms "arch" and "bridge" are inappropriate in the study of Colorado Plateau natural spans had no impact on their use by the public because the report was not widely circulated, or by the National Park Service, even though it had co-sponsored the study.

SOME SERIOUS ATTEMPTS

One of the earliest serious attempts to define and classify America's natural spans was made by geologist Herdman F. Cleland early in this century. The study may have been inspired by the publicity that attended the discovery of the great White Canyon bridges and Rainbow Bridge in Utah. It was first presented as a paper at a conference of the Geological Society of America on December 29, 1909, then published in that organization's Bulletin dated July 2, 1910.

The study, entitled "North American Natural Bridges, With a Discussion of Their Origin," was quite thorough in its description of all the known major natural bridges in America, and a few on other continents too, but was limited in that, other than the four big bridges in Utah, nothing was known at that time about the many other spans within the Colorado Plateau. Those in Desolation Canyon described by the members of the Powell expeditions were not even mentioned, probably because no geologist had studied them.

The report limited itself to natural bridges by establishing the definitions later adopted by the Society for "arch" and "bridge," then went on to say that many of the spans it described were "not strictly natural bridges." It defined fourteen types of natural bridges, based entirely on how they formed, and gave examples of each type. The report was fairly complete with respect to natural bridges, as the term was used at that time, but had only limited applicability to the spans within the Colorado Plateau.

Because of that limitation, this first scientific attempt to define and study natural stone spans fell considerably short of being useful in any study of Colorado Plateau spans, even though it was fairly comprehensive in its coverage of spans in the rest of America.

This immense opening in the rim of Taylor Canyon, in Canyonlands National Park, did not fit any definitions, so it was dubbed "Grand Central What?" For scale, note figure on left rim.

Professor Junius Henderson of the University of Colorado authored a study in 1933 that noted some inconsistencies in Cleland's 1910 report and added a few more spans to Cleland's state-by-state list, but was largely a summary of existing literature on the subject. It did, however, describe briefly the process of span forming that is unique to sandstone, the development of potholes and seep-caves. It contributed little to solving the problem of definitions.

In 1968, arch-hunter Dick Wilson of Moab, Utah, started his survey of southeastern Utah natural spans. The first thing he did was establish five types of spans, both arches and bridges. His types were quite subjective, perhaps in recognition of the fact that human factors must play a part in any system for classifying natural spans:

1. Large see-through arches and bridges exceeding 25 feet in span or height. Superior viewing and photography.

2. Large pothole arches, primarily with vertical openings exceeding 25 feet. Secondary horizontal opening must exceed 10 feet. Superior viewing and photography.

3. Small see-through arches less than 25 feet span and height. Good viewing and photography.

4. Small pothole arches less than 25 feet primary opening. Good viewing and photography.

5. Obstructed or insignificant arch, large or small. Difficult to photograph. May be immense but still classified insignificant due to obstructed viewing.

Wilson's survey was discontinued in 1971 without further refinement of his classification system. Unfortunately, while the system as started may have been of value to him in his survey, it was too ill-defined and subjective to serve as a basis for a set of definitions that could be applicable to all the natural spans in the Colorado Plateau.

As noted earlier, Dr. Dale Stevens, a geographer at Brigham Young University, published his system for classifying natural openings in 1974. He recognized that the terms normally applied to rock openings were inappropriate, so based his system strictly on the shape of the openings. He made no attempt to address the problem of establishing comprehensive definitions for the generally used terms:

> "The words arch, bridge and window will not be used in the traditional sense. Aperture shape will be the major object of examination."

> Further, his study was confined to:

> "-- that portion of the Colorado Plateau in Utah between the Utah-Colorado border and the Colorado River -- and all sizes and shapes are considered."

The area actually studied was largely restricted to one combination of geologic formations, thus severely limiting the applicability of the study's results to the rest of the Colorado Plateau. Stevens established, defined and illustrated two general classes of openings, with several types in each:

ARCHED APERTURES	HETEROMORPHIC APERTURES
Free standing arch	Free standing aperture
Cliff wall arch	Joint aperture
Pot hole arch	Bedding plane aperture
Bridging arch	Alcove beam aperture
Perforated alcove arch	Rock fall aperture
Jug handle arch	Natural tunnel
Alcove pillar arch	

Since the shape of the rock opening, whether arch-shaped or otherwise, is largely a matter of perception and personal judgment, this classification system is also subjective, making it difficult for others to apply, even though it was undoubtedly suitable for the purposes of the study. Stevens recognized the limitations of his study:

> "There is always difficulty in organizing such a group of landforms. There may be some other types which were overlooked, and placement within a specific group may seem arbitrary. It is assumed that further investigation could very possibly lead to a refinement of the present system. As with most other topics, there is ample room for further research."

The research Stevens started did, indeed, lead to further research, some of it sponsored by the National Park Service, as discussed later.

A group of seven geologists from Fort Lewis College at Durango, Colorado, presented a paper to the eighth field conference of the Four Corners Geological Society in 1975 entitled "Origin and Classification of Natural Arches in Southern Utah." The paper was based on field work that was largely confined to Arches and Canyonlands national parks, thus limiting its observations to a very few rock formations, and also limiting the applicability of its results. The study made no attempt to establish definitions for the terms "arch" and "bridge." It discussed formation modes, and set up two general classes, with three types of arches in each class:

JOINT CONTROLLED ARCHES WHICH HAVE BEEN DIRECTLY INFLUENCED BY PRE-EXISTING STRUCTURE

Open arch – opening through widely spaced joint controlled fins

Joint-wall arch – separated from adjacent wall by an open joint

Joint arch – arch developed along a pre-existing vertical joint

PROCESS CONTROLLED ARCHES DEVELOPED UNDER THE INFLUENCE OF EROSION AND STRESS RELEASE

Open arch – formed in a fin by adjacent entrenched drainages

Pothole arch – formed from the merging of alcove and solution pit

Pillar arch – rock column isolated by the joining of two alcoves

The limited nature of the study was demonstrated by one of the questions it posed: "-- why are arches rare in both the Wingate Sandstone and Navajo Sandstone?" While they are, indeed, rare in these formations within the study areas, they are not rare elsewhere. Further, there are several major spans within both formations, even in the vicinity of the two parks studied. Rainbow Bridge is in Navajo Sandstone, and is rare only with respect to size. Although the classes and types of arches established by the study provided no useful definitions of the terms arch and bridge, the study did make a start at classifying Colorado Plateau natural spans on the basis of formation mode. After describing several problems with this approach, the report noted that "it is hoped that these problems will stimulate further thought and study."

Mill Creek Arch is in Navajo Sandstone

Boulder Bridge is in Wingate Sandstone

In 1976 and 1977 Richard M. Pearl authored a six-part series of articles titled "Stone Rainbows" for a journal called Earth Science. As with Henderson's 1933 study, the articles were based on Cleland's 1910 report, with a few refinements and with more examples of Colorado Plateau spans. Pearl avoided the problem of definitions by placing all spans into nine groups:

A. By differential weathering

B. Stream-cut bridges:
   - B-1 Pothole arches
   - B-2 Incised meander bridges

C. Solution bridges

D. Wave-cut arches

E. Lava-tube arches

F. Petrified log bridges

G. Gravity arches:
   - G-1 By stone wedging
   - G-2 By rock slippage

H. Constructive arches

I. Ice and snow arches

DEFINITIONS

The report contained a number of errors. For example, it placed two significant spans within Canyonlands National Park that are not in the park, and completely overlooked one major type of gravity span, while emphasizing two other types that are not generally considered to be significant in the Colorado Plateau region. Openings formed by the "wedging" and "slippage" of large masses of rock are so extremely common there that they seldom get a second glance from serious students of natural spans, even though the openings may be enormous. The report also failed to mention the fairly common gravity openings formed in thin walls of fractured rock when a big chunk falls out, even though it depicted one -- Angels Window in the North Rim of the Grand Canyon. Again, the report was primarily a summary of earlier documents. It contributed very little new to the subject of defining and classifying Colorado Plateau natural spans.

It should be noted that up to this time, none of the serious studies of natural spans was based upon its author's broad personal knowledge of the wide variety of natural spans that actually exist within the Colorado Plateau, and that none of the study authors was satisfied with his definitions and classification system. Some studies even appeared to have been based largely, if not entirely, upon literature research, with little or no original field research. This lack of familiarity with existing variety led to the establishment of definitions and classifications of natural spans that were either simplistic, or of very limited usefulness.

In 1976 arch historian Robert Vreeland published a set of definitions and classes for all kinds of natural rock openings. He established six general classes of spans -- natural bridges, natural arches, natural windows, natural tunnels, sea arches and lava bridges -- with several sub-types of natural bridges and arches. On the whole, his system was the best produced up to that time, but it contained a number of errors in detail.

For example, he indicated that a natural bridge must span a water course that has "considerable length -- distinguishable banks -- and sand and gravel in its bed," without defining the term "considerable" and without regard for one major factor -- many sizable water courses within the Colorado Plateau that have produced obvious natural bridges could not meet the last two criteria because they run across solid slickrock for long distances.

In his definition of "natural arch," Vreeland noted that these spans are formed by "the erosive action of wind and rain plus freezing and thawing," thus perpetuating the myth that wind plays a significant part in span formation. He also ignored the major forming process of seeping water.

In his definition of "natural window," Vreeland noted that these openings form "by the action of freezing and thawing -- accelerated by the force of gravity," but left out the principal factor, that the rock must have been heavily jointed or cracked for these erosional forces to work.

For "natural tunnel," Vreeland failed to provide a unique definition, concluding that "-- in reality, it is a wide natural bridge."

Vreeland divided natural bridges into three sub-types -- meander, waterfall and siphon -- and natural arches into ten sub-types -- rockshelter, pothole, alcove, cave, bridging, buttress, fin, free-standing, arc and pillar. He defined each type in detail, and provided illustrations of the ten types of arches. His system, aside from its several errors, was probably as good as can be devised within the limits of using the inappropriate terms "arch" and "bridge." His observational base was certainly much greater than that of the authors of all earlier systems.

Vreeland also contributed to the subject by trying to establish a standard of "significance" from a rather judgmental combination of "size, name, history, density and appearance." As he applied this standard in his log of spans, however, he sometimes gave "significance" to quite insignificant spans, just because they had been named by someone. "Significance" is discussed further later in this chapter.

In a 1978 book, "Canyon Country Geology," this author noted the problems with establishing definitions for the various kinds of holes-in-rock, then went on to give working definitions for the four principal types:

> "There are innumerable natural holes in the rocks of canyon country. Many are small and insignificant. Some are larger and more aesthetically pleasing. A few are spectacular in size, shape and beauty. There is a human tendency to want to classify, label and name the larger and more interestingly shaped holes.
>
> "Unfortunately, no two classifiers of natural rock openings agree upon the same set of definitions, and the subject is not geologically important enough to warrant a standardization effort by national or international organizations of geologists.
>
> "For purposes of this book, natural rock openings have been grouped into four categories: arches, bridges, tunnels and windows. Defining each of these categories can be very complex."

DEFINITIONS

The book then assigned practical, but obviously not universal, definitions to each of these four categories of natural spans:

"ARCHES. In general, an arch is defined partly on the basis of what it is not. A natural bridge spans an existing or former watercourse. An arch does not. A tunnel is an elongated hole through a rock mass. An arch is not. A window is a hole through a thin wall of rock, so far from any wall edge that the hole does not give obvious shape to that edge. An arch opening does give shape to that edge.

"Arches are created by wind and water erosion, but not flowing water erosion. Arch openings start in thin walls of rock at points where the chemical composition of the rock allows faster erosion by rain and seeping moisture.

"Water is the main erosional force. Wind is minor, because most arches are too high above the ground for abrasion by wind-borne sand. Moisture uses both chemical and mechanical methods to enlarge arch openings. It dissolves the chemical cements that bind the rock particles, allowing them to slough away. It freezes and expands, fracturing the rock. In sum, when it comes to shaping arches, water does the work, and wind cleans up the mess, by blowing away the loosened rock particles.

"BRIDGES. Bridges are cut from either vertical or horizontal sections of relatively thin rock by the action of flowing water, either steady or intermittent. A stream may change its course and abandon its bridge, but the term still applies. After running water has started construction of a natural bridge, it may further widen the opening, but other sculpting is done by the same erosional forces that shape arches.

"TUNNELS. A tunnel is an elongated hole through a rock mass, generally through a thick, monolithic type of rock such as Entrada, Navajo, Wingate or Cedar Mesa sandstone. How short the tunnel can be before it becomes an arch or window is a matter of detailed definition, and will vary from expert to expert. Tunnel type openings that are clean holes through solid rock are rare, but another type that forms when two spring-seep caves meet is more common.

"WINDOWS. A window can roughly be defined as a hole through a relatively thin wall of rock, with the hole so far from any edge of the wall that it does not give obvious shape to the rock between the hole and edge. If the hole should lend shape to an edge, such as a recognizable arc, then the opening might be called an "arch" rather than a window. Or if the rock wall is thick enough, the opening might be classified as a "tunnel." The difficulty of objectively resolving such fine points of definition is one thing that prevents standardization of rock-opening nomenclature."

The author elaborated on these basic definitions and gave a number of examples of each. On the whole, these definitions were the best devised to that time, considering the inappropriateness of the four terms for this use.

The J. David Rogers 1979 report noted earlier summarized the results of a four year study of the spans in Arches National Park. In the report he established four types of sandstone arches, "on the basis of the mode of origin." These were pothole arches, exfoliation joint arches, progressive jointing arches, and bending-moment-induced arches. While these types served the purpose of the investigation -- to study how the spans within Arches National Park formed -- they were not broadly useful to the general public or even to the National Park Service. Rogers made this clear by noting that his classifications were "for demonstration and -- not for common usage." Their applicability to other spans within the Colorado Plateau was also limited because, as with most earlier research, the study's observational base was restricted to the very narrow range of rock formations found within Arches National Park.

John F. Hoffman's 1981 book, "Arches National Park," defined arches, bridges and windows in rather simple terms, and ignored tunnels altogether:

> "Bridges span present or former watercourses; arches and windows do not. Arches and windows are synonymous -- sculptured by weathering -- the chemical action of air and water, and the physical effects of temperature, frost wedging, and gravity."

On the whole, Hoffman's definitions were useful within Arches National Park, but were too limited for application in the rest of the Colorado Plateau, where there is much greater diversity of shape and formation mode.

In 1983, Arches National Park established a set of standards for its rangers to follow in documenting the rock openings in that park. The standard was entitled "How to Accession a Natural Arch or Natural Bridge." It was not intended to clarify the matter of definitions for public use, nor was it for application anywhere outside of the park, but it was an attempt to create uniformity in the reporting and classifying of the many natural spans within Arches National Park.

Although the standard used both "arch" and "bridge" in its title, it did not differentiate further between the two terms, but used the general term "arch" for all spans within the park. The standard based its definitions and classes of spans upon those established by Stevens, as noted earlier:

> "The present system of recording and cataloging has been changed substantially as to numbering, recording and supporting data. However, there was no change in the basic criteria used by Stevens -- since that would introduce a whole new set of circumstances."

ANALYSIS OF THE PROBLEM

Most of the definitions for the general terms applied to natural rock openings that were established early in this century were suitable for use with the spans then known, but were not suitable for application to the majority of the spans within the Colorado Plateau. Most of the definitions established after the Colorado Plateau spans were known to exist were based upon very limited knowledge of the great variety of spans in the region, and hence were not broadly applicable.

Only two attempts to establish definitions were based on wide knowledge of the immense diversity of Colorado Plateau spans. Of these, the one that attempted to be comprehensive contained significant errors and omissions, while the other made no attempt to be comprehensive. It simply discussed the problem and provided basic working definitions useful to amateur arch enthusiasts and tourists.

Serious students of the natural rock openings within the Colorado Plateau are just now beginning to study these spans in depth from an objective, scientific viewpoint. Even so, those doing the studying are not professional geologists, but are from other academic disciplines such as geography, history and engineering.

Regardless of academic background, anyone who attempts to establish broadly applicable definitions for Colorado Plateau natural rock openings must take into consideration the wide diversity of shape, geologic formations and erosional modes that occur in this vast region, and also realize that some of the formation modes common in other regions are non-existent within the Colorado Plateau. One such is the partial collapse of underground caverns of the kind that readily form in the limestone formations that are fairly common elsewhere. Within the Colorado Plateau, alcoves eroded into sandstone by seeping water rarely penetrate more than a few yards. There are also other major differences between the Colorado Plateau and other regions, such as precipitation rates and totals, vegetation cover and types, and the general topography of the land, all of which affect natural span formation.

There are many questions and problems associated with establishing objective definitions for the terms commonly applied to natural rock openings -- arch, bridge, window and tunnel. Here are some samples:

ARCH. Should the term apply to the opening exclusively, or to the isolated span of rock defined by the opening? Must the span be arch-shaped, or is an arching underside enough? How much intermittent water can flow through the span's opening without its becoming a bridge? How far can an opening be from the edge of the rock it is in before it becomes a window? How deep can the hole be before it becomes a tunnel? Can a span with a horizontal opening be an arch? Can a vertical column of rock between two connected alcoves be called an arch? Is an arch-shaped span formed by flowing water an arch or bridge? Is a span formed by the joining of a spring-seep alcove and a pothole an arch, even if considerable water flows through it now and then? Must an arch stand out from the rock in which it forms? How far? Or is it an arch if it is simply self-supporting, no matter how small or narrow its opening? Should the subjective aspect of being able to see sky through the opening be ignored? There are any number of such awkward questions that need to be addressed before an objective and consistent definition of "arch" can be established.

BRIDGE. How much water flow under a span does it take to make it a bridge? Is a span a bridge only if it was formed by flowing water? What about minor sporadic flows, such as off of a broad expanse of slickrock? Is a span formed by such a flow a bridge or arch? What about a bridge long since abandoned by the stream that formed it? Is it still a bridge? Can a water-formed span be arch-shaped on top and still be a bridge? Is a span that is truly bridge-shaped -- flat on top and arched beneath -- a bridge, even though it was not formed by flowing water? Conversely, can a stream-formed span that is definitely not bridge-shaped be a bridge? What about size -- is a small opening formed by stream action in a canyon wall a bridge, regardless of its size and shape? What about the kind of small arch-like or bridge-like openings so common in the Colorado Plateau that are carved out of solid slickrock directly within an intermittent stream course? Again, there are many such questions that must be answered before an objective definition of bridge can be created.

North and South Windows, in Arches National Park -- or are they arches?

WINDOW. What is the real, objective difference between window and an arch? Can a span with a horizontal opening be a window? Why not, if other criteria are met? Does being a window depend on the shape of the opening, its size, its height above the ground, its angle, or its proximity to an edge of rock? Is being a window constrained to particular formation processes? Can a span be too narrow to qualify as a window? Can an opening be arch-shaped and still be a window? How near to an edge of rock can a window be before it becomes an arch? At what-size-to-distance ratio? The questions are endless, and vary with any definition of arch already established.

TUNNEL. Defining tunnel depends upon defining the other three types of spans. The main problem is drawing a clean line between tunnel and other openings. How wide can a bridge or arch or window be before it becomes a tunnel? Can an arbitrary figure be set, or does it depend upon a ratio of opening size to width -- or length? Is an opening a tunnel, not a bridge or arch, if it is twenty feet long? Fifty feet? One hundred feet? Four hundred feet? Tonto Natural Bridge in Arizona is not within the Colorado Plateau but it can serve to illustrate the problem with length. Its opening is 400 feet long, but it is called a bridge. A subterranean stream-carved cavern near Lander, Wyoming, is about a mile long. Is the land above that cavern also a bridge? What if the cavern is 25 miles long, as with a river that flows between Yugoslavia and Italy? Or, within Utah, Dellenbaugh Tunnel is more than 100 feet long and Tusher Tunnel is more than 200 feet long. Are these bridges instead of tunnels? Perhaps the ratio of opening height or width to length should be the determining factor. If so, what ratio?

DEFINITIONS

SIZE. There are also problems with deciding what openings should be considered for inclusion, even after definitions have been established. Minimum size limits must be set for practical reasons. Small spans can be aesthetically pleasing, but can they be called arches, windows or bridges without destroying the carefully built definitions for these terms? Should the same term be applied to a span with a two-foot opening as to one with a two hundred foot opening? If a size minimum is established, how should this be measured? Maximum width or height? Total area of the smallest part of the opening? If, for example, a minimum opening area of thirty square feet were set, then a small pothole arch with a circular opening seven feet in diameter would qualify, while a much longer span with a narrow opening measuring twenty eight feet long and one foot wide would be excluded. Is there any way to establish size minimums for natural spans without using subjective criteria?

ROCKFALL OPENINGS. As noted earlier, openings formed by fallen rock are so common within the Colorado Plateau that they are not normally considered to be in the same class as openings in rock created by erosion. In the region's myriad deep, sheer-walled canyons carved out of sandstone formations it is commonplace to see enormous undercut slabs of rock that have sheared off along a near-vertical plane then slid down without breaking up further, to lean against the remaining cliff, thus forming a wedge-shaped gap. Some such gaps may even have water flowing through them. Similar gaps are produced when large rock masses are undercut by erosion and tilt against other rock, making windows of sorts. Large rocks also occasionally fall onto narrow cuts within sandstone, bridging them. These gravity constructions are not considered to be true arches, bridges or windows by students of Colorado Plateau natural spans, but are in a class by themselves having little relationship to the study of erosion-formed openings through native rock. For the purposes of this book, openings formed by fallen rock will not be considered further. Openings formed by the removal of rock by gravity-fall are, however, considered.

CHANGES. In establishing definitions for natural rock openings it must be borne in mind that a particular span may evolve through two or more phases during its existence. It may start out as a window, become an arch, then eventually a bridge, as the forces of erosion continue their work. Rainbow Bridge is an excellent example of such a three-phase span. Study of the many natural spans within the nearby Escalante Canyon system indicate that within the Navajo Sandstone in this region there is a tendency for streams to cut into narrow rock walls at the necks of meander loops, causing openings to form when the rock wall gets thin enough. The initial opening, however, does not always form at the base of the cliff and thus produce a natural bridge, with the stream flowing through the new opening. The new opening often starts farther up in the wall, thus creating either an arch or a window. Examples of these kinds of initial openings occur in the Escalante Canyon system, where in three different meanders the openings are high above the stream level -- La Gorce Arch, which is more properly a window, and Jacob Hamblin Arch and Broken Bow Arch, which have no water flow beneath them. Water does flow beneath Coyote Natural Bridge, and did beneath Gregory Natural Bridge until it was flooded by Lake Powell. All of these spans are within walls thinned by stream-flow undercutting, but only two of the openings are clearly bridges. Rainbow Bridge could easily have gone through three distinct stages in its development: first a high window through a sandstone fin, as with La Gorce "Arch"; then, after further erosion had enlarged the opening until it gave shape to a span of rock, an arch like Jacob Hamblin with no water flowing beneath it; and finally, after the stream had removed all the debris beneath the span and flowed through the opening, a natural bridge like Coyote Natural Bridge.

After taking all these many factors into consideration, it is easy to understand how the author of the newspaper article noted earlier finally grew disgusted and proposed a tongue-in-cheek set of definitions:

1. A window is a hole in a wall of rock.

2. A bridge is a hole through rock that has water running through it.

3. An arch is like a bridge, only without any water running under it.

4. Holes that don't fit these definitions have no right to exist.

A DEFINITION OF "SIGNIFICANT"

When is a natural span "significant"? Some writers use the term when deciding whether or not to measure or catalog a particular span. If "significant" is suitably defined, it can play a useful, objective part in the classification of natural rock openings.

As generally used in literature concerning natural rock openings, the term is either defined only within the mind of the writer, and is hence subjective and useless to others, or it is defined on the basis of size alone -- if an opening is smaller than an arbitrarily chosen size, it is "not significant."

In contrast to such subjective or arbitrary definitions of "significant," there is a way to make the term generally useful on an objective basis, by linking it to two readily observable factors. Both should be considered when deciding whether a span is "significant" for any purpose:

1. Within the Colorado Plateau, the majority of natural rock openings smaller than, say, three feet in their largest dimension are formed by a different combination of natural forces than those that have formed larger openings. As outlined in the chapter on formation modes, several factors contribute to the formation of each larger natural span, with the balance of these forces changing throughout the span's life. In contrast, most (but not all) of the literally millions of smaller rock openings within the Colorado Plateau have been formed by the combination of only two factors -- softer pockets of rock and surface water run-off -- and these factors do not change.

2. Probable life expectancy is another objective factor in determining the significance of a natural rock opening. An opening that, when enlarged, would soon destroy itself and the span of rock it has created -- as with all of the soft-pocket, water-runoff openings -- could be considered insignificant. An opening that has the potential for enlarging for centuries or even millennia, and thus creating a sizable span of rock, could be considered significant, even though it is small at present.

Thus, in determining the "significance" of any smaller rock opening, both its formation mode and probable life expectancy should be considered, rather than its size alone.

CONCLUSIONS

The earlier analysis of the problem should have demonstrated the impossibility of devising logical, consistent, objective definitions for the terms arch, bridge, window and tunnel when they are used to describe Colorado Plateau natural rock openings. Conversely, it is easy to devise a set of definitions that are based upon subjective, judgmental factors, but such definitions will probably be useful only to the person who devises them, and then only within narrow limits. The present situation is a reflection of this, with as many sets of definitions as there are individuals studying the region's natural spans.

This situation easily supports the conclusions noted earlier -- that the terms now used to label and define the natural rock openings within the Colorado Plateau are inappropriate, confusing, and often misleading.

What should be done about all this? That depends entirely upon the individual. This book provides no easy answers to the problem, but offers readers an analysis of the problem and a "menu" of four possible solutions:

1. If a person wishes to avoid contributing to the confusion, it is always possible to use generic terms, such as "natural span," or "natural rock opening," but even "natural span" can be misleading. There are rock openings that are too far from an edge of rock to produce a distinct "span" of rock. This leaves only "natural rock opening" as a universal generic term. That phrase is used throughout this book, although not exclusively.

2. If a person does not particularly care about the matter of precise definitions, then it is easy to ignore the problems and call all spans "arches," or use the terms "arch" and "bridge" interchangeably, leaving the challenge of distinguishing between types of spans to those who do care. Perhaps this is the best approach for the average tourist who has only a passing interest in natural rock openings.

3. Devise your own set of limited definitions, ignoring all misfits and inconsistencies. This approach has been used by virtually all "experts" in the field to date, so must have merit of some sort.

4. Devise a completely new set of logical, consistent, objective definitions for Colorado Plateau natural rock openings, avoiding the use of terms that are semantically confusing.

Would this last option be possible? Certainly, but probably not using the terms arch, bridge, window and tunnel, and probably not with results that would be acceptable to the general public and various authorities. Altogether, there are several possible approaches to establishing definitions. Each has its advantages and drawbacks. Definitions could be established on the basis of present opening geometry, geology, or predominant erosional force, or even upon a strictly anthropomorphic basis.

Stevens based his system upon opening geometry, or shape, and the authorities at Arches National Park have followed his example. While this system may be useful for certain purposes, it does not produce answers to the kind of questions commonly asked by the public, nor do the shapes of openings have any real, objective meaning from the scientific viewpoint. Such an approach may produce a very limited scientific tool, but not a set of universally useful definitions.

A system based upon geology, with span types keyed to the geologic formations in which they form, would also be of limited meaning and practical value. While there is a relationship between rock formations and opening shape and frequency of occurrence, definitions based upon this would have very limited applicability. Of what value would a system be that called an opening a Kayenta Span, or Navajo Span, or Entrada Span -- or perhaps Rainbow Navajo Span? Few people other than geologists would know what these cumbersome names meant, and even fewer would use the names.

Using a basis of predominant erosional mode, the definitions would be related to whether an opening was formed largely by flowing water, seeping water, wave action, or gravity, with other contributing factors being ignored. There would then be seep spans, flow spans, wave spans and gravity spans. But, as with other systems, what does the public call them? Distinguishing names are needed, suffixed by types that are objective and consistent. Should spans be called Sipapu Flow Span (now Sipapu Natural Bridge), or Delicate Seep Span (now Delicate Arch), or Angels Gravity Span (now Angels Window)?

Perhaps definitions based upon strictly anthropomorphic applications of the words arch, bridge, window and tunnel would be the least confusing. Several writers have suggested this. For example, a "bridge" would cross a barrier, as with constructed bridges. An "arch" would be like a doorway -- it would permit passage through a barrier, such as a rock wall. A "window" would permit viewing through a barrier, but not passage, just as with the windows in buildings. A "tunnel" would be like a human tunnel -- a long passage through or under a barrier. Such a system of definitions would permit easy classification of most natural rock openings, but would also require that practically all present span names be changed.

What is the bottom line? There are no generally accepted experts or authorities in the field of establishing definitions and types of natural rock openings. Few amateur enthusiasts are even trying to make sense out of the present confusion, and perhaps it is not really all that important. The field is wide open to anyone who wishes to try, with the best wishes of this author, who is satisfied with having defined the problem.

SERENDIPITY

The road turned and twisted as it gradually climbed through rolling terrain sparsely covered with gray-green desert vegetation. The arid land was tilted up toward the west and was set here and there with enormous domes and elongated masses of barren, salmon-hued Entrada sandstone that had been shaped and smoothed by eons of weathering.

As I drove slowly along, my wife compared the terrain with a topographic map of the area, seeking a certain rock mass that was shown on a crude map plotted in 1859 by the first U. S. Government exploring party to enter southeastern Utah after acquisition of the region from Mexico. We were trying to find the exact route traveled by that expedition as it left the Old Spanish Trail and descended into what is now the Needles District of Canyonlands National Park. That route had never been traced by historians because the terrain it penetrated was too wild and remote for access except on foot, and even that was difficult.

We soon spotted the historic landmark and, after a little searching, found a faint jeep trail that headed in the general direction that the early explorers had taken. The trail was difficult to follow as it wended its tortuous way through a broken wilderness of sand, rocky gullies, pinyon trees and immense masses of sandstone hundreds of feet high and miles long. After about a mile, the vehicle trail ended on a wooded rock-and-sand terrace at the base of one such monolith, overlooking the head of a canyon that plunged steeply into rock-layered terrain between a sheer sandstone wall and a wide peninsula of land that sloped upward into the distance.

We parked and started hiking out onto that peninsula, following the rim of the deepening canyon. After a mile or so the peninsula narrowed, as a developing canyon system to our right angled toward the one whose rim we were following. Then the slender strip of sand and slickrock terraced downward, forcing us to scramble down ledges and slopes of wildly eroded rock to continue.

By now, the canyon we were studying was deep and narrow, with its bottom boulder-strewn but still travelable by the mounted men and pack animals of an exploring party. Perhaps we had found the historic route. Again and again we descended to lower benches on the peninsula, as the two canyons that formed it approached each other. Finally, we could go no farther. We had reached the still-lofty tip of the peninsula and stood gazing down at the confluence of two appallingly deep gorges. Opposite us, a third canyon joined the one we had been following. Upper Harts Draw was a complex maze of tributary canyons.

As we studied the distant canyon bottoms with binoculars, it was immediately apparent that the 1859 expedition had not come this way. Just upcanyon of where the canyons joined, the bottom of each one had reached the top of the deeply buried Wingate Sandstone, then plunged abruptly downward. This formation, when undercut, collapses to form almost vertical cliffs. The entire inner gorge of this canyon system was sheer-walled Wingate Sandstone! Men could descend such cliffs by rappelling, but horses and mules certainly could not. The 1859 expedition had not come this way! We sighed in disappointment and marked our map. This was only the second of the five attempts it eventually took before we found that historic route.

Even though we had failed in our primary mission, the day had not been wasted. The canyon system was spectacular and breathtakingly beautiful. Intermittent streams, pools of clear water and twisting lines of cottonwoods and other water-loving vegetation traced the bottoms of each inner gorge. The canyon we had been following turned abruptly just after being joined by the canyon opposite us and there, where that canyon reached its sheer Wingate dropoff, was a strange sight!

I looked more closely at the distant inner gorge rim with our binoculars. The canyon's trickling stream didn't reach the rim, but instead disappeared into a big hole many yards back from the rim. It reappeared in a huge alcove below the rim, then plunged down an undercut cliff into an immense pool of water surrounded by trees, shrubs and grasses. Most certainly the isolated rim of rock was a natural bridge, one never before reported, but its shape and size could not be determined from that distance. Serendipity -- look for one thing in this fantastic land of canyons, and find another just as exciting!

Several weeks later we found a way to reach and hike down into the canyon with the strange natural bridge at its lower end. That was an adventure in itself. The meandering canyon exhibited no signs of prior human passage, nor even use by range cattle. Its walls were too steep for cattle to descend, although it contained an intermittent stream and countless lovely pools of water, always attractions to animals in this arid region.

After two or three miles of enchanted hiking, we reached the rim of the deep inner gorge and its unusual tunnel-like natural bridge. We studied it from several angles, then took pictures from a cliff-rim vantage point that looked back on the bridge and its plunge-pool far below.

It was, indeed, unusual. The span itself was about twenty feet long and almost as wide. Its rounded opening was twelve feet in diameter. Below that opening was a huge cavern in the solid rock some thirty feet in diameter. There was a shallow pool on the cavern floor beneath the bridge's opening. Water splashed into the pool, then ran down a short slope of slickrock before plunging into open air to drop about forty feet into a pool at least fifty feet in diameter and quite deep. It was late in the season and the pool was in deep shade, but the autumn-hued leaves of the surrounding cottonwood trees gave a faint golden glow to the dark depths, adding to its beauty and aura of mystery.

We searched in vain for a way down the inner Wingate cliffs to that enchanted pool but found none. Another trip, perhaps -- after we had found the elusive route into the canyon bottom used by that historic expedition.

We spent what was left of that Indian Summer afternoon returning up the canyon, admiring the backlit glow of autumn-amber foliage and talking about "our" unique and lovely natural bridge. What should we name it? We were certainly not going to call it just one more "unnamed span." It deserved better than that.

Eventually, we settled on a name with a conventional source -- that of the canyon system in which it was located -- but with a serendipitous twist. We named our remote and hidden span "Harts Draw Bridge."

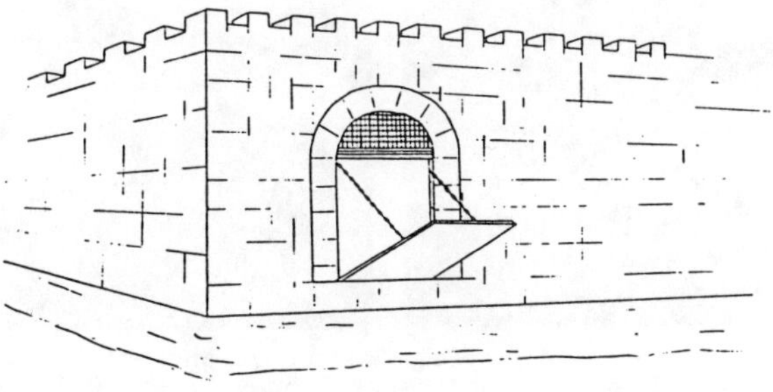

Harts Draw Bridge, from above and below.
Lower two photos by Rodney Greeno.

FORMATION MODES

INTRODUCTION

Early geologists and other scientists were entranced by the geology and topography of the Colorado Plateau as reported by educated and not-so-educated explorers, beginning in the mid-1800s. Some scientists explored parts of the Plateau themselves, generating detailed reports describing its unique stratigraphy, tectonics and erosional modes, and such unusual erosional features as deeply entrenched river meanders, sheer rock walls, canyon mazes, vast expanses of bare slickrock, gigantic alcoves and caves, and natural rock openings. Some even conjectured about how such openings formed. While some of these conjectures were remarkably astute, others were more dramatic than accurate, especially those formulated by individuals not well versed in basic geologic principles, the unusual nature of Colorado Plateau rock strata and climate, and how these interact.

Conversely, a few of the early scientists who were not geologists produced astonishingly detailed and accurate descriptions of the Colorado Plateau region. One such man, John S. Newberry, was a medical doctor who accompanied both the 1858 Ives Expedition up the lower Colorado River and across northern Arizona, and the 1859 Macomb Expedition into southeastern Utah, in each case as their all-purpose "scientist." Newberry's reports were rich in detail about Colorado Plateau geology, paleontology, botany and archaeology. He was probably the first scientist to attempt an overall description of the general Four Corners area as a geologically distinct region, and to apply the name "Colorado Plateau" to that region because it was essentially the drainage basin for the upper Colorado River.

Some of the descriptions of Colorado Plateau natural spans written by early explorers are sampled in the following pages. One such writer, a professor of geology at Marshall College, noted in a September 1943 article in The Scientific Monthly that interest in natural spans was not limited to geologists:

> "Natural bridges, not designed and fashioned by the hands of man, are particularly interesting. They, too, represent stages, not in human experience, but in the orderly progression of an ever-changing world. Our curiosity is at once aroused by the sight of a natural bridge, and we can not refrain from speculating about the mighty forces which brought it into existence."

This curiosity has doubtless been a part of humanity and its predecessors for as long as they have been intelligent enough to notice such natural phenomena. There is an old Chinese proverb, which may refer more to time than geology, that says "Time drills holes in stones," but the question remains -- how?

SOME EARLY DESCRIPTIONS

Although they had no written language with which to document their beliefs, the prehistoric and historic Amerind cultures that occupied the Colorado Plateau region had their traditions about how such gigantic natural spans as Rainbow Bridge came into existence. One such belief, taken from Navajo oral-history, was described by archaeologist Neil Judd in his 1967 Arizona Highways article about Rainbow Bridge:

> "Geologists tell us that Rainbow Bridge -- was carved by sand and water during immeasurable time. The Navajos have a different story. 'Long, long ago,' they say, 'one of their hero gods, while hunting in the canyon, was unexpectedly trapped by a rush of floodwaters. With escape cut off, death for the hunter seemed certain. But just then the great Sky Father cast a rainbow before the torrent, our hero god climbed to safety across the arch, the latter turned to stone, and is there today as evidence our great Sky Father constantly watches over and protects His earth children.' "

Alexander Ringhoffer, the miner whose report about the "Devil's Garden" eventually led to the establishment of Arches National Monument, had an idea about how such "Window Castles" came to exist that was no less fanciful. He conjectured that they were built by the region's prehistoric Indians:

> "The main castle is an immense sand rock with doors and windows. It has to be determined if it was formed by nature or built by hand and age wore it to seemingly a natural ruin. I suspect there are several closed-up caves which hold the mummies, as the earmarks show them to be such. I haven't had time enough to explore that part to open up the caves, but I am sure there is a great surprise for us closed up there."

Aerial view of Ringhoffer's "Window Castles," now the Windows Section of Arches National Park.

Frederick Gardiner, who first reported Arizona's Black Rock Natural Bridge in 1885, provided a description of how it formed that was substantiated by geologist H. E. Gregory in his 1917 paper, "Geology of the Navajo Country." Gardiner observed that:

> "The bridge is formed by a remnant of the overlying grit (coarse sandstone or conglomerate) which is continuous with it on both sides. The grit has every appearance of being a continuous cap over the ridge (and shows) an inclination to break into blocks, and it may be that a large crack, thus formed to the west of the present bridge, allowed the stream to reach the soft sandstone, and so cut it away beneath."

H. L. A. Culmer, the artist who visited the White Canyon natural bridges in 1905, gave a quite creditable description of how they formed:

> "I think I have hit on the -- causes producing the bridges. The canyons are all eroded through a series of sandstone strata. In some the cementing element is almost wholly lacking and they crumble away like brown sugar. In others the cement is lacking in the laminations and where they are undermined they fall in great flakes altho sometimes the chunks come down with a conchoidal fracture. (The bridges) are at the junction of streams. In each case, at the original base of the spring of the arch is this (soluble) clay bed, and it is evident that the streams attacking on each side of the wall that separated them, finally found a way through this clay seam, and the rest was merely erosion and successive fallings of the ceilings. The bridges will become higher all the time from these causes."

Archaeologist Byron Cummings, one of the "official" discoverers of Rainbow Bridge, provided a description in his 1910 report, "The Great Natural Bridges of Utah," of how Rainbow Bridge and other geologic features in the vicinity formed. His description of a region largely unknown and unstudied by geologists at the time was incorrect in both concept and detail:

> "Ages ago the great sandstone beds overlying this entire region must have been pushed upwards by the internal forces of the earth until in the places of their greatest elevation the various strata separated, mountains were formed, and large cracks opened up that extended in zigzag lines away through the slopes of this vast tableland. This process of elevation was undoubtedly a gradual one, and, as the waters of the mountains sought a lower level, they took their courses through these irregular crevices, searching for the ocean which was then not far away. Their rushing currents and surging eddies wore off the sharp corners, sought out the soft places in the yielding sandstone, dug out deep caverns and recesses in the cliffs, and left behind them a series of graceful curves and fantastic forms that amaze and delight the traveler at every turn."

Cummings did no better in describing how Pritchett Arch near Moab formed:

> "Caves have been hollowed out of these cliffs at various points and numerous natural reservoirs are found scattered on the surface of these bare rocks where soft places have

been found in the stone, or whirling eddies in former ages have ground out cisterns. Some of these are mere shallow tanks, while others reach down twenty feet and more through the solid sandstone. Some are irregular and winding in their course, while others look as though they had been sunk by some titanic drill when the gods were playing with the earth's crust. A few drain considerable areas of the cliff, and in time of storm many a rushing torrent loses itself in their depths. In a few instances such a reservoir has been formed directly behind a cave that was being hollowed out of the side of the cliff. As the walls of the cave gradually extended backward farther and farther into the cliff, the reservoir was sunk deeper and enlarged little by little until its bottom broke through into the back of the cave. Then the waters formerly gathered into the reservoir and held, surged through the cave and lost themselves in the valley below. Every downpour of rain and every driving wind carried the work a little farther until the former roof of the cave became an arch. When the reservoir held the waters until its depth about equalled that of the cave, then the gracefully curving arch of the cave became a real bridge as in the case of the fine arch already mentioned, which we have christened Pikyabo (Pee-kya-bo), the Ute name for water tank."

Cummings was incorrect in almost every detail of his descriptions, but his biggest error was in assuming that it took violently moving water to form the potholes and caves that so often join to create natural rock spans. Many other writers, some of them professional geologists, also completely overlooked the fact that in this semi-arid country slowly seeping water is responsible for the creation of far more natural spans than flowing water, and even plays an important part in the initial formation of the "natural bridges" that span water courses.

Herbert Gregory, a geologist for the U.S. Geological Survey, recognized the importance of seeping water and stress relief exfoliation, although he did not use these terms. He described these processes, and provided a succinct summary of how many Colorado Plateau spans form, in his 1917 report on the geology of Navajo country:

"The conditions that permit the formation of rock shelters in massive beds of sandstone -- namely, tangential cross-bedding and porosity in rock of uniform composition -- also favor the production of windows and natural bridges. The arch is the dominating architectural form. Blind arches and curved panels are recessed in the cliff faces, and the roofs of niches and rock shelters are vaulted. Detached arches of sandstone rest against cliffs; here and there they swing free and rise above rock slopes in a fantastic manner.

"Erosion of these massive cross-bedded sandstones involves not only disintegration and decomposition but also the removal of disks of rock of varying thickness and width. All cliff faces are subject to attack and are forced to retreat at varying rates. If the wall is thin -- a cusp of an entrenched meander or a dike-like buttress -- the removal of successive shells from opposite sides may result in a perforation, or window. Windows thus formed may be seen at a number of points in the Navajo country. Some, like the Defiance Window, are set high on the wall; others are near

> the surface of the ground; a few are cluttered with debris, but most of them are swept clean by rain and wind. These perforated walls are not a distinct species of erosion form; they represent merely a stage in an erosion process -- one term in a series that includes niches, rock shelters, blind fan windows, open windows, and the arches of natural bridges."

In his 1929 book about Rainbow Bridge, explorer Charles Bernheimer offered his own "amateur theory" as to how Rainbow Bridge and the surrounding territory formed:

> "Over bed-rock a layer of limestone and hard sandstone was spread by an ancient ocean. This ocean bottom was lifted up some 5,000 to 6,000 feet. The loose sands, which were hundreds of feet thick, dried and became the playthings of wind and weather. Like the sands of the beaches they deposited themselves in cross-bedded, curving strata, to be blown off, deposited, and re-deposited, for hundreds of years. This age-long grinding of the wind gave the sands a uniform structure and size. Their final resting place was provided by some unknown force; possibly a thorough wetting and covering by a flow of lava kept them in leash. This plateau-like crust of cross-bedded sandstone, with its layer of lava as a lid, now formed the level into which, in later geologic times, the present network of canyons was cut.
>
> "Somewhat more complicated were the causes which produced the Rainbow Bridge. The streamlet in the canyon called Bridge Creek had two channels, its present one under the arch and an ancient one some hundred feet higher. The old channel, extremely tortuous and meandering, was twisted in sharp and frequent loops from one side of the canyon to the other. Where it played long enough against an obstacle it formed caves. In one instance two such caves collided, their backs, or rear walls, broke through, collapsed, and offered a new and more direct channel for the waters. These widened out the hole and, in the course of time, even dug into the next lower stratum, the lime and hard sandstone underlying the softer, cross-bedded sandstone from which the arch is formed. Its peculiar chipping characteristic helped by frost, sun, heat, and wind blast, completed the process."

Bernheimer's ideas about the general terrain were correct in concept, even though wrong in detail. The "bedrock" is not "limestone" but formed from highly mineralized freshwater sediments. The land was not "uplifted" until megayears later. The sands of the next layer were not "sands of the beaches" but those of a vast Sahara-like desert, and they were not later covered by a "flow of lava" but by more freshwater sediments. His description of how Rainbow Bridge formed was as correct as the descriptions of modern geologists, all of whom omit mention of one important factor. This factor -- how such openings first form -- is discussed later in this chapter, in the section describing how flowing water forms natural spans.

Robert Frothingham, author of the 1932 book "Trails Through the Golden West," provided an almost poetic but fairly accurate explanation of the formation of Rainbow Bridge. His only mistake was the assumption that the Navajo sandstone from which Rainbow Bridge formed was deposited by a "vast inland sea." The real sea came millions of years later.

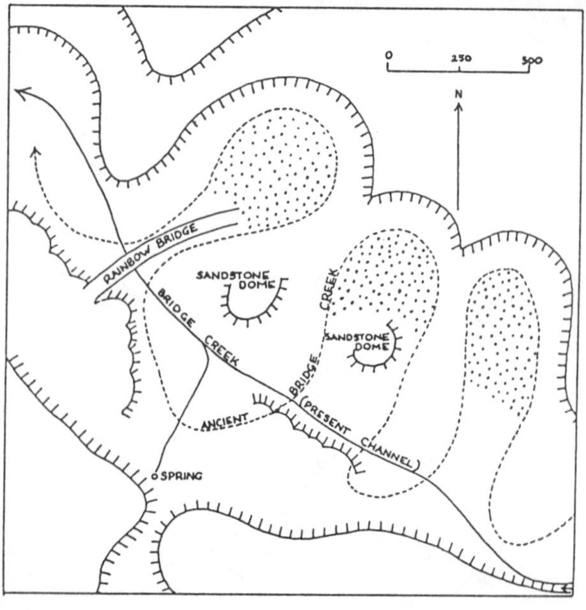

Drawing based on a Bernheimer sketch, with a similar aerial view, showing the abandoned meanders near Rainbow Bridge.

"Nature has outdone herself in the gigantic scale upon which she has fashioned this masterpiece. Then think of the antiquity of this marvel and the almost unbelievable fact that there was a time when Non-ne-zo-she Na-gee-lid (Rainbow Bridge) and the mountain peaks surrounding it were buried beneath the waves of a vast inland sea. In considering the glacier-like progress of erosion, one vainly tries to visualize the fact that the prehistoric rocky river-bed beneath, now dry, was at one time 150 feet higher than the present one and that the original stream that raged through that canyon centuries agone flowed around one end of that prodigious flying buttress before the inexorable combination of frost, wind, rain, sand and sun chiseled out the center and wrought such an amazing transformation."

In his 1938 report on the geology of San Juan County, Utah, geologist Gregory provided a description of how the White Canyon natural bridges formed that substantiated and added technical detail to Culmer's 1905 conjectures, then discussed how the arches and windows in that region formed:

"In rocks that are evenly bedded or angularly cross-bedded blocks pried loose by frost or by undermining commonly have angular outlines. In rocks with curved cross-bedding the detailed sculpture is characterized by curved surfaces in both vertical and horizontal directions. Thus on many bare exposures of Navajo sandstone, also in places on the Cedar Mesa, Entrada, Wingate, and Bluff sandstones, cross-bedding outlines arches. As erosion continues most of the arched structures are detached from the walls and, lacking adequate support, collapse in response to gravity, exposing new curves and arches behind them. Some arches retain their individuality for long periods, and as the surrounding rock is worn away they stand like hoops with ends firmly attached to the ledge high above present stream beds. Thus in Dark Canyon an arch with a space of about 100 feet clings to the sandstone wall, and arches in various stages of development stand here and there on Grand Gulch and Red Rock Plateaus -- the free-swinging arches develop most readily at the heads of short box canyons, where erosion is most vigorous and continuous.

"Arches and windows seem to have the same physiographic history as rock shelters, bridges, and the multiform cavities in walls of porous sandstone. In early life they express the work of ground water; with their later development all agencies of erosion are concerned; at all stages rock structure is an influential control."

In this description, Gregory clearly assigned a major role to seeping "ground water" in the initial formation of natural rock openings, but he overlooked the importance of stress relief exfoliation in span formation. Further, he overemphasized the importance of cross-bedding in the shaping of arches. While exfoliation due to seeping water is influenced by bedding planes, later geologic scientists reported that stress pattern exfoliation beneath arched openings generally ignores bedding planes.

Unfortunately, not all the writers who have published descriptions of how Colorado Plateau natural spans form have been even as knowledgeable and observant as Gregory. One major fallacy about natural span formation -- that wind plays a major part -- got an early start and still persists in modern literature and thinking.

THE MYTH OF WIND FORMATION

The most durable myth about how the natural spans within the Colorado Plateau form first appeared in print in the late 1800s. It was immediately refuted in the same publication by another writer, but seemed to take on a life of its own, a ghostly misconception that has haunted otherwise rational writers ever since, and by now seems to have become embedded in public awareness as "folk knowledge." It is high time this ghost was put to rest, if this is possible.

In what may have been the first written report of the natural spans in Arches National Park, a Mr. Arthur Winslow noted in the April 22, 1898, issue of Science that the magnificent span now called Landscape Arch was:

> "-- in all probability, a monstrous product of wind erosion. The rock appears to be one of the friable Mesozoic sandstones which are widely exposed in this region. Other examples of wind action were seen by me while traveling through the country, so located that no other cause could be assigned. Strong and prolonged winds are frequent here, as any one who has sojourned in that country can testify to his misery. The sands carried by these winds are whirled about in the depressions of the rocks, and excavate wind pot-holes in the friable sandstones with great rapidity. A wall or slab of such rock is by degrees entirely penetrated, giving rise to the so-called window rocks which are frequently seen in isolated buttes high above the surrounding level. Our natural bridge, I conclude, is simply an extreme or abnormal enlargement of such a 'window.' Possibly some water channel may have assisted in the process, but the view (photograph) does not indicate this, but shows the bridge to be high above the main water course. The dimensions of the bridge, or rather the shape of the space covered by it, are also against this idea, as the ordinary channel cut by a stream through rock is deep and narrow."

In the May 20, 1898, issue of Science, Frederick Dellenbaugh, a member of Powell's earlier explorations of the Green and Colorado river gorges who had noted numerous natural spans there, took strong exception to Winslow's concept of wind-formed spans:

> "-- there are in the great arid region a large number of these natural arches. In the Canyon of Desolation, on Green River, they are particularly common, and from the surface of the river some of them seemed of huge proportions. All I have seen occur in formations exactly similar in kind -- homogeneous sandstones with tendencies toward conchoidal fracture -- and my observations are against the wind erosion theory as a prime factor.
>
> "The beginning appears to be in some natural crevice or cleft on the face of the bare cliff wall, where water is able to penetrate and allow frost to start operations by throwing out a fragment that leaves a cavity almost a miniature of the final perforation which marks one further

period in the demolition of the cliff. This fragment is followed by many others, till the cavity presents the appearance of an alcove with arched top, and a talus floor. The arch gradually deepens into the cliff, and I have seen one so deep that its floor was a lake, with a grove of trees at the opening. Frequently, if not generally, the deepening is assisted by water percolating from above.

"At a certain depth, if the cliff is a thick one, the arch begins to protect itself, and the excavation proceeds more slowly. It becomes a cave with floors of various character according to circumstances that vary with other conditions. But if the cliff is comparatively thin the wearing finally cuts through to the opposite side, and then wind erosion becomes a more potent factor. I have seen many examples of every stage of progress, and I have seen at least one beginning where a rain torrent was in active operation, and made a sketch of it. Frost, and the disintegrating and dissolving power of water combined with structural tendencies, appear therefore to be the chief causes of these natural arch forms."

While Dellenbaugh's ideas about span formation were somewhat incomplete, he recognized the importance of seeping water, and his explanation was certainly far more correct than Winslow's.

In his article, "The Natural Bridges of Utah" in the May 11, 1907 issue of The Mining World, professor of geology Arthur Lakes described how these phenomena form:

"The rocks of the Jura-Trias (Jurassic-Triassic), consisting of beds of coarse sandstone and conglomerates, alternating with finer grained and harder sandstone layers, are particularly susceptible to erosion. Water getting into the joints and between the softer rocks and pebbles of the conglomerates finds a point of attack. This weak spot, long after, when the stream has passed away or subsided to its present diminutive size in the depths of the canyon, is constantly being enlarged by the agency of wind blown sand."

The eminent professor's description was far from complete, and was dead wrong about the importance of "wind blown sand" in the enlargement of the opening.

In his November 1911 National Geographic article about Rainbow Bridge, Joseph Pogue made the same error of assigning importance to wind erosion:

"The hole, once made, has been enlarged and given its present shape by the combined action of weathering, expansion and contraction due to changes in temperature, and the carving effects of wind-blown sand --"

Another geology professor, Frederick J. Pack, took an opposite but equally incorrect position in a book, "Natural Bridging in High Plateaus of Utah," dated 1922:

"Notwithstanding the fact that wind action on the under side of the bridges is at present so slight as to be almost negligible, yet, on the upper surface, it is of far more than passing importance. Circular wind-pits, ranging from one to four feet in diameter, are very abundant, particularly on the top of the Edwin Bridge, where they comprise an intimate mass of shallow borings covering half the entire surface.

It should be mentioned here that the "wind-pits" noted by Pack and Winslow are formed primarily by water, not wind. Small amounts of rainwater or snow-melt standing on an exposed sandstone surface dissolve or loosen the cementing between sand granules. Later, these loose particles may be removed either by rain or wind, enlarging the depression slightly. The process then repeats over the years, making the "pothole" larger each time. The critical factor is that water alone can make such potholes, but wind alone cannot.

Moab Rim Arch, Moab Valley. For scale, note figures.

In his description of the 1921 Bernheimer Expedition to Rainbow Bridge, writer Earl Morris noted incorrectly that the surrounding country was "a hades of aolian erosion." He wrote correctly that the Navajo Sandstone there was "solidified sand dunes," but again incorrectly that these petrified dunes had been "resculptured by the agency that formed them," thus giving full credit to wind for creating and shaping the region. In fact, water has done virtually all the shaping, with wind only a minor factor. The mistake was, perhaps, understandable, because the expedition encountered so much wind that they even named one place "Sirocco Pass," after the hot desert winds of North Africa.

In his 1925 book, "Mesa, Canyon and Pueblo," writer Charles F. Lummis described a span near Fort Defiance:

> "There is a curious natural bridge near Fort Defiance, New Mexico. It has an arch of only about sixty feet, but is remarkable because it was carved not by water but by sand-laden winds, as are some of the most beautiful and fantastic erosions of the dry Southwest."

Lummis was wrong on all counts, but the myth of wind formation was seldom challenged. It continued to persist, even among professional geologists who should have known better. The myth even became "official" when Arches National Monument was established in 1929. The proclamation signed by President Herbert Hoover, stated that:

> "-- these areas contain extraordinary examples of wind erosion in the shape of gigantic arches, natural bridges, 'windows,' spires, balanced rocks, and other unique wind-worn sandstone formations,"

The 1941 WPA guidebook to the state of Utah spread the misconception still further in its description of Arches:

> "In the Monument proper, the wind has carved these canyon walls into forms that, even in a region noted for spectacular erosion, are remarkable. The wind, driving crumbled sand, has found weak spots and pecked windows through, chiseled out pinnacles, and chipped the contours of a stubborn piece of stone into life forms and ragged battlements. The wind is tireless -- cowboys say it is useless to build a pine board fence in the desert. If the knots loosen, the wind blows through the holes and wears the boards out, 'which,' they add, 'is how come an arch.'"

Not only did the anonymous WPA writer perpetuate the wind formation myth, he also apparently "bit" on a local cowboy tall-tale. The book also quoted National Park Service literature which repeated the wind myth by noting that the Fiery Furnace in the monument "-- contained many arches and windows and 'other formations that are among the finest examples of wind erosion in the world.'"

In the September 1943 issue of The Scientific Monthly, geology professor Raymond E. Jensen of Marshall College published an article about "Nature's Bridges" that gave "scientific" weight to the wind formation myth:

> "Although Utah has become noted for its natural spans cut by running water, it also has some which were not water-carved but were produced by the blasting action of strong desert winds. Wind-made arches exist in Arches National Monument -- these openings range from small windows a foot or two in diameter to great natural archways more than a hundred feet in diameter. Eight of these are of enormous size, one being two hundred feet high and nearly circular in shape. In this region strong prevailing winds pick up loose grains of sand from the desert floor, carry them for a distance, and hurl them with tremendous force against the sandstone walls, effecting a natural sandblast. During long ages the walls have been worn through at intervals, forming the many natural windows and arches."

The good professor had obviously never been to Arches National Monument and was simply repeating "facts" he had acquired from questionable sources. He certainly did no original research or thinking about his subject, and the editor of the distinguished scientific journal obviously ran the article solely on the basis of its author's otherwise creditable background.

In his September 1949 article in National Geographic about exploring the Escalante area, Jack Breed described the discovery of Grosvenor Arch, then added:

> "Actually, it is a double arch, with the larger span on the end of a buttress that juts from the main sandstone butte. Near the anchor end wind has blasted a smaller hole through the buttress."

With even the respected National Geographic continuing to support the wind-formation myth, it is easy to see why that misconception has persisted.

In the May 1959 edition of Natural History, Richard L. Threet attempted to correct the popular fallacy in an article titled "The Wind At Work." He subtitled his article "The powers of wind erosion are less than is commonly supposed." Threet began by discussing the mechanics of wind movement in general, then went on to add evidence from his own observations of Colorado Plateau erosional features:

> "As the wind streams over the landscape, it is generally at a serious disadvantage in abrasion of hard, rocky surfaces, because the strongest winds are usually found high above the ground. Only where the air is funneled through local constrictions will the velocity of wind-driven particles be high enough for significant abrasion. (Further) the flow of air at the earth's surface is generally turbulent, sucking up loose particles of rock into the stronger currents and carrying them for considerable distances. Because large and heavy rock fragments settle rapidly, usually only fine sand and dust rise more than a few inches off the ground, even in very severe windstorms.
>
> "Pedestal rocks, pinnacles and arches may develop in humid regions, but they are most conspicuous in desert regions, where winds rip across the barren landscape. Hence the common misconception that wind abrasion has been the chief process in shaping these bizarre forms.
>
> "Consider, in conclusion, additional evidence that casts doubt on wind abrasion as a significant factor in producing arches and similar features. First, leeward and windward surfaces are usually carved and smoothed alike. Second, the necks of pedestal rocks and the surfaces of arches are usually soft and powdery (while) wind abraded surfaces are hard and polished. Finally, many arches and other overhangs are found high above the level at which winds should ordinarily carry the gritty particles that are required for abrasion."

Threet then quoted a paper on this subject by Herbert E. Gregory, then a professor of geology at Yale University and long-time geologist for the U.S.G.S., which concluded:

> "In general, it may be said that direct abrasive work of the wind is not a prominent feature (in shaping sandstone) but as an agent that removes the products of decomposition -- the wind has high value."

Threet then went on to describe how seeping water, rather than wind, did most of the active work in natural span formation within the Colorado Plateau region, using quotes from a paper by another geologist, a Professor Kirk Bryan of Harvard University, to support his own observations and analysis.

Threet's 1959 article in Natural History should have put the matter to rest, at least within the scientific community, but did not. Another repetition of the myth appeared in the July 1964 edition of the same respected publication, in an article authored by a Willard Luce. The article, entitled "Arches and Bridges of Stone," noted while describing how the spans in Arches National Monument formed, that while openings within Entrada Sandstone were created by the combined action of "water, wind and frost," openings in the formation beneath the Entrada:

> "-- happened far more rapidly. Because the wind is seldom still on the high Colorado Plateau country, sandblasting became a continuous process. Slowly, holes were worn through the fins, and arches were born."

Eagle Arch, Johnson Canyon. Bell Brothers photo. Wind plays only a minor part, even in the formation of such low spans as this. Note the sharp edges in the opening. Wind erosion would leave smooth, abraded surfaces.

In 1976 and 1977, a Richard M. Pearl authored a six-part series of articles entitled "Stone Rainbows" that appeared in the journal, Earth Science. In the second article of the series Pearl put the matter of wind-formation into the proper perspective:

> "Wind-blown sand acts as an abrasive, although the present tendency is to regard it as less effective than we used to. Even the one yearly rain is believed to accomplish more erosion and real transportation (of material) in a desert than the perpetual blowing of the wind, which does as the maid who comes in once a week and 'rearranges the dust.'"

FORMATION MODES

While factual, scientific discussion of the matter did eventually put the ghost of wind-formation to rest within the scientific community, the idea persisted with others who wrote about the natural spans of the Colorado Plateau. In his log of spans in the region, arch historian Vreeland gave new life to the myth. In his basic set of definitions he stated that "A natural arch was formed by the erosive action of wind --," then compounded this error in some of his descriptions of specific spans. For example, in his description of Phipps Arch he notes that:

> "It is a very young, rock shelter type natural arch, developed by erosion of the wind blasting through an elongated dome of Navajo sandstone."

In his 1981 book on Arches National Park, geographer John Hoffman emphasized that "-- wind is NOT a major agent of erosion," but respected Utah author Ward J. Roylance kept the myth alive in his otherwise excellent 1982 rewrite of the 1941 WPA guidebook to the state of Utah. In describing how the numerous spans in Arches National Park formed, Roylance retained most of the original author's erroneous wording:

> "The wind, driving crumbled sand, has found weak spots and pecked windows through, chiseled out pinnacles, and chipped the contours of a stubborn piece of stone into life forms and ragged battlements."

Broken Arch, Arches National Park

Thus, to this day, the myth of wind forming natural spans holds onto its ghostly, insubstantial life, despite a total lack of evidence to support the idea, and in spite of all that has been written to disprove the concept. There is little doubt that this myth will continue into the far future, even after this book has been widely distributed. But perhaps the readers of this book will come to understand that when it comes to the formation of the natural spans within the Colorado Plateau geologic province --

WATER DOES THE WORK -- WIND JUST HELPS CLEAN UP THE MESS

GENERAL FACTORS

CONTINENTAL DRIFT. As with the Colorado Plateau itself, and so many of the unique geologic features within it, Colorado Plateau natural rock openings are the products of a rare combination of factors directly or indirectly related to continental drift -- the ancient and stately dance of continental land masses upon the face of the planet.

Rock formations favorable to the formation of natural spans were deposited layer after layer during a 100-million year period long before the present North American continent was thrust away from the single super-continent Pangea, during and even before the period when that enormous land mass was slowly drifting northward from the south polar region. For most of that 100 megayears, the Colorado Plateau was on Pangea's west coast and was either equatorial or near-equatorial, with a corresponding climate. The land was either broad coastal plains hundreds of miles wide, or equally extensive coastal shallows. Climate and sea level variations influenced by continental drift factors alternately covered these vast lowlands with shallow marine sediments, then freshwater deposits brought from the Ancestral Rocky Mountains to the east, in particular the immense Uncompahgre Range that then dominated what is now western Colorado.

By the end of the Pennsylvanian period, some 225 million years ago, the sea lost the ages-long battle for supremacy, and for a very long time land deposits predominated within the Colorado Plateau. These were either freshwater sediments spread across the vast floodplains from the Uncompahgre highlands or, during more arid times, were enormous deserts of dune sand, with more sand constantly being carried into the region from the east and south by the prevailing winds to create an immense Sahara-like desert that persisted for most of 50 million years. The climate was arid, and the "aeolian" deposits, sand dunes, were perennially worried into new and different contours by the strong desert winds.

The little precipitation that fell was not enough to form a system of streams and rivers that eventually reached the sea. Instead, as is common in many desert areas, the scant rainwater accumulated in broad ephemeral "playas," or very shallow temporary lakes with no outlets. As the water evaporated, it deposited layers of silts and soluble minerals, which were eventually covered by the shifting dunes. These thin "lenses" of highly mineralized, water-deposited sediments ultimately became the impervious layers of rock within the massive aeolian sandstones of the Colorado Plateau that are otherwise fairly permeable to seeping water. Such "petrified playas" within aeolian sandstones are now major factors in the formation of natural spans in such rock. Similar lenses of relatively impervious rock also formed within geologic strata, such as Cedar Mesa Sandstone, that are composed of shoreline sands. These lenses formed from the minerals and fine sediments deposited by the sea in tide pools, and by freshwater streams and flooding in shoreline swamps.

About 150 million years ago, the land mass that became North America became detached from Pangea and thrust westward. This was the beginning of a series of continental drift events that, in time, brought dramatic changes to the hitherto calm lowlands of the Colorado Plateau. Changes in global sea level brought back invading seas. As these seas advanced, then later retreated, they left behind thick deposits of beach sands that later became such sandstones as the Cedar Mesa, Entrada (Moab member), Dakota and Mesa Verde. The shifting continental land masses changed major ocean currents, which in turn caused major changes in global climate.

During much of this period, new tectonic activity -- physical geologic events -- also caused by continental drift, thrust up mountain ranges all around and even within the Colorado Plateau, causing still more climate variations within the region. Continental crustal pressures and tensions caused by North America's westward drift violently distorted the Colorado Plateau with folds, uplifts, basins, and overthrusts. The Plateau resisted the forming of volcanoes, confining such forces to the creation of the eight "intrusive" mountain ranges within the region, but active volcanoes ringed the Plateau and added highly mineralized ash to freshwater deposits there that later became the Chinle and Morrison formations.

One enormous continental drift event -- a long period of exceptional sea-floor spreading activity -- brought a drastic rise in sea level that inundated much of the Colorado Plateau for some 45 million years under a deep but stagnant inland sea. The invasion and later retreat of this sea left beach deposits that later became the Dakota and Mesa Verde sandstones, and the intervening Mancos and Lewis shales.

Subsequent climatic changes, also influenced or created by continental drift events, brought on long glacial epochs and trapped vast lakes within or adjacent to the Colorado Plateau. Sediments deposited within these lakes later became the rock formations that are presently exposed within the Bryce-Cedar Breaks area, along the Book Cliffs, in the Uinta Basin and within Desolation Canyon and its numerous tributaries.

Fairly recent continental drift events uplifted the whole western half of North America, causing another great climatic change and beginning the present 10 million year erosional cycle that has removed about a mile of newer sediments and rock from much of the Colorado Plateau. This has exposed the older rock that is suitable for natural span formation, and provided the semi-arid climate also essential to most span-forming processes.

Thus, while it is clear that continental drift set the stage for natural span formation within the Colorado Plateau, for full understanding of its impact it is necessary to examine other general factors.

BASIC STRATIGRAPHY. The exposed rock strata in the Colorado Plateau are generally ideal for the formation of natural spans because they were deposited in level, fairly thick layers, then distorted only minimally by later tectonic activity. Further, most span-forming Plateau rock strata are composed of sediments that were well sorted by wind or water, and thus fairly uniform in texture and chemical content. There are few igneous formations within the Plateau, except within the several laccolithic mountain ranges, and no metamorphic strata except where erosion had cut deeply into really ancient rock within the river gorges. These few igneous and metamorphic strata do not form natural spans of any significance.

Certain Colorado Plateau rock strata and groups of strata tend to form spans more easily than others. The individual sandstones common to the Plateau in which spans are more likely to form are the Cedar Mesa, De Chelly, Wingate, Navajo and Entrada. Some of the combinations which enhance span production are Navajo-Carmel-Entrada or, in some areas, Navajo-Dewey Bridge-Entrada; Kayenta-Navajo; and Wingate-Kayenta. There are other combinations of lesser importance. Some writers have claimed that natural spans are rare within Wingate Sandstone, but this is not true. There are many outstanding spans in this formation within the Colorado Plateau in Arizona and Utah, and a few in Colorado and New Mexico.

Within the Plateau there are also several areas that have relatively high concentrations of natural spans because of unusual local stratigraphy and topography. Among these are Monument Valley; Canyonlands, Arches, and Bryce Canyon national parks; Natural Bridges and Cedar Breaks national monuments; the regions surrounding Navajo Mountain and the Abajo Mountains; the Escalante drainage basin; the Rattlesnake Canyon area in Colorado; and the vast desert-canyon area to the northwest of Moab, Utah, that is wedged between the Colorado and Green rivers. In most cases, the exceptional density of spans is not confined to the park areas noted, but includes adjacent areas.

Typical windows, Bryce Canyon National Park

REGIONAL TECTONICS. There are massive overthrusts and other crustal distortions on the border of the Colorado Plateau, but within its boundaries rock strata have been bent, in some cases into almost 90 degree monoclines, but usually with relatively gentle radii. Monoclines, synclines, uplifts, and intrusive mountain ranges that are many miles long seldom produced sharply distorted rock strata as they formed. This has left the distorted rock layers heavily jointed in many areas, often in immense parallel or cross-hatched patterns, but not severely fractured. Even after such bending, the rock strata still retain enough integrity to permit the formation of natural spans. In most cases, the jointing occurred while the strata were still deeply buried, and seeping mineral-laden water cemented the joints long before they were exposed by the current erosion cycle.

In sum, the numerous rock joints formed in Colorado Plateau rock strata by past tectonic activity often play an important part in span formation by promoting the formation of relatively thin sandstone "fins," by channeling water seepage, and by helping to define the shape of spans as they form. These processes are examined in more detail later in this chapter.

REGIONAL CLIMATE. As noted earlier, the Colorado Plateau for the last ten million years or so has been experiencing a major erosion cycle due to having been uplifted about a mile by a continental drift process. The mountain ranges that surround the Plateau on all sides have contributed to this cycle in two ways. Ranges to the west have caused the Plateau region to be largely arid by squeezing out much of the moisture from passing clouds, thus creating huge "rain shadows" within the Plateau. Mountains all around the Plateau have promoted erosion within the Plateau as their precipitation has drained across its arid reaches, often with devastating fury.

Over the last ten million years, several major and minor glacial periods have come and gone. Each has played a major part in shaping the present Plateau landscape. The most recent mini-glacial period to affect the Plateau ended in the mid-1800s, and was a major factor in the formation of most existing natural spans in the region. The glaciation itself played no part in span formation, but the cooler, wetter climate during that period did, as did the glacial melt-water to some extent. The more arid phase that now prevails is slowing down the aging of existing spans.

There is plentiful evidence of the growing aridity of the region. In places there are stands of Ponderosa pine trees, remnants still clinging to life in environments too arid in which to prosper as a species. There are still a few stands of aspen trees at around 6000 feet elevation, too low for them to thrive in this region except during wetter, cooler periods. Tree ring studies also show the drying trend, as do a half-century of flow records for the region's major rivers. Early reports about the region indicate that much of it was far more heavily vegetated even sixty years ago. The first reports about the Rainbow Bridge area noted a profusion of plant life there that no longer exists. Bernheimer's 1920 report in Natural History described a canyon near the bridge this way:

> "In this canyon there are more varieties of plants and flowers than are to be found between Asbury Park and New York City, and their blossoms are exquisite."

He went on to describe "acres" of paintbrush, columbine, daisies, asters, sunflowers, yuccas, blooming cacti, sego lily, purple sage and other flowering annuals and perennials. Other historic reports describe vast meadows waist-high in grasses and other annual plants, where none exist now, indicating that such lush vegetation was normal, and not due to a single exceptionally wet spring.

PRECIPITATION. Within the Colorado Plateau the total amount of precipitation is fairly small, although somewhat greater at higher elevations. In addition to the total precipitation, there are several factors related to regional precipitation that affect natural span formation and life expectancy. The rate of rainfall governs how much is absorbed and how much runs off in flash floods. Snow and ice generally melt slowly and hence are absorbed by the ground, so the ratio of rain to snow is important. Winter temperatures control the amount of freezing, an important factor in the formation of some kinds of spans. The amounts, patterns, ratios and rates of precipitation in turn control the amounts and types of vegetation that grow in the region, and that affects water runoff and span formation. Plant roots also play a small part in this process. Thus, directly or indirectly, precipitation is a major factor in natural span formation, but too much precipitation reduces the life expectancy of spans in sandstones. This is discussed more fully in a later chapter.

TOPOGRAPHY. The present and geologically recent topography of the Colorado Plateau also plays a major part in span formation. Before the general uplift of the western part of the continent, the surface of the Plateau was fairly level and low in elevation. The small amounts of water falling on the region and draining into it from the surrounding highlands simply disappeared in vast expanses of arid land. But as the land slowly rose, and local and intruding runoff increased, meandering rivers reached farther and farther into the Plateau, slowly cutting their twisting channels deeper and deeper. At first, a

much smaller Colorado River did not even reach the sea, but drained into a vast inland desert lake, there to evaporate. Later, due to a complex process called "stream piracy," the Colorado was diverted through the present Grand Canyon and on to the Gulf of California.

Geological researchers have proven by experiment that water flowing across gently sloped land shapes its stream into looping meanders. The more gentle the slope, the wider and more frequent the loops. As the slope increases, the meanders straighten out and become less frequent. From this it is clear that when Plateau rivers first formed, the land was fairly flat and the rivers dropped very little per mile. Now, after some ten million years of deepening and realigning their courses, the rivers drop more rapidly, and the increased runoff from the mountains surrounding the Plateau continues the entrenchment process.

The deeply entrenched meanders of the Colorado Plateau, with their many loops, sheer cliffs and innumerable tributary canyons play important parts, not only in the formation of large natural bridges, but in cutting down through multiple layers of rock and exposing vast reaches of the jointed massive sandstones in which other types of natural spans can form. Thus, the shape of the land itself, past and present, has been critical to span-forming processes.

THE HUMAN FACTOR. As noted earlier, human activities affect natural processes that in turn affect natural span formation. Weather modification affects precipitation amounts, rates and ratios. The destruction of natural surfaces for agriculture, mining and other developments, affects runoff and erosion rates. Blasting and sonic booms can cause direct physical damage to existing spans. All things considered, humanity is the single most violent "geological" activity on the planet at this time, and such activity is bound to affect the formation and destruction of natural spans.

SPECIFIC FACTORS

A number of specific factors can enter into the process of natural span formation, in various combinations. These factors are rock type, composition, mechanics and thickness; rock strata relationship; tectonic faulting and jointing; exposure geometry; exfoliation processes; gravity; flowing water; seeping water; differential erosion; and, occasionally, such processes as travertine deposition and wood petrification. Not all of these factors are involved with any particular span. Each span has its own unique set of factors, and balances between factors, even spans just a few feet from each other, and these factors and balances change during the life of each span.

In the following pages, each of the specific factors is discussed in detail, but it must be borne in mind that in the formation of natural spans, these factors do not work separately. Each factor always works with several others, with the balance between them changing constantly. For example, running water may be cutting through the neck of a meander loop, but at the same time seeping water, exfoliation, gravity, rock characteristics, and strata relationships will be affecting how fast the water cuts, where the opening first appears, and the shape of the opening.

ROCK TYPE. Each of the several types of rock found in the Colorado Plateau -- aeolian, fluvial, marine, igneous and metamorphic -- has its own distinct characteristics. Some of these are related to the way in which the original material was deposited, some to the way in which the resulting rock erodes upon exposure. As dune-sand rocks (aeolian) and water-deposited rocks (fluvial and marine) erode, the erosion tends to follow their original bedding planes. It does not with igneous and metamorphic rocks, because their internal structure is different. Thus, an arch forming in an aeolian sandstone such as Navajo may have the shape of its top or opening influenced by the preserved curve of an ancient sand dune. A natural bridge in a water-deposited rock such as White Rim sandstone may be flat topped because the original bedding planes were flat. An opening within an igneous lava dike may be rather shapeless because of the somewhat random, patternless nature of cooled molten rock.

ROCK COMPOSITION. The chemical composition of rock affects how it weathers and erodes. Most sandstones are silicon dioxide granules, with the sand particles cemented together by water soluble minerals containing various oxides, hydroxides, carbonates and other compounds of iron, magnesium, sodium, potassium and aluminum, possibly with trace impurities of many other minerals. Some sandstones even contain relatively insoluble silicon dioxide cementing. The porosity and mechanical characteristics of a sandstone depends upon the size of its granules and its chemical content. Chemical exfoliation, discussed later, also depends upon the water soluble mineral content of a sandstone. Anomalies within rock -- impurities and irregularities -- also affect rock weathering, erosion and other span formation processes.

ROCK MECHANICS. The tensile, compressive and shear strength of rock is a function of its composition, and these factors affect how it erodes, exfoliates and collapses during span formation. Most if not all Colorado Plateau massive sandstones are not strong enough in shear or tensile to support their own weight, although they are quite strong under compression. Thus, when the base of a sandstone cliff is undercut, it shears away in a near-vertical plane. The shape of the resulting wall will be affected by other factors, such as its original curvature, adjacent formations, and any anomalies within the rock. In general, if the original wall was straight, the shear plane will be near vertical; if concave, the shear plane will slope inward at the bottom; and if convex, the shear plane will slope outward. It should be understood, however, that massive slabs of sandstone can also shear away by stress relief exfoliation, without undercutting. Further, seeping or running water can undercut a cliff in such a manner that the weight above the void is redistributed slowly over an arching opening, so the rock above does not shear off.

ROCK THICKNESS. The thickness of a sandstone formation affects span forming processes in several ways. Relatively thin rock layers do not respond to seeping water in the same manner as more massive sandstones. Thin, laminated sandstones may be relatively impervious to downward seeping, yet be undercut by horizontal seepage and either collapse or form natural bridges. Thicker sandstone is essential to spans that are formed principally by seeping water, and to certain kinds of natural bridges formed by running water. A Rainbow Bridge could not take shape within a sandstone formation less than hundreds of feet thick. With few exceptions, spans form within single rock

El Morro Arch
Verne Huser photo.

Arch 117, as seen from NM 117
Verne Huser photo.

Stimper Window,
Dana Ertel photo.

Dolores River Arch,
Verne Huser photo.

formations, even though they may start at junctions between formations. Thus the total thickness of a rock layer limits and controls the kinds of spans that can form within it.

STRATA RELATIONSHIPS. Where two adjacent rock strata have drastically different characterstics, erosion can undermine the more resistant, possibly creating a natural opening. If a more erodable layer is removed, exposing the top of a more resistant rock, this can lead to pothole formation, seeping water and span formation. If a fairly permeable rock layer has one less permeable below it, seeping water can form spring-seep alcoves at the interface, which can in turn create natural spans. Where a meandering water course has been cut into a rock like Navajo sandstone, when the water course reaches the harder Kayenta layer below, undercutting of the Navajo will begin, accelerating the collapse of Navajo walls and straightening the water course. This process is clearly visible in aerial views of Rainbow Bridge. As noted earlier, many of the largest concentrations of natural spans in the Plateau owe their existence to particular combinations of rock strata.

TECTONIC FAULTING AND JOINTING. The difference between a fault and a joint is that with a fault the faces of rock on each side of the crack are displaced to some extent, while with a joint they are not. As far as natural span formation is concerned, there is little difference between the two, as long as the displacement was not very great, and the resulting crack was later cemented shut by seeping, mineral-laden water. In a few instances, however, faults are so great that they leave whole cliffs of arch-forming rock exposed, while the matching rock on the other side of the fault may be far underground. This can enhance the chance of span formation within the exposed cliff line.

Both tectonic faulting and jointing within the Plateau took place ages ago, but they still affect how natural spans form. Even well cemented joints, once exposed by erosion, can guide water flow and seepage, affect exfoliation patterns, and be re-opened to create vertical fins of rock in which spans can form. Within the jointed, tilted strata of monoclines, joints that are close together can contribute to the formation of double spans. Double Arch in Upper Muley Twist Canyon of Capitol Reef National Park formed this way.

Areas of rock that have been heavily jointed in their geologic past are much more apt to form natural spans once exposed to erosion than the same kind of rock that was not jointed. This explains the proliferation of natural spans in areas that have been deformed by past tectonic activity. One writer has claimed that natural spans do not form in the absence of jointing, but this is not true. Spring-seep spans can and do form in unjointed sandstone, as do other types. There are also many cases where the only part played by jointing in span formation was the creation of rock fins, with the actual opening made by other processes.

EXPOSURE GEOMETRY. The present physical shape of exposed rock affects and limits other natural span-forming processes. Within the Colorado Plateau region rock is exposed in several general forms, such as cliffs, terraced ledges, and open expanses of rock that varies from almost flat to mazes of rounded fins and domes. The shapes of cliffs can also vary. Some slope inward at the bottom, others outward. Cliff curvature can vary from gentle convex or concave sweeps to sharp turns. Protruding abutments and the slashing cuts of sidecanyons can complicate the geometry of canyon walls. Minor drainage lines in the rock formations above canyon walls can channel water flow and seepage in ways favorable to the formation of alcoves and openings in the cliffs below.

In open expanses of flat or rounded rock masses, the flow, entrapment and seepage of water is dictated by the surface geometry. This in turn affects whether natural openings form in the rock, and how. Patches of relatively impervious "caprock" left perched on older mounds or ridges of massive sandstone can lead to span formation by erosion of the older rock from beneath the caprock. This effect is quite noticeable in some areas, for example to the west of the Warm Creek Arm of Lake Powell. Terraced rock layers affect the flow of water and can lead to the formation of small to moderate sized natural bridges. The Kayenta formation is especially susceptible to this.

In sum, present rock shape controls exfoliation processes and the flow, entrapment and seepage of water -- a major span forming agent -- and to that extent affects other span-forming processes.

Skylight Arch, in Wahweap Basin,
formed in a hard, impervious layer.

EXFOLIATION PROCESSES. There are three types of exfoliation -- chemical, heat and stress -- and there are two types of stress exfoliation, stress relief and stress pattern. Chemical exfoliation is brought on by the reaction of moisture with the mineral content of rock in such a way as to weaken surface layers of rock. These then fall away as granules of sand or thin sheets or plates of rock. Within massive aeolian sandstones, chemical exfoliation may be influenced by the curvature of bedding planes and by such bedding irregularities as varying sand grain sizes, limited wetting of the original sand dunes, and slight differences in mineral content.

Chemical exfoliation effects can be observed in spring-seep caves throughout the Colorado Plateau. The moisture can be surface water from rain runoff, but is usually seeping water that causes chemical exfoliation when it finally reaches a surface. This slow but continual scaling away of moist rock can create massive alcoves and caves in time. Such cavities are common within the Plateau. The plant life that usually thrives on the moisture in such caves contributes to the process.

Chemical exfoliation is also a major factor in the forming of potholes in sandstone. Even a tiny amount of water standing on a sandstone surface dissolves enough cementing that after the puddle dries, a few more grains of sand have been liberated for the wind and later rain to remove, thus deepening the depression. Each succeeding cycle of chemical exfoliation deepens the depression still further, until eventually the pothole may be deep enough to hold water for months, or even year-around. If such a pothole is near a cliff rim, it may become deep enough to join the alcove that its seeping water has formed in the cliff below the pothole, thus creating a pothole arch. Chemical exfoliation may appear slow to human perception but as a geologic process it is fairly rapid, and it plays an important part in the formation and shaping of most Colorado Plateau natural spans.

Heat exfoliation is similar to chemical exfoliation in that it causes rock to flake away in fairly thin slabs, but it is not nearly as common. Stresses caused by extreme surface heating of rock generally occur on relatively level expanses of rock, and the resulting jointing may correspond to weaknesses in bedding planes, as in the curving layers deposited in aeolian sandstones. Humped slabs of heat-exfoliated rock that are still attached in several places are common on large exposures of Navajo sandstone. Some such slabs that are still connected all around, like big unbroken blisters, can be detected only by the hollow sound they make when walked on. While heat exfoliation contributes little to the initial formation of natural spans, it can be one of the processes that contribute to the subsequent shaping of the upper surface of an existing span.

Stress relief exfoliation is another major factor in span formation. The Colorado Plateau rock presently exposed formed when it was beneath a mile or more of newer sediments. The weight of this enormous overburden caused the deeper rock to form under great pressure. In many areas additional pressures were added later by crustal deformations from tectonic activities. Such pressures are still within the rock as it is exposed on the surface, especially in massively thick sandstones, and are released in the form of stress relief jointing. Such joints form roughly parallel to the exposed surface, without regard to bedding planes or even older tectonic-formed joints within the rock. They may form anywhere from a few inches to many feet in from the surface. Research has indicated that in Plateau massive sandstones, most stress relief joints form within twenty feet of an exposed surface, with none deeper than about thirty feet. Beyond this depth, the strength of the rock is great enough to contain its internal stresses and resist jointing.

In contrast to joints created by ancient tectonic activities, stress relief joints are relatively recent and have thus not been well cemented by seeping, mineral-laden water. They are usually tight cracks, but do permit water channelization. This flow often contributes to natural span formation.

If a cliff or rock slope that has a stress relief joint a few feet in from its surface is undercut either by flowing water, or by water seeping down through the joint, it may start collapsing in a complex pattern dictated by the changed internal stress patterns of the loosened slab of rock. This pattern usually makes a shallow arcing alcove at first, then, as more rock collapses above the arc, it rounds toward an "ideal" shape which is approximately, but not exactly, a half-circle. Such arch-topped alcoves shaped by stress pattern exfoliation are common within Colorado Plateau sandstone cliffs. Some of them, such as the Great Arch of Zion, reach enormous proportions. With some, the stress relief joints eventually erode, chemically exfoliate or selectively collapse in such a manner as to create sizable openings behind the arc of rock, thus creating clearly visible, free-standing natural spans such as Kolob Arch in Zion National Park.

In outward sloping walls of sandstone, stress relief exfoliation sometimes creates another unique type of natural span. If an arcing slab of rock is not undercut, or is undercut at each end rather than the middle, it may collapse in such a way as to leave a remaining, relatively slender arc of rock that is attached at top and bottom, but with a sizable opening where the original joint was, thus making a jughandle-type arch. Such stress relief exfoliation jointing may even form double-handled spans, such as the one in Four Arches Canyon in southeastern Utah. There are many jughandle spans in the Colorado Plateau. They are a distinct type of span that is rarely if ever described by writers in this field of study. They are unique in that they are formed almost exclusively by just two factors -- stress relief exfoliation and gravity.

Human activities within the Colorado Plateau directly affect stress pattern and chemical exfoliation processes. Small amounts of moisture cause chemical exfoliation. Weather modification programs can increase total precipitation or change the proportions of rain and snow. Further, both stress pattern and chemical exfoliation are drastically accelerated by the upward seepage of standing water in such reservoirs as Lake Powell.

Standing water also brings the premature collapse of stress relief jointed slabs of rock by undercutting them. Many such collapses can be seen in the sandstone cliffs that bound Lake Powell. Big construction projects that cut into natural sandstone cliffs, such as the building of Glen Canyon Dam and the highway cut through Comb Ridge in southern Utah, can cause new stress relief jointing and collapse. This has been a problem at these two sites, as well as others.

In sum, two of the three types of exfoliation -- stress and chemical -- play critical parts in the formation of most Colorado Plateau natural spans. Without them, there would be very few spans in the region.

GRAVITY. Gravity plays an inconspicuous but still very important part in natural span formation. None of the processes that shape rock by loosening rock particles of various sizes could work effectively unless gravity aided in the removal of the freed material. Gravity also works directly by pulling downward constantly on rock masses, thus creating internal stress patterns that shape the rock above openings that have started at the bases of cliffs and elsewhere. Gravity works by directing the flow of surface water and, to a much lesser extent, the direction of flow of seeping water, both of which play critical parts in the formation of natural spans. In special circumstances, gravity can form natural spans virtually by itself. If a protruding buttress of rock is jointed or otherwise fractured in just the right way, gravity, with very little help from water, can cause pieces of rock to fall out, thus creating an opening in the buttress. There are more than a few such "gravity spans" in the Plateau. Two are Hunter Arch and Window Arch, both in canyons near Moab, Utah. Others are Keyhole Arch in Wetherill Canyon, now a tributary of Lake Powell, and Angels Window, on Cape Royal in the Grand Canyon.

Hunter Arch, Hunter Canyon

FLOWING WATER. Flowing water can contribute toward natural span formation in several ways, whether the flow is continual or intermittent. It can undercut cliffs by eroding away softer material and causing collapses. In the narrow necks of entrenched meanders, this can create openings through the cliff. Flowing water can abrade cliffs and slopes of rock with the rock and other debris it carries along. It can contribute to chemical exfoliation by wetting higher rock surfaces by capillary action, by this plus splashing at the bases of pouroffs, or by direct wetting during flash flooding. It can create series of cascade potholes on sloping rock surfaces, some of which may become undercut and form spans. Rain water running down cliff faces can remove softer inclusions in the cliff, forming numerous small natural openings. These are common in the walls of some Plateau canyons.

In rare cases, intermittent water flows can grind potholes in their solid rock beds by twisting and turning harder "pestle rocks" that are too heavy to carry downstream. This is fairly rare in sandstone country because most potential pestle rocks are also sandstone and disintegrate too quickly to do much work. Further, in this land of violent flash floods, few of the suitable pestle rocks are heavy enough, or their potholes ground deeply enough, to prevent them from being washed away. Such mortar-and-pestle hole grinding is more common in country with harder rock and more gentle waters.

Flowing water within entrenched meanders often cuts through narrow meander necks to form natural openings but, because of the mechanics of such cutting, the opening may not at first be a natural bridge with the stream flowing through it. The odds are that with a stream cutting sporadically at both sides of a narrow neck of rock, at any particular time one side will be further along than the other. When this occurs, the initial rock opening will probably be well above stream level. Even if the first opening is at stream level, debris on the other side of the new opening in the rock wall may prevent the stream from going through it. Thus, in most cases, openings cut through meander necks go through at least two and possibly three phases -- a window opening above the stream level, an opening at stream level but blocked on the other side and, finally, an opening the stream can flow through. The various stages of this sequence can be observed in the Escalante canyon system and elsewhere. As noted earlier, it is quite likely that famous Rainbow Bridge began as a window high above the water flow. Then, as the water cut down to the top of the Kayenta sandstone, the opening was enlarged by faster undercutting and collapse until the water could flow through the opening.

Most water-cut natural bridges form as a stream cuts through one of its own meanders, but there are several exceptions to this within the Plateau. In Natural Bridges National Monument, with two of the three big bridges -- Kachina and Owachomo -- tributary canyons cut through into adjacent canyons just above their confluences. In Capitol Reef National Park, Hickman Natural Bridge formed when the water flowing in one drainage cut through a rock wall separating it from a lower drainage system.

In sum, flowing water plays an important and complex part in the formation of many Plateau natural spans in addition to the obvious "natural bridges."

SEEPING WATER. Seeping water works far more slowly than flowing water, but is an even more important factor in natural span formation. Any precipitation that falls within the Colorado Plateau that is trapped and does not run off immediately becomes a part of a process common within the region, but quite uncommon elsewhere. Water can be caught many places -- in sands and soils, in porous rock, in cracks and crevices, in underground cisterns and aquifers and, more obviously, in the rock potholes that are so plentiful in intermittent water courses and wherever expanses of bare sandstone rock exist. Any water so held continues to flow, however slowly, toward a lower level. As long as this seeping water remains below ground, it plays no part in span formation. But whenever it reaches the surface, whether on a lower level, a slope or a cliff face, if that surface is rock the seeping water weakens it by chemical exfoliation and by

A. SIPAPU NATURAL BRIDGE.

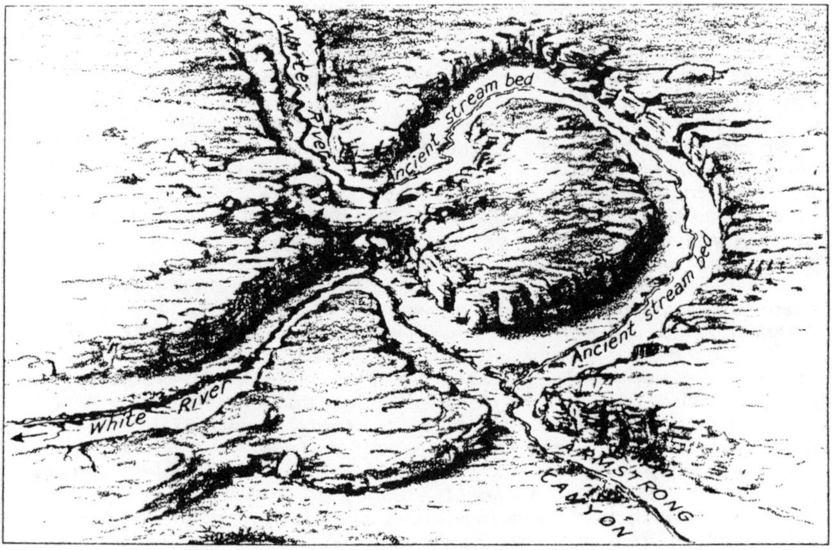

B. KACHINA NATURAL BRIDGE.

DIAGRAMS SHOWING FORMATION OF NATURAL BRIDGES.

Reproductions of illustrations from a 1938 U. S. Geological Survey report, showing how flowing water cuts through canyon meanders.

dissolving and removing some of the rock's cementing material. This can eventually form natural openings.

Surface potholes in sandstone are very common in the Plateau region. Viewed from above after a rain, thousands can be seen in large expanses of slickrock. The Waterpocket Fold in Capitol Reef National Park received its name from the myriad water-catching potholes on its surface. When such potholes form in open rock, the water slowly seeps into the rock or evaporates. Potholes near cliff edges, however, can readily form natural spans. The water seeping into the rock from the pothole travels downward through the permeable rock. Some of it reaches the cliff face and promotes chemical exfoliation, creating a seep-cave.

Depending on how near the pothole is to the cliff face, its water may go downward until it reaches the next formation below. If this is less permeable, the water will then travel to the cliff face at the base of the permeable formation, creating a spring-seep cave there. The seeping water may, however, encounter one of the lenses of impervious rock that so often occur within aeolian and beach-sand sandstones, and thus be channeled toward the cliff face, to form an alcove above the impervious layer. The seeping water can also be directed by cracks, faults, joints, bedding planes and other anomalies within the permeable rock, and these will each affect how and where the moisture reaches the cliff face and the shape of the resulting alcove. If the pothole is near enough to the cliff face, an alcove will form below it even though the rock through which the water is seeping is completely uniform, with nothing in it to channel the water. Such alcoves are elongated -- deep at the top, but tapering to nothing toward the bottom as the water seepage diminishes. There are many such alcoves that have been formed almost exclusively by seeping water within uniform, massive, permeable sandstones.

As noted earlier, potholes deepen continuously, and every pothole near a cliff face that is deep enough to hold much water will have a seep-cave below it. The rate of seepage will depend upon the capacity of the pothole, the porosity of the rock, and the local supply of water. The size of the alcove will vary with the rate of seepage, the capacity of its parent pothole, and the distance between the pothole and cliff edge. Alcove size may also be affected by rock characteristics which determine how far below the cliff rim the alcove forms. Many such pothole-alcove combinations eventually join, to form "pothole arches." Such spans are probably the most common type within the Plateau, although they are very rare elsewhere because they only form in massive permeable sandstones. Certain rock formations in the Plateau are especially suitable for the formation of pothole arches. These are the Navajo, Entrada, Wingate, De Chelly and Cedar Mesa sandstones, plus related strata under different names in western areas of the Plateau.

Seeping-water alcoves can also form another kind of natural opening. Occasionally, adjacent alcoves join at the rear, to form columnar spans. Seeping water also plays a part in the formation of some natural spans other than through chemical exfoliation. During the colder season, seeping water can freeze and thus expand. This can accelerate the deepening of potholes and alcoves, and expanding ice can easily wedge apart rock cracks and stress relief joints, causing rock falls and massive gravity collapses. Freezing seep-water can also accelerate the collapse of rock weakened by stress pattern exfoliation.

Certain types of natural spans are formed almost exclusively by seeping water. Water seeping downward from potholes on top of a ridge, fin or buttress of sandstone can form a span by creating seep-caves on each side of the ridge. Sometimes embedded lenses of impervious rock within the ridge help in the process by forcing the water to travel laterally. Massive natural spans are created when such opposite alcoves join. There are many such spans. Good examples can be found within Arches National Park. While it may seem unlikely to those unfamiliar with Colorado Plateau topography, potholes on top of narrow sandstone fins and ridges are very common. Some may reach depths of fifteen to twenty feet. Viewed from below, a series of high, narrow slickrock fins in such sandstones as Navajo, Entrada, Cedar Mesa or De Chelly may seem to be so rounded on top that they could not possibly be potholed, but a hike across the tops of such fins will reveal that many do have potholes, some deep enough to hold water perennially.

Potholes are not, however, absolutely essential to the formation of double-alcove spans in narrow walls of sandstone. Even without potholes, water from rain and melting snow can penetrate a fin's upper surfaces, then seep downward, or a sand dune against one side of a fin may contribute moisture to its base. In such cases the process is the same as with fin-top potholes, but slower.

In such non-uniform rock as the Wasatch formation, which is exposed in the Bryce Canyon-Cedar Breaks area, some of the work of forming the many windows found in exposed narrow ridges of rock is done by the surface weathering of softer layers, which leaves holes between the irregular harder layers. However, seeping water captured from rain or melting snow helps in this process. It also plays a major part in the formation of the large alcoves that are so plentiful just below the rims of canyons, some of which eventually break through at their tops to create large but relatively short-lived natural spans.

Wall of Windows, Bryce Canyon National Park. N.P.S. photo.

FORMATION MODES

DIFFERENTIAL EROSION. Once an opening has formed in rock, the growth and shaping of that opening is governed by many factors, one of which is the fact that weathering, erosion and other destructive processes do not work at uniform rates on all rock surfaces. Upper surfaces are subject to heat exfoliation, and a surface exposed to direct sun and rain will weather (decompose) more rapidly and thus lose material faster than a vertical surface that receives far less sun and rain. An underside may, however, decompose faster because of seeping water and the resulting chemical exfoliation, or it may be very stable if no seeping water reaches its surface. Other factors, such as stress pattern exfoliation, also work on the undersides of free-standing spans of rock, but not upon other surfaces. Differential erosion plays an important part in natural span formation, shaping and life expectancy. It is discussed in more detail later in this chapter and in the next chapter.

TRAVERTINE DEPOSITION. Although there are no travertine bridges within the Colorado Plateau, Tonto Natural Bridge in Arizona is on the borderland of the Plateau and is mentioned here because of the rare but not totally unique manner in which it formed. Ages ago, a mineral spring in the side of a narrow valley slowly deposited a growing terrace of travertine. As the terrace advanced across the valley, it forced the valley's small freshwater stream toward the opposite side of the valley. In time, a thick travertine deposit dammed the valley, seepage from the valley stream cut a channel under the travertine dam, and the entire stream diverted into this channel. This flow, plus seepage from the surface, eventually created a large tunnel beneath the huge travertine mass. There is only one such span in the United States, but others have been reported in Italy, Turkey, Africa and Chile.

WOOD PETRIFICATION. The fact that petrified wood is generally harder and more durable than the deposits in which it occurs contributes to the formation of one type of Colorado Plateau natural span. The erosion of softer rock or other deposits by rain and runoff sometimes leaves petrified logs bridging narrow canyons, dry water courses or minor gullies. There are several such pertrified log bridges in Petrified Forest National Park in Arizona, and in a few remote locations in Utah. One of these is within the inner gorge of the Colorado River and its logs are remnants of the oldest true trees on Earth.

Historic N.P.S. photo of Onyx Bridge,
Petrified Forest National Park.

OPENING ENLARGEMENT PROCESSES

Once a natural rock opening has been formed, the various factors that contributed to its formation continue to work at enlarging the opening, but with a different balance between those factors. Following are examples of how this balance would change with three types of natural openings.

Once a pothole has broken through into its seep-cave, the action of seeping water will be drastically diminished because there will no longer be a pothole full of water feeding the process. Opening enlargement from chemical exfoliation will depend solely upon water from rain and snow-melt.

Once a jughandle type opening has been formed, it may in a few cases be enlarged by the subsequent collapse of still another stress relief exfoliation slab behind the opening, but more probably the opening will be enlarged very slowly by upper-surface weathering, by under-surface chemical exfoliation from slight amounts of seeping water from rain and snow-melt, and by freeze-expansion of moisture in the winter, assisted by gravity.

Once an opening has been formed by seeping or flowing water at the base of a stress relief exfoliation slab, most opening enlargement will be from stress pattern exfoliation, perhaps with some chemical exfoliation from seeping water.

Stress pattern exfoliation works only where the weight of a mass of rock is redistributed above an opening by its loss of load-bearing capability. The concept of this kind of exfoliation in relationship to natural spans was recognized early in the history of Colorado Plateau spans, although not fully described until much later. In his 1911 National Geographic description of Rainbow Bridge, Pogue noted that the span had reached a "curvature of maximum stability," but he did not go on to explain his remark.

In his 1953 report on the geology of the Henry Mountains region, Hunt partially described the process in relation to the formation of alcoves, but did not extend this to natural openings:

> "When an (alcove) has been started -- if the strength of the rock permits, it may grow to the limit of the thickness of the formation by solution of the cementing matrix along joint planes permitting blocks to fall from inside the (alcove). Joint planes within the roofs of the (alcoves) are conspicuous, but in the adjoining cliffs few can be found. They may be due to the opening of incipient, or tight, joints by the new stress relations that are set up in the roof as the supporting rock is removed."

Hunt was correct in his suspicion that the joints only found within the alcoves were caused by "new stress relations," but incorrect in assuming that the openings were enlarged by "solution of the cementing matrix" along such joint planes. Most such exfoliation joint planes are fairly new. Any cementing in them got there after the joints formed, and merely delayed very slightly the gravity-collapse of the rock plates defined by the joints.

In a 1970 report in the Geographical Society of America Bulletin, Edwin S. Robinson of the Virginia Polytechnic Institute described the "mechanical disintegration" of massive sandstones, using the Navajo sandstone in Zion Canyon as an example. The report described the process of gravity-collapse within arch-roofed alcoves formed in the massive slabs of rock created by

La Ventana, near New Mexico 117, exhibits typical stress-relief joint planes. Verne Huser photo.

joint planes parallel to cliff faces, but attributed all such joints to ancient tectonic stresses rather than attributing some to recent stress relief exfoliation:

> "Extensive regional joint systems have developed in the Zion Canyon region during and possibly prior to the uplift of the Colorado Plateau."

While tectonic stress joints do often parallel canyon wall surfaces, stress relief joints always do. Within most Colorado Plateau massive sandstones, stress relief jointing dominates except in specific areas distorted by tectonics. In Zion Canyon both types occur, but Robinson overlooked this fact by noting that big slabs of rock created by jointing formed arch-topped alcoves when undercut by seeps or flowing water:

> "As support is removed from beneath the plate and cementing material is dissolved and washed from the joint plane --"

Old joints have cementing, but stress relief joints have little if any. What counted, however, was that the writer noted that such uncemented joints leave their isolated slabs of rock supported largely at their bases, and that as openings develop under them, their weight is redistributed to each side of the opening. The writer then went on to describe how this affected subsequent opening enlargement. He drew sketches showing how the "curved stress trajectories" (patterns) all around the opening affected exfoliation within the opening. He noted that rock above Pogue's "curvature of maximum stability" was under compression, while rock below such an ideal arc was under tension and thus prone to relatively rapid failure "along arctuate surfaces" by the collapse of the curved exfoliation plates. He also noted that this process was affected by rock composition:

> "Arches would not be expected to maintain an ideal symmetry during development owing to imperfections in the rock and pre-existing weakness along bedding planes --"

More recently, geographer Dale J. Stevens noted in his 1974 report about southeastern Utah arches that:

> "The form of the opening is most frequently an arching shape which results from either prolonged differential sapping (seeping water affecting the underside of the opening) or failure along arching tension planes of the rock. As stress changes on the surface of a large rock, mass rupture or deformation may occur with failure following along shear or tensile planes. As chunks of rock fall from the rock mass, stresses change and the arching span of rock creates new stresses which tend to favor and sustain an arch."

One abutment of Covert Arch, in Wintercamp Ridge, shows typical stress-pattern exfoliation striations.

FORMATION MODES

To summarize this important process by which rock openings enlarge:

1. The weight of a relatively thin slab or fin of massive sandstone is supported primarily on its base, whether it is a free-standing wall, or a slab separated from the main rock mass by a stress relief exfoliation joint and still leaning against the parent mass with the joint tightly closed.

2. When either flowing or seeping water undercuts such a base-supported wall of rock, an arching opening develops, and a new stress pattern is created in the rock above the opening.

3. Inherent within this stress pattern is an "ideal arc" that is almost, but not exactly, a half-circle. The shape of this arc is defined by gravitational pull and internal rock strength, and its "legs" are just outside of the existing opening.

4. All rock above the ideal arc is under compression, while the rock below the arc is under tension. Rock under compression is very strong. Rock under tension is fairly weak and tends to fail quickly, in this case by the collapse of stress-defined curving plates.

5. Since the rock below the arc is under tension, the opening will tend to develop toward the shape of the ideal arc. Further undercutting of the base by any means will make the ideal arc larger. Thus the opening may grow laterally as well as upward, but it will still tend toward the ideal arc shape. As noted earlier, imperfections in the rock may affect this trend.

6. As the upper contour of an opening approaches the ideal arc shape, stress pattern exfoliation will slow. This in turn will slow opening enlargement, although rock composition and other factors may complicate or even overcome this trend.

7. Once the opening reaches the ideal arc shape it may enlarge further, but will tend to enlarge uniformly under the entire opening, retaining its ideal shape, unless other factors interfere. The enlargement will be by the collapse of plates of rock parallel to the opening surface, along planes defined by internal stress patterns. Such curved plates of rock that have not yet fallen can be seen on the undersides of many spans, including Rainbow Bridge.

8. With such spans as pothole arches, where the opening is horizontal, stress pattern exfoliation will already have played a part in the formation of the seep-cave below the span, and will continue to affect the shaping and enlargement of the cave, although the internal stress patterns in such arches are more complex than those in openings that develop in free-standing vertical walls or fins of rock.

9. In the heavily jointed and cracked walls in which gravity windows form by the falling away of blocks of rock, stress pattern jointing plays a very small part if any, because the rock is too fractured for the usual arcing pattern of internal stress to form.

10. With jughandle type arches, the arc of rock is usually already a segment of a very large ideal arc, and the span is not

bearing the load of an overhead mass of rock. It is thus little affected by arc-forming internal stress patterns.

11. Natural bridges formed in harder, thinner layers of water-deposited rock, such as Kayenta sandstone, are less subject to arc-forming internal stress patterns because they are much stronger than massive aeolian sandstones. Such bridges may be fairly flat on top, and even for small lengths on their undersides, but if they are to last very long, their abutments must provide a certain amount of arching support. This support can be seen under such flat-topped spans as Owachomo Natural Bridge, Morning Glory Natural Bridge, the Gemini Bridges, Musselman Arch and Uranium Arch. These last two "arches" are actually natural bridges. The length of a bridge that can be flat on both top and bottom is a function of rock strength and the area of the bridge's cross-section. The longer such unsupported flat lengths, the more their undersides are subject to tensional stresses that can cause their collapse.

12. With such spans as Landscape Arch in Arches National Park, where the entire span is below the ideal stress arc, the upper part of the span is under extreme compression. Such an arch can last only as long as its abutments give adequate support to that lateral force, and nothing else causes its collapse. With Landscape, the abutments are more than adequate, but the span itself is so thin that it could collapse at any time, either from internal failure or from the shock of seismic mineral search explosions or aircraft sonic booms. Earthquakes of any significant intensity are very rare within the Colorado Plateau.

With natural bridges formed in canyon meanders by flowing water, opening enlargement will be relatively rapid as long as the flowing water is widening the base of the opening either by abrasion from flooding, or by continual wetting. This rapid widening will also accelerate upward enlargement by making the internal stress pattern ideal arc larger and thus placing more of the underside of the span under tension.

Some natural bridges, however, eventually reach a stage of development where the streams that formed them no longer attack their bases. The development of such spans then slows, with only the processes of chemical and stress pattern exfoliation continuing to enlarge their openings, although internal compositional anomalies may also play a part.

The White Canyon bridges represent three different stages in this rate-of-opening-enlargement process. Kachina is in an early stage of development, with intermittent stream flow still actively eroding the base of the opening. Sipapu's opening base is now well above stream action, while Owachomo has reached a stage where the stream that formed it no longer even flows beneath it. The small amounts of rainwater and seeping water that flow under the span are not enough to enlarge its opening.

Opening enlargement processes are somewhat different where the opening is within a single rock formation than when the opening is at the interface between two formations. Generally, openings develop at formation interfaces when the lower formation is softer, or in some way less resistant than the upper formation. Thus, the mode and rate of opening enlargement will mostly be a function of the nature of the lower formation, with the span itself shaped, enlarged and perhaps prematurely destroyed by what happens to the less resistant rock below it.

With another type of opening that occurs at the interface

between two formations, the opening forms within the exposed top of a harder formation once the softer formation above it has eroded away. In such cases, the softer formation may play an active part by absorbing water and releasing it in such a manner as to help opening formation in the harder rock below, or it may play a passive role by simply being absent and allowing the elements full access to the top of the harder formation, thus allowing the creation of arch-forming potholes. This has occurred in Rattlesnake Canyon.

Many free-standing spans, such as Rainbow Bridge, Delicate Arch, Corona Arch and Angel Arch, ultimately reach a state of relative equilibrium, where the faster processes that formed and shaped them have slowed almost to nothing, and only the slower processes continue to work. At that point, such spans can last for a long time, unless there are drastic changes in climate, or unless human activities change local environmental conditions.

Geographer Stephen C. Jett noted in his 1980 preface to Kluckhohn's book about his 1925 trip to Rainbow Bridge that there is evidence that the span's opening enlargement process was greatly accelerated, even before it formed, as the sandstone fin split:

> "Rock remnants adjacent to either end of the Bridge suggest that the (original meander) wall was further thinned by collapse of the upstream half -- from which the downstream half was separated by a vertical crack or joint running through the sandstone."

With the rock fin from which Rainbow Bridge was to form made thinner by this collapse, the original opening may have formed sooner and, once formed, it probably enlarged more rapidly.

The east abutment of Rainbow Bridge shows fallen rock that has exfoliated along internal stress lines parallel to the inner arc of the span. For scale, note figures, lower right.

Natural rock openings of the same type in all stages of enlargement can readily be observed by studying areas that contain many openings within the same rock formation, or the same combination of formations. Some good areas for this are:

Arches National Park, where there are many spans entirely within the Entrada Slickrock member, many at the interface between the Entrada Slickrock and Dewey Bridge members, and a few between the Entrada Slickrock and Moab members.

Bryce Canyon National Park and Cedar Breaks National Monument, where there are many natural openings within the Wasatch formation.

Canyonlands National Park, where there are many arches within the Cedar Mesa formation in the Needles district to the north of the Abajo Mountains, and in the Maze district in the western part of the park. There are many others within this same formation to the west and south of the Abajos.

The Rattlesnake Canyon vicinity in western Colorado, where there is a concentration of arches along the rim of Rattlesnake and nearby canyons, between the Moab and Slickrock members of the Entrada Formation.

The Escalante Canyon system, where most of the natural spans have formed within Navajo Sandstone.

The Mineral, Bull, Spring, Hellroaring, Taylor, Sevenmile and Day canyon complex in southeastern Utah, where most of the spans are in Wingate Sandstone.

FORMATION MODES

SPAN DEVELOPMENT RATES

Of all the specific factors that contribute to the formation of natural spans in the Colorado Plateau, the only one that is both active and variable is water, with the variables being the amount and form that falls within the region, its rate of fall, and the amount that flows into and through the region from the surrounding mountains. All the other factors are constant, and either do not change or change very slowly. A secondary but related variable, of course, is temperature. A cooler climate brings more freezing, which plays a part in the formation, shaping and ultimate destruction of natural spans. But without water to freeze, low temperatures have even less effect on span-forming processes than the hot desert sun.

To understand how major variations in regional precipitation affect span-forming rates, it is necessary to know how the climate of the Colorado Plateau region has varied during recent millennia, when most of the present spans actually formed.

As noted earlier, a major erosional cycle was triggered about ten million years ago by a continental drift event, the uplifting of the western part of the North American continent. This provided a general change of climate within the Plateau, but during this ten megayears other climatic changes have also produced varying precipitation in the Plateau. While scientists do not know precisely why these changes have occurred, they do know when they occurred, that each change produced a wetter and cooler period that lasted from a few hundred to many thousands of years, and that each was characterized by glaciation.

A major glacial period began about 125,000 years ago, peaked about 18,000 years ago and is still in its receding phase. This major continental glacier covered most of what is now Canada and Alaska and the northern parts of the lower 48 States. It was responsible for the creation of the present Mississippi River system, as well as the Great Lakes that lie between the United States and Canada, neither of which existed before. It is quite probable that most if not all of the presently existing natural spans within the Colorado Plateau have formed since this glacial period peaked.

During this long major glacial period, however, two other climate fluctuations have produced "mini-glacial periods" in the Colorado Plateau region. Each of these was brought on by a prolonged stretch of wetter, cooler climate. This in turn meant more precipitation, more freezing and hence faster span formation rates.

One such period that put active glaciers in all the mountains within and surrounding the Plateau lasted from about 4000 to 900 years ago. More recently, the mountains surrounding the Plateau, and some of the higher ranges within it, experienced still another mini-glacial period. That one lasted from about 800 years ago to as recent as 1850. Remnants of the glaciers formed during this period were seen by the region's early pioneers. Fresh-looking glacial moraine can still be seen on some slopes of the La Sal Mountains. Other mountain ranges within the Plateau were not high enough to form glaciers during this most recent period.

Glaciers themselves have little if any direct effect upon natural span formation, although they do have an indirect effect in that as the climate warms and they melt, water flow within the region increases, cutting existing canyons deeper. It is possible, of course, that such increased flow may even have played a part in the creation of a few of the Plateau's existing natural bridges.

In general, however, glaciers are simply indicators of periods of wetter, cooler climate, and these increases in moisture and freezing did directly affect natural span formation rates. Conversely, between such wetter, cooler cycles span formation rates slow, as moisture totals drop and the balance between rain and snow changes. The snow and ice produced during winters and periods of cooler climate have more effect on span formation than the same volume of rain water. Most rain water runs off quickly, causing a certain amount of direct "mechanical" erosion, but snow and ice generally melt slowly, so more moisture sinks into the ground and rock to seep out later. Seeping water is the single largest active factor in the formation of Colorado Plateau natural spans.

Evidence for a present trend toward a drier climate has been presented earlier in this book, but it is interesting to note that virtually all of the major stream-formed natural bridges in the region span watercourses that are now intermittent. Since it is unlikely that such intermittent, seasonal streams could have formed the spans, it is safe to conclude that most of the shaping of these bridges took place earlier, when the streams were far more active.

It could be argued that during drier climatic cycles violent flash floods are more likely to occur, and that these might compensate for the diminished regular flow. Against this argument is the fact that the kind of continual wetting of canyon walls provided by perennial streams is very important to the creation of openings. It causes fairly rapid chemical exfoliation and weakens sandstone cementing, thus hastening its disintegration. It is this perpetual wetting that makes the standing water in reservoirs so devastating to sandstone. In one sense, it could be said that sandstone is one kind of rock that is not waterproof. Thus, while flash floods do some mechanical damage, they usually do not last long enough to provide more than shallow surface wetting of canyon walls.

To summarize, within the Colorado Plateau natural span formation rates vary in general with the amount and type of precipitation that falls on the region. A period of wetter, cooler climate accelerates formation rates. Rates slow during more arid periods, although the increase of flash flooding during drier periods compensates to some extent for the diminished total precipitation. The Plateau is presently on a general trend toward a more arid climate, although this trend will continue to exhibit minor fluctuations from time to time on the human scale of affairs.

FORMATION MODES

SUMMARY AND EXAMPLES

Within the Colorado Plateau geologic region the stage was set for the formation of natural rock openings by the general factors of continental drift, stratigraphy, past tectonics, climate and topography. More recently, human activities are playing a part. The actual creation and shaping of such openings involves a complex and interrelated blend of two or more of many specific factors, some variable and some constant. With each opening, the blend of these factors differs and constantly changes throughout the life of the span. Because of this complexity, most authors either present simplistic explanations as to how natural spans form, or no explanations at all. While it may be difficult to explain in detail how any particular natural span formed, all of the various factors that could be involved are described within scientific literature and are summarized within this book.

There are minimum conditions for the formation of a natural rock opening. The presence of water is among these conditions with most, but not all, openings. If not enough factors are present, or the right combination of them, openings will not form, or will quickly collapse if they do. Conversely, if the minimum number or combination of factors is present, openings will form. Murphy's Law -- if anything can happen it will, given enough time. Within the vast scope of geologic time, given the geologic composition, structure and climate of the Colorado Plateau, the only questions are -- how many will form, how fast, and how long will they last. In an area that is favorable to the formation of natural rock openings, at any one time there will probably be openings in all stages of formation, from just short of opening, to maturity, to recent collapse.

The general rates of formation of natural rock openings vary with major climatic shifts, but are also affected to a lesser extent by year-to-year variations in precipitation and temperature. Some human activities affect precipitation totals, rates of fall and form, and thus affect rock opening formation rates. There are two types of natural rock openings -- gravity-formed windows and stress relief exfoliation arches -- that are less subject to climate variations.

With any particular natural rock opening, a full understanding of all the factors that could be involved can lead to a good estimation of how the opening was created and shaped, but such a description would probably apply exactly to few if any other openings. With many openings it is possible to determine the "principal factor" in their formation, although there are always several minor factors as well. To give some examples:

With pothole arches, the principal active factor is seeping water, but the permeability of the rock is also vital.

With certain types of windows, gravity is the principal factor, but jointing or fracturing within the rock is also essential.

With stream-formed natural bridges, flowing water is the principal factor, but the canyon must also be an entrenched meander.

With jughandle type spans, the principal factor is stress relief exfoliation, but gravity and the selective undercutting of the joint-formed slab are also vital.

It would be possible to establish general classes or types of natural rock openings on the basis of their principal formation factors, but this could only be done by highly knowledgeable scientists. The results would be of interest to other scientists and serious students of the subject, but would be of little value to the general public.

Most of the processes that contribute to the formation and shaping of natural rock openings are quite slow by human standards. Although a few of us have had the dubious privilege of witnessing the inundation and destruction of a number of splendid natural spans by the rising waters of Lake Powell, the natural processes that form rock openings are generally too slow for human observation. One exception to this is Skyline Arch in Arches National Park. Early photographs, such as the one in the 1928 General Land Office survey made just before designation of the area as a national monument, depicted a large, well-defined opening, but with almost half of that opening still occupied by one of the blocks of rock that defined it. This boulder finally fell from the opening in 1940, giving short-lived humanity a chance to observe the before-and-after of one tiny phase of a geologic process that is almost imperceptible to humans.

Following are some examples of how several specific natural rock openings have formed:

Angels Window, on the north rim of Grand Canyon National Park, formed primarily as gravity pulled blocks of rock from a jointed, layered wall of rock, but rain runoff, seeping water and freezing probably played a part in loosening the blocks.

Cedar Tree Arch, in Rattlesnake Canyon, Colorado, was formed in the top of a massive, permeable sandstone formation by seeping water gathered, held and delivered slowly from an overlying softer formation, then enlarged by chemical and stress pattern exfoliation.

Cleft Arch, in Canyonlands National Park, was formed in a massive sandstone fin by water seeping down through a longitudinal joint to an impervious layer, then horizontally in both directions to create seep-caves on both sides of the fin. Eventually, the caves joined, and the longitudinal joint opened slightly to form the present "cleft." Subsequent opening enlargement was by chemical and stress pattern exfoliation.

Sipapu Natural Bridge, in Natural Bridges National Monument, was formed by an entrenched stream creating an opening through one of its own meander loops. Eventually, the stream enlarged the opening so that the stream was diverted through the opening. The stream continued to contribute to the enlargement of the opening until its bed was well below the opening's abutments. From then on, the opening was enlarged by the slower exfoliation processes.

Bridge Mountain Arch, in Zion National Park, is the remains of a stress relief exfoliation slab that was selectively undercut so that most of the slab collapsed, leaving the present span. There may have been additional exfoliation behind the span that totally collapsed, thus enlarging the original opening. The span is presently subject only to very slow weathering on its surface.

Angels Window

Cleft Arch

Sipapu Natural Bridge

Cedar Tree Arch

Gemini Bridges

Longbow Arch

FORMATION MODES

Hickman Natural Bridge, in Capitol Reef National Park, formed when an intermittent canyon stream cut through its sandstone wall into an adjacent lower canyon. Subsequent enlargement has been by the slower exfoliation processes, perhaps aided occasionally by stormwater runoff.

The Gemini Bridges, in Bull Canyon, Utah, were formed by water seeping down through and enlarging two joints parallel to a cliff face until it met a less permeable layer and was forced to the cliff face. The seep-cave thus formed eventually undercut the two slabs of rock defined by the enlarged joint planes, leaving two parallel spans. The spans may have formed at about the same time, but it is more probable that one is older than the other.

Little Arch, beside Utah 279 in the Colorado River gorge, is a typical pothole arch. It was formed when a large cliff-rim pothole joined the alcove its seeping water had formed. Subsequent opening enlargement has been mainly by weathering and the slower exfoliation processes.

Longbow Arch, on Poison Spider Mesa near Moab, Utah, was formed by water seeping downward from a cliff-rim pothole through a joint in aeolian sandstone until it was forced to the cliff face by a thick "desert playa" impervious layer. The opening has since been enlarged by the slower exfoliation processes.

Dellenbaugh Tunnel, in the rim of upper Spring Canyon, Utah, was also formed by water seeping along a thick, impervious "desert playa" layer within desert-dune sandstone, but its water did not originate from a cliff-rim pothole. Instead, intermittent runoff dug a pothole in a shallow water course well back from a canyon rim. Water held in this depression seeped downward to the impervious layer, then horizontally toward the cliff rim. In time, this undercutting seepage caused the rock to collapse above it, forming a broad terrace of the impervious rock with a sizable tunnel between the pothole and the terrace. So little water flows through the tunnel during the present arid period that much of the exfoliation debris from its recent enlargement still remains on the floor of the tunnel.

Dellenbaugh Tunnel

Hickman Natural Bridge

Tusher Tunnel, in the Tusher canyon complex to the northwest of Moab, Utah, formed in an entirely different manner. There, rainwater seeping down through a joint in a narrow peninsula of Moab member Entrada sandstone widened the joint at the base of the 75-foot-thick rock layer to form a straight opening some 250 feet long. This tunnel is fairly uniform, with its inverted V cross-section averaging about 12 feet wide and 20 feet high. Rain water entering the tunnel through the joint flows its length. Its flat sandy floor thus contains relatively little collapse debris. The Joint Trail, between Chesler Park and Chesler Canyon in the Needles district of Canyonlands National Park, goes through an even larger such joint tunnel.

Tusher Tunnel

FORMATION MODES

Double Arch, in Capitol Reef National Park, formed in the extremely tilted and tectonic-jointed sandstone of a monocline, as water seeped down through and enlarged two parallel joints, until it surfaced in one wall of Muley Twist Canyon. The original openings in the joints were then enlarged by the slower exfoliation processes.

Owachomo Natural Bridge, in Natural Bridges National Monument, had an unusual beginning. The stream in Armstrong Canyon had cut down into a layer of impervious rock, but the less active stream in one of its tributaries had not, leaving the tributary floor somewhat elevated. Eventually, the tributary stream, with its lateral cutting enhanced by the impervious layer, cut through its wall into Armstrong Canyon just above their confluence, forming and flowing through the original opening that was to become Owachomo. Some time later, the tributary stream again cut through the same wall, but even farther above the confluence, thus diverting its flow into Armstrong above the original opening. Since then, the abandoned stream bed behind Owachomo has filled with collapse debris, Armstrong Canyon has been cut more deeply, and the original opening has been widened and deepened by erosion and exfoliation, leaving the present very old but still stable natural bridge.

DELLENBAUGH'S TUNNEL

A friend had told me about a very rare phenomenon -- a natural tunnel through solid sandstone -- and had given me an approximate location. I was on the way there, to see if I could find this strange opening. It was late winter, but the day was bright and warm. With me were two young men, a geologist and an historian. Both were as eager as I was to find the tunnel, and to explore the remote canyon in which it was located.

Access to the area of our search from the nearest paved road was by several miles of rough dirt road, another mile or so on a four wheel drive trail, then the rest of the way on foot. The challenge began as I left the dirt road. After traveling for several hundred yards through very broken terrain, the trail reached the bottom of a broad lowland and shortly entered a dry, sandy wash several yards wide. I drove down this wash between sand dunes and low rock outcrops.

After less than a mile, the sand in the wash gave way to water-sculptured solid sandstone that was scored too deeply for wheels. I slowed our pace and steered up out of the wash bottom onto a higher bench of sandstone, then stopped. My passengers quickly scrambled out of the vehicle to peer excitedly down at the deep gorge into which our shallow wash had abruptly plunged, just a few feet away. We knew such pour-offs were common within this region, but the reality of the spectacular drop was still breathtaking.

Within minutes we were on our way along the rim of the canyon, cameras in hand and eyes peeled for signs of the tunnel. It was supposed to be somewhere along the mile of rim between the pour-off and a spur canyon. For the first hundred yards or so we hiked together, staying as close to the rim as possible, but soon that route became too hazardous for my conservative tastes, so I took safer routes a bit back from the rim, while my companions continued to follow it.

As I studied the depths below us I soon noticed that the 200-foot-deep canyon had cut through one of the "petrified playas" that occur so often in Colorado Plateau aeolian sandstones. I could see each end of the "lens" of harder water-deposited rock. It was visible in both walls of the canyon. The original ephemeral pool that formed it must have been about a quarter of a mile long and at least as wide. Its sediments were about three feet deep over most of its length, but tapered to nothing at each end. Fairly uniform dune-sand rock dominated above and below the lens of harder grayish rock.

The petrified playa affected the canyon rim in a strange way, since it was only about fifteen feet below the rim we were following. At intervals, the impervious layer of rock had channeled seeping water toward the rim. This had caused the softer aeolian sandstone above it to collapse, leaving crescent-shaped alcoves in the rim, floored by the gray rock of the lens, and walled by the aeolian sandstone. At each end of some of these crescent-shaped terraces, the harder rock of the lens floor would continue along the face of the sheer cliff as a narrow, crumbling ledge of harder rock. As the rim became more difficult and hazardous to follow, we took alternative paths more frequently. After slipping on a patch of snow while descending into one of the alcoves, and sliding for a few feet on my posterior and camera pack, I went my way while the other two kept as close to the rim as they could. They were looking for the tunnel mouth, which was supposed to dump into the canyon depths, but I decided to look for its upper end instead. Surely there would be visible clues to a tunnel mouth, if you knew this kind of terrain as I did from having hiked it for two decades.

After another ten minutes of search, with our trails

touching and parting now and again, I saw a strange depression in the terrain ahead. It was walled on all sides, with no drainage apparent from a distance. I think I said something like "AHA!" As I approached, I could see that the hole was in a shallow water-drainage line, and that it was deeper than a pothole should be in such a small watercourse. Then I reached the hole's rim -- and saw that it wasn't just a hole. It was the mouth of the tunnel!

I scrambled down the slope, shouting to my companions, hoping they were near enough to hear me. As I disappeared down the hole they came into view and quickly joined me. Together, we stood looking the length of the tunnel. Its floor was mounded with collapse debris, wind-borne sand and dust, because the water flow through the tunnel was not enough to keep it cleared. We could still walk the tunnel with plenty of head room by following the tiny dry watercourse that wound its length.

At its lower end the tunnel opened out onto one of the crescent-shaped terraces, walled along its curved side, and edged by a sheer drop into the canyon. This terrace was about 150 feet wide and perhaps 100 feet from the canyon rim to the tunnel mouth. It had its own isolated plant community of pinyon and juniper trees, desert shrubs and an assortment of winter-dormant grasses and dry seasonal plants. It also had its own sand dunes. One of these at the mouth of the tunnel forced the tiny water course to travel along one rock wall before plunging into the depths of the canyon.

After photographing this strange and beautiful natural tunnel, we explored the terrace further and discovered another curious thing. Wherever the edge of the hard petrified playa that floored the terrace was exposed, we could see that it contained its own even harder tiny "lenses" of red-brown jasper. On vertical edges these layers of jasper showed as thick lines of color in the gray rock matrix. Where the layered rock had exfoliated along horizontal planes, the jasper was irregular patches of rich color, with each one several square inches in size. Strange! How could this lovely and colorful mineral have formed within a matrix of gray desert-playa sediments? Perhaps some day I will find a geologist who can tell me.

As we headed on from the tunnel to continue our explorations of this wonderful peninsula of rock between two canyon spurs, we discussed names for the tunnel. Surely, such a rare feature deserved a name! Finally my historian friend suggested a name we all thought fitting -- Dellenbaugh Tunnel. Dellenbaugh was one of the hardy adventurers who accompanied Major John Wesley Powell on his early explorations of the Colorado Plateau, and who also documented much of what he saw. Dellenbaugh had never, of course, seen "his" tunnel, but had nonetheless been fairly close to it as the Powell expeditions had boated down the nearby Green River gorge.

And what could be more fitting? A long natural rock opening named after a Colorado Plateau explorer with an equally long name.

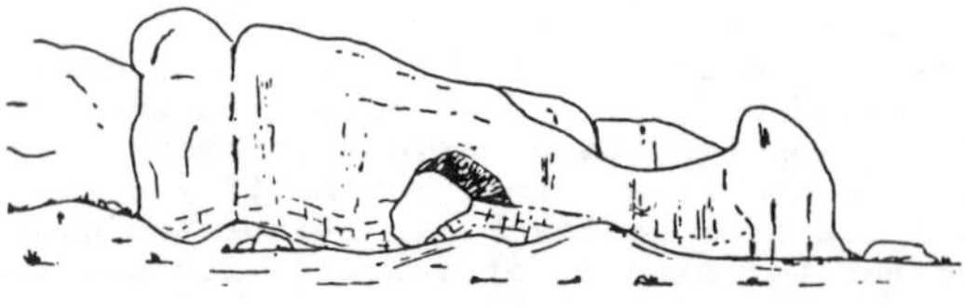

AGE AND LIFE EXPECTANCY

HOW OLD ARE THEY?

How old are the natural spans in the Colorado Plateau? This question is possible to answer only in a general way, by placing existing spans within a high-probability time bracket. Of course, no two spans within this vast region are exactly the same age. If their ages could be determined with accuracy, it would be seen that they range from a few years for spans with smaller openings, to thousands of years for many, and perhaps tens of thousands of years for a very few spans.

There are two principal factors that prevent accurate measurement of the age of any particlar natural span. One is that there is no way to measure how fast a particular mass of rock will erode, even under uniform climatic conditions. The other is that climatic conditions have not been uniform during the lives of most natural spans in the Colorado Plateau.

Even within a fairly short time, span-shaping rates may vary drastically as the more rapid shaping forces, such as stress exfoliation or flash-flooding, remove big chunks of rock within moments, then for years, decades or even centuries the weathering and chemical exfoliation processes continue working very slowly.

Thus, because of drastic variations in weather, major changes of climate, and the different rates at which various span-shaping forces work, it is simply not possible to determine the ages of particular spans with any reasonable accuracy. There are no "age-dating" tests for natural spans, as there are for prehistoric human artifacts and other remnants of life.

It is possible, however, to place most existing spans within general age brackets. As noted earlier, the climate of the Colorado Plateau region has varied over the last few million years. Glacial periods and changes in vegetation are clues to these variations. Tree ring studies give quite accurate pictures of year-to-year climate variations for the past several thousand years. Examination of these clues can lead to at least a general idea as to the ages of the region's natural spans.

One parameter can be established immediately. In a 1975 report to the Four Corners Geological Society on the development of the Colorado River system, E. E. Larson, et al, note that:

> "From 10 megayears on, the existence of the Colorado River system is easy to document -- by contrast, unequivocal evidence for a Colorado River older than 10 megayears has yet to be produced."

Since the Colorado Plateau is essentially the upper drainage basin for the Colorado River and its many tributaries, by implication the region was relatively flat and featureless prior to 10 million years ago, and all of its present fantastic erosional features have been shaped since then.

Geologists know that during this time, erosion has removed at least 5000 feet of material from the Plateau, most of it during the last 5 million years. In a May 1963 report in the Geological Society of America Bulletin, geologist W. C. Bradley bracketed the ages of Colorado Plateau natural spans more closely by noting that stress relief exfoliation within Plateau sandstones is "recent," and that canyon cutting there "-- has been accomplished since late in the Tertiary and probably largely during the Pleistocene."

The Pleistocene period began about two million years ago. If most of the Plateau's canyon cutting occurred during that period, then certainly even the oldest existing natural spans would be only a small fraction of that age. It becomes obvious, then, that present Plateau topography has endured not for ages, but for a relatively short time. To humanity, Plateau landscapes seem unchanging, and they are, indeed, changing quite slowly during the present arid climatic phase. But the loss of thousands of feet of rock indicates that in the past erosion has often worked much faster.

As noted in the last chapter, glacial periods have indicated fairly recent wetter-cooler times. The retreat of the last major continental glacier left much of western Utah and Nevada inundated under immense lakes, because the region has no drainage to the sea, and melting glacial waters simply accumulated to as deep as a thousand feet.

During that same period of time, much glacial melt from the Rocky Mountains drained westward across the Colorado Plateau, carving its canyons deeper and carrying countless megatons of rock and sediment to the sea. Later mini-glacial periods did much the same. The wetter-cooler climate that brought the glaciers increased Plateau precipitation, and that is what accelerated the formation, shaping and destruction of its natural spans. The wet climate phase of 4000 to 900 years ago probably started and shaped many if not most of the larger present spans, while the one that began 800 years ago and that is still phasing out, probably formed many more. A few of the larger existing natural bridges may have been formed by glacial runoff and the concurrent heavier precipitation before these two recent wetter periods. Smaller spans that existed that long ago would have been destroyed by the more recent wetter periods.

To understand how natural spans age, it might be useful to compress geologic time to a scale more conceivable to the human mind. If the life of a span were compressed to just a few hours, it could be seen that its formation and shaping would in general be rapid during the hours of wetter-cooler climate, and would then slow drastically during more arid times. The shaping of some spans would also seem "jerky," as the more rapid shaping processes did their occasional work. Few spans would form, then smoothly "melt" away to their final collapse. Some spans might even be destroyed by the first wetter period encountered, while the life expectancies of those that survived such periods would be extended by the succeeding dry spells.

From what is presently known, it is safe to conclude that most if not all existing Colorado Plateau natural spans are less than some 18,000 years old, and that only the largest and most stable spans are older than 5,000 years. Attempting to judge span age by extrapolating from present relatively slow formation and shaping rates will give ages much too old, because these rates have varied considerably during the lives of most spans.

It must be concluded, then, that most spans are younger than many writers have believed, and that while natural spans may appear ancient to mankind, they are short-lived phenomena on the geologic scale, just as most geologic dictionaries and encyclopedias contend. Following are some ideas about Colorado Plateau natural span ages that have appeared in various publications since these spans become widely known early in this century.

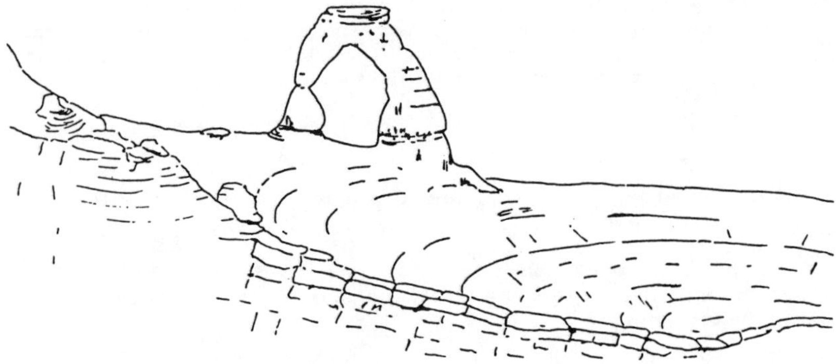

WHAT OTHER WRITERS HAD TO SAY.

In his 1902 book about the second Powell expedition down the Colorado River, Dellenbaugh noted in a photo caption that it depicted "The beginning of a natural arch," thus showing his awareness of the growth-aging process. This contrasted sharply with the philosophy that "everything has been here forever, unchanged" so widely held then, and even now to some extent.

In his March 1907 National Geographic article about the White Canyon bridges, E. F. Holmes made a shaky case for their antiquity by judging their ages on the basis of the human time scale:

> "These bridges, composed as they are of light sandstone, might seem to be wearing away very rapidly. Such, however, is not the case, as is evidenced in the caves beneath the Caroline (Kachina) abutments -- where was found what may have been a workshop of the ancients, for evidences of their pottery work were scattered around, and at one place a number of fiber sandals were found in a fair state of preservation."

Artist H. L. A. Culmer offered his subjective feelings about the ages of the White Canyon bridges in his September 1908 article in Technical World Magazine:

> "A sense of enormous strength pervades (Augusta natural bridge), a sense that it has endured for ages, and will endure for ages yet to come."

H. F. Cleland reached two conclusions about the ages of natural bridges in his July 1910 report in the Bulletin of the Geological Society of America:

> "1. (They) occur in glaciated as well as unglaciated regions, and that the former, at least, have required a short time, geologically speaking, for their formation.
>
> "2. (They) must necessarily be short-lived structures, the life depending upon several factors -- the character and composition of the rock, the climate conditions, etcetera."

In J. E. Pogue's November 1911 National Geographic article about Rainbow Bridge he noted that its shape had the "curvature of maximum stability," but added:

AGE AND LIFE EXPECTANCY

> "Though doubtless requiring many years for its formation, the arch is nevertheless a very recent geological feature, and destined to withstand the forces that gave it being for only a brief period as geologic time is reckoned."

Clyde Kluckhohn, who made a trip to Rainbow Bridge in 1925 wrote that:

> "Unmarred by the water, ungrieved by the wind, it will endure forever."

In his 1929 book about his 1921 trip to Rainbow Bridge, C. L. Bernheimer also noted the stable shape of the span, but in more poetic terms:

> "How has this gem in high relief -- withstood destruction by the very forces that reared it? Imagine a structure so massive that the evolutions of the ages have merely brought to the surface its muscular structure, divesting it of weak and useless particles. How long will it stand?"

The answer to that today, of course, is -- not nearly as long as it would have without reservoir water standing beneath it. Geologist F. J. Pack, in his 1922 book about natural bridges, mentioned that in White Canyon:

> "Edwin bridge (Owachomo) is the oldest of the three structures. The bridge itself is chiseled down to a mere skeleton of its former massiveness. The whole structure is so slender as to give one the impression of great insecurity, but, of course, barring some unlooked for cataclysm, it will probably not collapse for many centuries. The Caroline Bridge (Kachina) is still in its early formative period. The sublime Augusta Bridge (Sipapu) is at its prime. The Edwin Bridge already reaches its old age (and) it cannot be denied that the centuries of its existence are rapidly drawing to a close."

Western author Zane Grey apparently held the conventional view that land forms remained unchanging over long periods of time. In his 1922 book, "Tales of Lonely Trails," when describing his trip to Rainbow Bridge, he commented on the age and life expectancy of the span:

> "I realized that long before life had evolved upon the earth this bridge had spread its grand arch from wall to wall, black and mystic at night, transparent and rosy in the sunrise, at sunset a flaming curve limned against the heavens. When the race of man had passed it would, perhaps, stand there still."

More to date, the question is -- which will fall first, man or span? In 1923, U. S. Geological Survey writers Miser, Trimble and Paige commented more realistically on the probable age and life expectancy of Rainbow Bridge. They even included a qualifying phrase about climatic conditions, but did not dream that human activities would jeopardize this most magnificent of all natural spans:

> "The bridge may have first come into existence during the Glacial period. At any rate the event took place many

thousands of years ago, as measured by the probable time required for the eroding of the present inner canyon. From time to time sheets or slabs of sandstone have peeled from its surface and fallen to the bottom of the canyon, and sand grains that have become loosened by weathering have been carried away by wind and rain. Although these processes have gradually increased the span of the bridge and diminished the diameter of the immense sandstone cylinder that forms it, the bridge is still of substantial proportions and will, under present climatic conditions, stand for many centuries."

In a 1980 preface to a reprint of Kluckhohn's 1927 book, geographer Stephen C. Jett summed up the matter of Rainbow Bridge's life expectancy:

"-- whether one perceives the great Bridge as having been debased by the reservoir water under it, or enhanced, there is no question that the stone rainbow will someday fall. Even if those who believe that the lake water will hasten the collapse of the bridge are proven wrong, the very forces which created the stone span as it is today will inexorably wear away at the great arch, grain by grain, until it is too diminished to support its own weight. Then it will give way, leaving, when the dust has cleared, only the remains of its abutments and a pile of rocky rubble."

Neil Judd, archaeologist and member of the 1909 expedition to Rainbow Bridge, commented about the ephemeral nature of natural spans in his March 1924 National Geographic article about later explorations in that region:

"Two days later we stood beneath the graceful arch of the colossal Rainbow Natural Bridge, marveling at the stupendous folly of Nature, who builds temples to herself and then, in changeful mood, tears them down again."

In the 1941 WPA guidebook to Utah, its anonymous author quoted Zeke Johnson, then custodian of Natural Bridges National Monument, as saying:

"-- one of the most noted geologists said in his opinion Owachomo, which is considered the oldest bridge, was at least ten million years old."

If so, it was under the mile of rock and sediments that covered the region at that time. In his October 1942 article "Rainbows of Rock" in Natural History, paleontologist H. E. Vokes commented on the ages of natural spans:

"But while the bridges are thus, in terms of geological time, only ephemeral phenomena, in terms of human existence they probably achieve hoary antiquity. The extremity of old age is exceedingly well shown by the unusual Landscape Arch bridge of the Arches National Monument. In terms of geologic antiquity, it is living out its last days, soon to collapse and to be remembered only by remnant buttresses on the canyon wall."

Once again, who could have dreamed in those days that such human activities as aircraft sonic booms, seismic mineral surveys and mine blasting would in the future be allowed to jeopardize this rare and delicate phenomenon of nature?

In his September 1943 article about "Nature's Bridges" in The Scientific Monthly, geologist R. E. Janssen commented on the temporary nature of natural spans:

> "Like most scenic formations, natural bridges are ephemeral features in the ever-changing landscape. The same agencies which fashioned them -- are relentlessly wearing away the rock grains of which they are built. Many, which existed in the past, have disappeared forever. Those which exist today will sometime disappear. But others, still in the making, will loom in their stead. Nature, the great builder and leveler of all material things, working hand in hand with time, is continually busy, designing and redesigning new structures in an ancient world."

More recently, in his 1969 book "Scenes of the Plateau Lands," geologist W. L. Stokes said much the same thing, but in less poetic terms:

> "A natural bridge is temporary and insecure; the falling of blocks from its under side and the general weathering of all its surfaces will weaken it and cause an eventual fall. But new ones appear to be forming to replace those that are lost in the ever changing landscapes of the plateaus."

In an Empire Magazine article that was reprinted in Our Public Lands in 1981, writer Zeke Scher observed that:

> "Another million years and Arches National Park can pack it in."

While it is doubtful that humanity will be around then to care, there will still be plenty of natural spans in that vicinity a million years from now, although it is certain that none of the present spans will survive that long. Barring massive human destruction, or some major geologic cataclysm, new generations of natural spans will continue to develop within the newly exposed Entrada sandstone flanks of the Salt Valley anticline and the underlying Navajo sandstone. Thus, there will probably be as many spans there in the future as there are now.

The 1976 "Illustrated Encyclopedia of the Earth Sciences" puts it all into perspective and makes it official by noting that natural spans:

> "-- are of little geologic importance -- because they are short-lived, incidental features doomed by the processes that created them."

Elephant Rock, Valley of Fire State Park, Nevada

FACTORS IN JUDGING AGE AND LIFE EXPECTANCY

Although it is not possible to judge the absolute age of an individual natural span with accuracy, there are usually clues that permit a general estimation, within the established limits for Colorado Plateau spans. Such clues can also lead to an estimation of a span's relative age, that is, its degree of development for the type of span it is, and its probable remaining life.

These clues can be found by examining the span and its immediate vicinity. As noted earlier, no two spans are identical in how they formed, even those within a few feet of each other. Thus, the factors that contributed to these differences also determine their ages to the present, their degree of development and, of course, their future life expectancies.

Even so, the most careful study of a particular span can lead to little more than the conclusion that it is hundreds, thousands, or tens of thousands of years old, that it is young, mature, or old for its kind of span, and that its future is probably short or long. Following are some of the kinds of clues to look for in trying to judge a span's absolute or relative age and its life expectancy, and what these clues indicate. It should be understood that not all of these factors apply to each kind of span.

AGE AND LIFE EXPECTANCY

THE NATURE OF THE ROCK. Water-deposited rock that contains a wide range of particle sizes, with these particles well cemented with minerals, is more durable than rock formed from wind-deposited sand that has a narrow range of particle sizes. Lava and metamorphic rock are also durable. In general, the harder the rock, the longer natural spans in it will last.

UNIFORMITY OF THE ROCK. Defects and anomalies within rock where a span forms can severely reduce its total life expectancy. Thus, spans in such rock are usually not very old. There can be exceptions to this, however, such as the impervious layers often found in some fairly porous sandstones. These anomalies may not affect a span's age at all.

THE SIZE OF AN OPENING VERSUS HOW IT FORMED. A large opening behind a pothole arch usually indicates considerable age, but a large opening created by stress relief exfoliation may be fairly young, depending on other factors. An opening formed by stream flow that is no longer affected by the stream is probably quite old. A similar opening still being undercut by the stream may be much younger.

THE TOTAL GROUND AREA THAT DRAINS THROUGH OR UNDER A SPAN. Since water in one form or another is the principal forming agent for natural spans, the aging of a span is in general related to the amount of water directed toward it. For example, a pothole arch with an opening ten feet in diameter that drains a large area is probably younger than one of similar size that drains a smaller area. Natural bridges follow much the same rule. Spans that have no water flow through their openings are still affected by what falls on them directly, but such spans generally last longer and are hence older.

ELEVATION ABOVE SEA LEVEL. This to some extent controls how much precipitation a span receives, in what form, and how much freezing occurs. As noted earlier, spans age more rapidly in wetter climates, so a span at 6000 feet elevation may be younger than a similar span at a lower elevation.

THE SIZE OF THE HOLE BEHIND OR BELOW A SPAN. Generally, the forces that create a span continue to work. Thus, the openings behind and below certain types of spans continue to enlarge. The nature of the rock affects this process, of course, but in general larger alcove openings indicate that the associated spans are quite old.

HIGHER ABANDONED-MEANDER BEDS. To some extent, the age of a natural bridge in a meander loop can be judged by the height of the abandoned stream bed above its present bed. This can be misleading, however, because some meander loop openings form long before the stream actually diverts through them. Even so, a bridge whose present stream level is below that of its abandoned meander bed is probably quite old. The more the difference, the older the bridge.

DEPTH OF CUT INTO THE NEXT LOWER ROCK FORMATION. With spans near the bottom of such formations as Navajo sandstone, if the opening then cuts into a harder formation below and the cut is fairly deep, the span is probably quite old. If the next formation down is less resistant to erosion and the span is still standing, it is probably not very old.

NEARBY JOINTING, FRACTURING AND COLLAPSE. Such rock defects near a span can indicate that its age and life expectancy are both short, although this is not always the case. The span may be in an unusually durable section of the formation. Even so, if the rock around a span collapses, this can also destroy the span. In general, spans in badly fractured rock are not very old and have short life expectancies.

SURFACE WEATHERABILITY. Rock with little mineral cementing will erode away quickly, as surface moisture removes that cementing and allows gravity, wind and rain to remove the sand granules. A hand rubbed across a vertical surface will give an indication of this. If sand rubs away freely, any span in that rock is probably not very old and will not last long.

SHAPE OF SPAN'S UNDERSIDE. If the opening shape is at or near the ideal arc, the span is probably older than if it is not. It takes time to reach this ideal arc, and many spans are destroyed by other factors before they reach this stable configuration. This clue does not apply to most pothole arches, however, because the alcoves below these spans often acquire an arching shape from chemical and stress exfoliation long before the pothole breaks through and forms a natural span.

ROUNDNESS OF UPPER SPAN SURFACES. Rock surfaces that are well-rounded from weathering are usually much older than those that have sharper edges, although this does not apply to the undersides of spans, which may be subject to stress exfoliation, or to spans in rock with exceptionally high or low resistance to weathering.

Gothic Arch, Canyonlands National Park

AGE AND LIFE EXPECTANCY

DESERT VARNISH BUILDUP. A span in harder rock may accumulate a surface coating of desert varnish. Although the buildup rates for such coatings are not uniform, a dark coating generally means long exposure to the atmosphere. A span with a dark coating on its sides will thus be older than one in the same rock formation with a lighter coating, perhaps by many centuries.

ACCESS OF FLOWING WATER TO THE SPAN'S ABUTMENTS. A span that has developed to the stage where its supporting abutments have not been affected by water flowing beneath it for a long time is probably quite old, and may have a long life expectancy. This is especially true of stream-formed natural bridges, but also applies to types of spans that are subject to transient water runoff.

SIZABLE POTHOLE ON TOP OF SPAN. The presence of such a pothole indicates that the span is probably not as old as a similar span without a pothole on its top. Such water catchments promote more rapid exfoliation on the undersides of spans, and are often what formed and shaped the opening.

INTERMITTENT STREAM BELOW A LARGE NATURAL BRIDGE. In some cases, such a stream indicates that the span was formed and shaped during a wetter climate phase, when the stream was much more active, making the span at least several thousand years old.

OPENING IN A MEANDER LOOP, BUT NO STREAM FLOW THROUGH IT. This is generally an indication of a younger span, although there are other factors to consider in such cases. If the meander wall is exceptionally thick, or its rock unusually resistant, the opening may still be quite old. Other factors should then provide indications of age.

RELATIVELY THIN SPANS THAT HAVE LESS THAN IDEAL-ARC UNDERSIDES. Such spans could be almost any age, but will probably not last much longer, although there are exceptions to this. Thin spans in fairly durable water deposited rock are more resistant to aging forces and may endure for a long time if their undersides are adequately supported.

TOP OF SPAN CLOSE TO IDEAL ARC SHAPE. Since upper rock surfaces wear away much more slowly than under surfaces, if the top of a span is now close to the ideal arc shape, it is more likely to last a longer time than one whose top is more level, especially if the opening is fairly well centered below the rounded upper surface.

CLIMATE AND WEATHER VARIABLES. These can severely affect the life expectancies of existing natural spans. Human activities such as weather modification programs and air pollution affect precipitation form and amount, and even its chemical nature (acid rain), and hence affect span life.

SONIC SHOCK. As noted earlier, human activities that produce blasts of powerful sound can also shorten the lives of fragile natural spans.

EARTHQUAKES. Natural earth shocks can, of course, shorten the lives of rock spans or even destroy them, but geologists have long noted that the Colorado Plateau is exceptionally resistant to earthquakes of any magnitude. It receives many small tremors, but has had no large quakes for many centuries, as proven by the existence of Landscape Arch.

There are countless other clues that can be used to estimate a span's absolute or relative age, all of them based upon detailed knowledge of Colorado Plateau geology, weather and climate, past and present. Many clues to age can also lead to estimates of remaining life expectancies of spans.

SOME EXAMPLES

Rainbow Bridge is an excellent example of a very stable natural span, with almost everything indicating that it is quite old, with a long natural life expectancy. Its upper surface was originally the rounded edge of a large sandstone fin, and the original opening broke through in a location that allowed its later rounding toward the ideal arc to be closely parallel to the slowly weathering outer edge. The stream below it had also cut deeply into the harder Kayenta sandstone, protecting the span's abutments from significant wetting and from flowing-water erosion. The rock in the span is quite uniform. The only process presently affecting the span other than slow weathering, is some continuing stress pattern exfoliation on the underside of the span, but the span is so massive that this would not cause its collapse for thousands of years yet. Unfortunately, the reservoir water now beneath the span has changed several of these factors, and the span's life expectancy is now much less than it was before.

Gregory Natural Bridge was, before its inundation by Lake Powell, a good example of a natural bridge whose stream had abandoned its meander a long time ago. The abandoned meander bed was perhaps a hundred feet higher than the stream bed under the bridge, although some of this was due to sand deposited in the meander by wind. This much higher meander produced a curious effect as Lake Powell waters rose. At first, as the reservoir water penetrated the canyon, it was possible to boat beneath the span and on up the canyon. Then, when the water level reached the top of the bridge's opening, access beyond the span by boat was cut off, because the meander that looped around the span was still high above the water level. The water was then far up the canyon, but boats could not go there until the reservoir level rose still higher and flooded the abandoned meander.

Mineral Canyon Arch in Mineral Canyon near Moab, Utah, is a good example of a canyon-rim pothole arch with an enormous alcove behind and below the span, indicating considerable age. The remaining span of rock is arched on its underside, and hence stable, and its top is very narrow with no potholes to catch water and promote underside chemical exfoliation. It is old, and should last a long time.

Lopez Arch, visible from U.S. 191 south of Moab, is an example of a very young span. The opening is in a rounded fin of fairly uniform rock and is small, but the rock all around the opening is quite thin so it should develop fairly rapidly into a graceful long-lasting span. Lopez Arch is an excellent example of one type of span-in-the-making.

Musselman Arch in Canyonlands National Park can serve as an example of a flat-topped span in rather durable rock with a stretch in its center that is also flat on the underside. Its abutments, however, are well supported by arching rock. The alcove below the span is huge, because it has been cut from a softer underlying formation, but the rock in which the span formed is massive and uniform enough to protect the span from this undercutting for a long time.

Window Arch, in Pritchett Canyon near Moab, is in a badly fractured wall of rock and represents a span without much future. Angels Window, on the North Rim of the Grand Canyon, is also in fractured rock but it will last longer because of its immense size.

SOME SPANS THAT COULD COLLAPSE ANY TIME

Landscape Arch, in Arches National Park. This magnificent natural span is still intact only because of an extremely rare combination of circumstances. It is in danger from human-generated sonic impact, as well as from natural forces. In 1972, the U.S. Air Force announced that by Executive Order supersonic flights over or near National Park areas were no longer permitted. For years this presidential order was obeyed, but in the 1980s it has been ignored countless times. Danger to Landscape Arch has been further increased by mineral search activities near the park.

Aqueduct Arch, in a tributary of Harts Draw, to the east of Canyonlands National Park, is quite mature and badly fractured. It could collapse at any time from natural causes.

Bridge Mountain Arch, near the entrance to Zion National Park, is a relatively stable type of span, but is so long and slender that it is in danger from both man and nature.

As noted earlier, Window Arch and many similar spans are not due for long lives. Any of them could collapse at any time.

Rainbow Bridge is the most famous span in danger. Its life expectancy is even the subject of a federal law suit. A summary of the controversy appears in the later chapter on protection. The most detailed scientific case for jeopardy to the span from Lake Powell water appeared in Landslide Magazine in 1973. In brief, the article pointed out that when wetted, the softer layers of the laminated Kayenta sandstone that support the bridge abutments deteriorate to the point that they cannot support the harder layers. This allows these to collapse, even of their own weight, let alone the massive weight of Rainbow Bridge. The author provided photographs to support his contention, some of them taken just downstream of Rainbow Bridge. The author's position was supported by geologists from several major universities and research organizations. Despite this, the U.S. Geological Survey geologist who originally claimed that reservoir water beneath it posed no danger to Rainbow Bridge, reiterated his contention.

Rainbow Bridge

SOME SPANS THAT HAVE ALREADY FALLEN

Although there were reports of collapsed natural spans in the West as early as 1890, Frederick J. Pack made what was probably the first report of a fallen span within the Colorado Plateau region, following his 1921 explorations in the White Canyon area:

> "Actuated by the hope that somewhere in the vicinity of the present bridges a structure still older than the Edwin Bridge (Owachomo) might be found, the writer recently made an investigation of the area and was rewarded by finding the remnants of a collapsed bridge, situated in White Canyon about two miles above the Augusta Bridge (Sipapu). The abutments of this bridge are still standing, one of them in an excellent state of preservation, and the other badly disintegrated and largely removed. The greater part of the collapsed arch has been carried away by the stream across the channel of which it had fallen. In point of size this bridge had a span as great as that of the other bridges, but its height was somewhat less. The same evidences of origin, in the form of undercutting stream-action and the abandonment of certain parts of the old channel, are present here as elsewhere. As a name for the remnants of this mighty structure is suggested the 'Fallen Monarch.'"

In a 1983 personal note to this author, National Park Service ranger Douglas Crispin added to this by observing that:

> "-- at least two sites upstream from Sipapu suggest locations of former natural bridges with abandoned meanders."

In his 1938 report on the geology of San Juan County, geologist H. E. Gregory noted two collapsed natural spans in Grand Gulch:

> "Prospectors report that about 2 miles below Collins (a tributary canyon) the buttress of a ruined bridge stands like a tower on the canyon floor between the present channel and the abandoned channel, and farther downstream the stub end of a broken bridge girder projects almost horizontally from the wall. Bridges in process of formation and others in ruins seem to be characteristic of intricately meandering, swift streams deeply entrenched in canyons."

In his September 1955 article about exploring the Escalante River system, National Geographic writer W. Robert Moore reported finding a fallen span in the Coyote Canyon arm of the Escalante:

> "At one of the kinks in the Coyote, a short distance downstream from Jacob Hamblin Arch, we saw what remained of the pedestal and the broken arm of a once mighty arch that had collapsed. As time is measured in geological formations, it had fallen only recently, for the broken edges were little eroded. Numerous large rocks that once formed its span lay strewn over the talus slopes and on the canyon floor."

In a 1975 newspaper article, arch historian R. H. Vreeland reported the "recent" collapse of an arch formerly on the rim of Mineral Canyon, to the west of Moab, Utah. In 1977 Vreeland reported a fallen alcove-type arch near the head of Explorer Canyon, another Escalante tributary.

In his 1981 book on Arches National Monument, geographer J. F. Hoffman speculated about the earlier existence of a double opening in the park:

> "In prehistoric times, two large arches -- may have stood in Courthouse Towers. Sheep Rock, and the fin pierced by Hole-in-the-Wall, may be remnants of the abutments of such arches."

In his March 21, 1984, status report on the inventorying of spans in Arches National Park, ranger Ed McCarrick noted that the figures given excluded three known spans that had fallen, two of them since the inventorying had begun.

Perhaps the most dramatic and fascinating case for a fallen western natural span was made in a book first published in 1890, then reprinted in 1902, entitled "The Bridge of the Gods." The book explored Indian traditions of a natural bridge that once spanned the Columbia River gorge. In a note in the 1902 edition, the publisher had this to say about the book's author:

> "By personal inquiry among old natives, he learned that the Bridge which suggested the title of his (book) was no fabric of the imagination, but was a great natural bridge that in early days spanned the Columbia, and later, according to tradition, was destroyed by an earthquake."

In the book, its author, F. H. Balch, discussed the authenticity of the legendary bridge:

> "It may be asked if there was ever a great natural bridge over the Columbia, a 'Bridge of the Gods,' such as the legend describes. The answer is emphatically, 'yes.' Everywhere along the mid-Columbia the Indians tell of a great bridge that once spanned the river where the cascades now are, but where at that time the placid current flowed under an arch of stone; that this bridge was 'tomanowos,' built by the gods; that the Great Spirit shook the earth, and the bridge crashed down into the river, forming the present obstruction of the cascades. All of the Columbian tribes tell this story, in different versions and in different dialects, but all agreeing upon its essential features as one of the great facts of their past history."

The author then went on to quote several recorded interviews with older local Indians about this story, and to discuss available proof of its authenticity:

> "'My father talk one time,' said an old Klickitat to a pioneer at White Salmon, Washington; 'long time ago liddle boy, him in canoe, his mother paddle, paddle up Columbia, then come to tomanowos bridge. Squaw paddle canoe under; all dark under bridge. He look up, all like one big roof, shut out sky, no see um sun. Indian afraid, paddle quick, get past soon, no good. Liddle boy no forget how bridge look.'
>
> "Local proof is not wanting. In the fall, when the freshets are over and the waters of the Columbia are clear, one going out in a small boat just above the cascades and looking down into the transparent depths can see submerged forest trees beneath him, still standing upright as they stood before the bridge fell in and the river was raised above them. Scientific men who have examined the field with a view to determining the credibility of the legend about the bridge are convinced that it is essentially true. Believed in by many tribes, attested by the appearance of the locality, and confirmed by geological investigation, it is surely entitled to be received as a historic fact."

In a 1909 guidebook to the Columbia River, its author, W. D. Lyman, described this cascades area, confirmed Balch's observations of a submerged forest, then added other observations that supported the idea of a fairly recent damming of the cascades:

> "As we near the Cascades we note another curious phenomenon. This is the sunken forest on either side. At low water these old tree trunks become very observable, and their general appearance suggests at once that they are the remains of a former forest submerged by a permanent rise in the river. This explanation is confirmed by the fact that from The Dalles to the Cascades the river is very deep and sluggish. When we reach the Cascades, a third fact is revealed and that is that at the chief cataract the river bank is continually sliding into the river. Trees are thrown down by this slow sliding process, railroad tracks require frequent adjustment, and on clear, still nights there is

sometimes heard a grinding sound, while a tremor from the subterranean regions seems to indicate that the upper stratum is sliding over the lower toward the river. In fact, the mighty force of the stream is all the time eating into the bank and gradually drawing it down.

"From those and other indications the conclusion has been drawn that some prodigious avalanche of rock at a not long distant time dammed the river at this point, creating the present Cascades and raising the water above so as to submerge the forest, whose remains now attract the attention of the observer at the low stage of water. To confirm this theory we have the Indian story of the 'tomanowos bridge,' the quaintest and most interesting of the long list of native myths."

In his 1956 book about the Columbia River, S. H. Holbrook noted that the construction of the Bonneville Dam on the river had flooded the Cascades, or the "Great Shute":

"Not only the Great Shute disappeared in the impounded waters. With it went the physical evidences of the Bridge of the Gods which most condescending white people referred to as a 'favorite myth' of the Indians. This was a natural stone arch under which, so the natives told the pioneers, the canoes of their ancestors had floated in the days before the various gods who inhabited the Cascade Mountains had got to warring and started throwing smoke, fire and boulders, to the end that the bridge collapsed and forced the river into a new channel. There is considerable geologic evidence that something like this did happen. The dam waters also covered an extensive submerged forest, the stumps of which were occasionally twenty five feet tall, well preserved, and included both fir and pine trees."

If this great natural bridge did exist, as the evidence seems to indicate, then it probably formed when an ancient lava flow from one of the nearby volcanoes dammed the river, then the river undercut the barrier leaving a wide bridge of hard lava spanning the entire river gorge. Hundreds, perhaps thousands of years later, the same volcanoes then destroyed the span with eruptions and accompanying earthquakes -- the legendary "battle of the gods" -- probably temporarily damming the river from the bridge's collapse and also causing numerous local landslides.

There are other hypotheses that attempt to explain the Bridge of the Gods legend, but with critical evidence now lost, the truth will probably never be known.

YOU CAN'T GET THERE FROM HERE

No doubt a few prospectors and cattlemen had seen it, and Indians before that, but a discovery that dwells only in the mind of the discoverer means nothing to the rest of the world. The arch had never been reported, nor a picture of it published.

A couple of BLM specialists spotted it from the air, while surveying some public land by helicopter. I heard about it from them a few days later, and saw their photographs taken from the canyon rim opposite the span, the only place in the vicinity a chopper could land. They showed me the arch's location on a topographic map.

The following weekend, my wife and I started out to reach the arch by land. The route was paved, then graded dirt, then very eroded seismograph trail and, finally, a trackless route winding among trees and boulders down a steep slope toward an unseen canyon rim. Upon reaching a ledge too steep to descend, we parked and walked a few yards to get an unobstructed view of the narrow inner canyon still farther below us.

I could tell where we were on our topo map, and we should be within sight of the arch, but it wasn't where it was supposed to be. We walked a few yards farther, to where we could see more of the canyon, and there it was -- about a quarter mile from where the BLM had placed it. No matter. The span was ancient and fragile looking, but still graceful. It was shaped like a section of an historic Roman aqueduct, with high, arching supports -- a very old arch in the top of the Wingate sandstone inner gorge.

With some difficulty we scrambled down to the inner rim of the deep gorge, took a few pictures, then looked for a way to descend to the bottom of the gorge so I could get pictures of the lovely span against the sky. No luck! And I could see that the tentative BLM plan to establish a hiking trail up the canyon to the span was impractical. A hundred yards downcanyon from the arch we could see a vertical drop in the canyon floor that made access from that direction impossible without the use of climbing gear.

Although there was no way down the walls of the sheer inner gorge that we could see, I still wanted to get below the span we were now calling "Aqueduct Arch." Maybe by hiking down the drainage from its head? As far as we could see upcanyon there was nothing to prevent this, but a sharp bend kept us from seeing very far. I was pessimistic, however. Canyons in Wingate sandstone are noted for their sheer walls and canyon-floor drops.

Even so, weeks later we tried descending the canyon, after figuring out how to reach and identify the inconspicuous wash that became this spur of the immense Harts Draw canyon complex. After careful study, I drove to the head of that wash and parked. We then hiked down the dry drainage.

It started in a grassy meadow, but soon began plunging steeply down terraced steps of sandstone through a dense pinyon-juniper forest. We slid down some of these water-smoothed drops, but had to work around others. After a mile or so, the dry watercourse crossed another stretch of slickrock then plunged abruptly a hundred feet or more into a spectacular canyon. Beneath the pouroff, an immense alcove undercut the rim. An even larger alcove set with aspen trees marked another pouroff across the canyon. The aspens and other vegetation there told of seeping springs.

We hiked around the sheer rim, seeking a way down, and soon found a talus slope we could descend with care. The canyon floor was a maze of rubble slopes, sand dunes, rock shelves and massive rock falls, and was heavily vegetated. The canyon-bottom watercourse twisted and turned and was alternately dry, then a series of crystal-clear pools connected by tiny streamlets of sparkling water. And wherever there was water, there were also water-loving grasses, reeds, shrubs and trees. These wetter areas hosted thriving colonies of insects, amphibians, birds and wildflowers. Large dragonflies patrolled each microswamp on wings of iridescent blue.

We hiked on, enchanted by the sight of this unspoiled linear oasis rampant with life at the bottom of an arid, rock-walled desert canyon.

After a mile or so, our enchantment was interrupted. The trickling stream flowed across a stretch of smooth slickrock, then plunged abruptly downward for perhaps seventy feet, there to flow into a series of large, deep pools in solid rock. Our eyes followed the flowing water beyond the pools to still another sheer drop. Far below, the water splashed into another pool that we could not see.

We examined the canyon rim. It would be possible to get down the first drop at one point, with a little rappelling -- if we had the rope -- but the canyon beyond the second drop was protected by the same sheer, unrelieved inner walls of Wingate sandstone that had frustrated us in the vicinity of Aqueduct Arch. We could continue along the canyon rim with considerable difficulty and walk across the arch. It should be only a few hundred feet ahead, around the next bend. But that was not our goal and would be hazardous, so we headed back, frustrated again. As we left, my wife expressed both our sentiments by paraphrasing the trite but appropriate old saying --

"You can't get there from here -- or from anywhere else!"

To the best of my knowledge, the canyon directly below Aqueduct Arch has yet to be trod by the feet of man, either ancient or modern. I would like to see the soaring, graceful beauty of that ancient span set against the blue high-desert sky, like the equally graceful human construct for which we named it -- and I may yet, if it will just endure a bit longer.

Aqueduct Arch

NAMING

QUESTIONS

Most people who visit the national park areas in the Colorado Plateau that are famous for natural spans give little thought to the matter of the names these spans bear. Each popular span seems to have a name, just like every person has a name. So where are the problems?

The problems soon become apparent when the more perceptive visitor begins to think a bit and ask questions, such as:

> "Are all the known natural spans named? If not, why not? Shouldn't they all have names? If not, why not?"
>
> "Who has the right to name natural spans? Is there an official agency responsible for naming spans?"
>
> "Is there a widely known and accepted system for naming natural spans? If not, what are the common practices for naming them?"
>
> "When a particular span has several names, which one is official, who determines this, and on what basis?"
>
> "What do arch historians have to say about the span-naming questions? Don't they clarify things?"
>
> "Why do some spans have more than one name? Who provided the names for Colorado Plateau natural spans that are named?"

And -- does it all really matter? Did Shakespeare put it into proper perspective in his play, Romeo and Juliet, when he had one of his characters ask:

> "What's in a name?
> That which we call a rose
> By any other name
> Would smell as sweet."

Yet those who have more than a casual interest in the fascinating subject of Colorado Plateau natural spans deserve to have some answers to the perplexing questions about names that are bound to occur to them.

The answers to most of the questions above are not simple. Like most matters involving human communications, the answers to questions and problems are complex. Semantics also provides an element of confusion. Perhaps the following analyses of problems highlighted by the foregoing questions will clarify this confusion to some extent.

ARE ALL NATURAL SPANS NAMED?

There are many presently known natural spans that do not bear "official" names, and some that do not even have locally used names. Further, there are beyond doubt many spans yet to be found, and others that are known to a few cattlemen or prospectors but not reported or named. Why are there known spans that have no names? The answer to this question is a matter of personal opinion. If spans are known, and have thus appeared in pictures and print, is there any reason why they should not have names? Some writers say names must be "official" or "accepted." Others are less stuffy about the matter and are willing to assign names to spans, if no one else has, particularly if they are the first to report them in print.

Is there any practical reason for not naming a known span, other than to avoid the possible confusion of several names for the same span, or several spans with the same name? None that are apparent, and there are definite advantages to assigning names to things people wish to talk or write about. Children and pets are named at birth. Towns and highways and a host of other individual things are named when established or built, all for ease of communications.

It needlessly complicates human communications for a father to say to his wife "What is our third kid up to, the boy with the blond hair?" It is much easier to say "What's Jerry up to?", especially if he needs to ask this question ten times a day. It is much simpler for a businessman to tell his secretary "I am going to Seattle tomorrow," than to say "I am going to that unnamed city by that big unnamed bay in the unnamed northwesternmost state of our unnamed nation tomorrow." Names enhance communications, and help avoid semantic confusion. This principle applies to whenever two or more people wish to communicate -- verbally or in writing -- about the same specific person, item, place or event.

It also applies to natural spans. If a span is worth noticing and reporting, it is worthy of having a name, for clarity of communications if nothing else. It is irrational to ask someone "Have you seen unnamed arch?" Which one? There are presently hundreds of unnamed arches. It is awkward and not much better to ask "Have you seen the unnamed arch that is quite a piece up a hanging canyon that drains into the Colorado River gorge several miles southeast of Moab?" It is unlikely that your listener will have enough knowledge of that area to respond. But if the arch in question has a name, the listener will know immediately whether or not he has seen it. The question "Have you seen Longbow Arch?" will produce an instant reply.

To summarize -- if a natural span is worthy of note, it should be given a name for the sake of practical communications.

WHO HAS THE RIGHT TO NAME NATURAL SPANS?

Is there an official agency responsible for naming natural spans? This question does not have a clean answer. There is an official federal agency responsible for geographic names. Many states also have such organizations, but most of the land in the Colorado Plateau is under federal rather than state administration. Thus the federal agency has jurisdiction.

In most cases, however, such agencies merely serve as arbiters. They make decisions when naming problems are referred to them, but do not play active roles in seeking out and naming geographic features, especially ones as relatively unimportant as natural spans.

Within the U. S. Department of the Interior, the Board on Geographic Names handles naming problems, primarily for purposes of uniformity of nomenclature within the federal government, but also as a service to the general public. This board summarizes its purposes as follows:

> "The United States Board on Geographic Names is a Federal body created in 1890 and established in its present form by Public Law in 1947. Comprised of representatives of several Federal agencies, appointed for a two year term, the Board is authorized to establish and maintain uniform geographic name usage throughout the Federal Government. Sharing its responsibilities with the Secretary of the Interior, the Board has developed principles, policies, and procedures governing the use of both domestic and foreign geographic names as well as underseas and extraterrestrial feature names. Although established to serve the Federal Government as a central authority to which all name problems, name inquiries, and new name proposals can be directed, the Board also plays a similar role for the general public.
>
> "Any person or organization, public or private, may make inquiries or request the Board on Geographic Names to render formal decisions on proposed new names, proposed name changes, or names that are in conflict."

Does this system work? For the most part, yes, but not always as efficiently as it should. For example, after providing the first published documentation of a pair of twin natural bridges in southeastern Utah in the late 1960s, this author proposed an official name for the magnificent spans by applying to the Board on Geographic Names. The application was dated September 16, 1968 and included exact details concerning the twin spans' location, dimensions and discovery, as well as an 8 by 10 inch photograph of them and a map showing their location. Acknowledgment of the application was received within three weeks, but the final disposition was dated March 7, 1969, almost six months after application was made. The proposed name of "Gemini Bridges" was accepted and listed in the Federal Register, making it official for all federal use.

Was this official name then used by everyone? For the most part yes, perhaps largely because it was subsequently used by this author in widely distributed publications, but the federal agency that administers the land where the spans are located has been somewhat less than consistent in using the official name. For years, Bureau of Land Management road signs pointing the way to these popular natural features said "Gemini Arch."

It should be noted that the federal Board on Geographic Names does not claim the exclusive right to create geographic names or make them official. It only speaks for federal agencies, and is available for public use. Thus, legally, anyone can create names for geographic features. But for that name to stick, to become widely accepted and used and thus become "official" by default, several factors are involved. The assigned name must appear in printed publications that are widely circulated, it must be accepted by most published writers, there must be few if any other published names for the same span and, eventually, the name must be accepted by various levels of public officialdom, such as the agency that administers the land on which the span is located and, perhaps, by the Board on Geographic Names.

Unfortunately, even if all this happens, as with the Gemini Bridges, the chosen name may not be accepted by everyone. A writer or a person in a responsible position may be ignorant of the generally accepted name, or may not like it and assign one he prefers. Many other strange things can happen to names honestly assigned to natural spans. Some of these are described later in this chapter.

To summarize, anyone has the right to name a natural span, but that person has a certain obligation to determine that the span does not already have a name. And there is no guarantee that the assigned name will stick. A great many have not.

Gemini Bridges -- a photograph taken on the original documenting trip.

IS THERE A SYSTEM FOR NAMING NATURAL SPANS?

Is there a consistent, widely known and accepted system for naming natural spans? Are there common practices that most people follow?

The federal Board on Geographic Names has a published general policy for naming geographic features but, unfortunately, that policy is neither widely known nor accepted by everyone familiar with it. To quote the Board:

> "With respect to domestic names, it is the policy of the Board to recognize present-day local usage or preferences when possible. To implement this policy, there is close cooperation with State geographic boards, State and local governments, and with the general public. When there is confusing duplication of local names or when a local name is derogatory to a particular person, race or religion the Board may disapprove such names and seek alternate local names for the features. In cases where local usage is conflicting or weak, well established documented names and names with historical significance are given strong consideration. The Board also has a policy of not approving new domestic geographic names that commemorate or may be construed to commemorate living persons."

Among individuals not familiar with this official policy, or those who choose to ignore it, the common practice is to give a span their own name, in direct conflict with the Board policy against this. The tendency is only human, but can lead to the kind of conflict the Board policies were established to avoid or resolve. Examples of this are the recently documented "Lin-Lynne Arch" and "Jack's Arch." These names are used in this book, in spite of their conflict with sound naming policy, because they have appeared in public print.

Among more responsible, conscientious private citizens, more thought and objectivity are generally put into the naming of natural spans. An unnamed span may be named after some local feature. For example, a span might be dubbed "Mineral Canyon Arch," or "Harts Draw Bridge," if it is the first one discovered in a canyon with that name. Or, if appropriate, it might be assigned a name based on its appearance or nature, such as "Wedding Ring Arch," "Rainbow Bridge," or "Covert Arch," or its fancied resemblance to entities from religions, folklore or legends, such as "Angel Arch," "Druid Arch," and "Paul Bunyans Potty."

A span might be named after some early explorer who ventured into the general area, such as "Armijo Arch," after an early Spanish explorer, or "Newberry Window," for the first American to describe the area that is now the southern part of Canyonlands National Park. On occasion, natural spans have been given native Indian names, but most of these are fictitious, not actually used by historic local Indians.

In a holdover from earlier decades, when only a few natural rock openings were known to exist in America, some spans were given names that were purely descriptive, such as "Natural Bridge" or "Double Arch" or "Window Arch," but such "generic" names usually lead to duplication and conflict, and have rarely been applied to Colorado Plateau natural spans.

A span could even be named out of sheer whimsy, such as "Fraggle Rock," an odd triple arch with a boulder perched on it, "Cow," "Calf" and "Steer" arches in Bull Canyon, and "Burrito Bridge." There is certainly room for alliteration, humor and whimsy in the naming of natural features.

To summarize, there is an official federal policy for naming geographic features, but it is not universally used. It is also doubtful whether the Board on Geographic Names would care to become involved in the naming of every Colorado Plateau natural span. There are simply too many of them.

There are also some published ground rules for naming natural spans and other geographic features, although these are not universally known and applied. The game is open to all, and

Angel Arch

Druid Arch

the rules are flexible, but winning the "name game" is not guaranteed.

WHICH NAMES ARE OFFICIAL?

When a natural span has been given more than one name, which one is official? Are state and federal geographic name boards the only ones who have the right to make such decisions?

There are many Colorado Plateau natural spans that have more than one name. A few have acquired many names. Which one is "official" depends to some extent upon where a span is located. While the federal Board on Geographic Names has the final word in cases of conflict, or if an official name is important, the primary responsibility for assigning names to geographic features belongs to the agencies that administer the lands where the features are located. That would be the National Park Service with spans in National Park areas; the U. S. Forest Service has priority when it comes to naming features within National Forests; and the Bureau of Land Management has primary responsibility for those within its jurisdiction. The same would apply to other federal and state agencies that administer areas of public land. On private land, the matter is less clear, but the ultimate authority should probably rest with the state.

Of course, not all public land administration agencies care to become involved in such matters, and interest may vary from one region to another, depending upon the views of agency administrators. However, individuals who are seeking names for natural features, or wishing to assign names themselves, should first check with such agencies. If a particular agency does not wish to assign a name, the individual is then free to choose, but should still follow the general federal ground rules for naming, and get that agency's informal approval of the chosen name.

In the preparation of this book, several federal agencies were contacted regarding natural spans on lands under their administration. In some cases little-known names existed and were used. In others, names were suggested, informally approved by the agency, and then used in this book. With several of the less spectacular spans, names were simply assigned on the basis of the common ground rules described earlier, with the full understanding that they were not "official," but also with the understanding that names are needed for communications and that "well established documented names" do have a certain priority with the Board on Geographic Names. To all objective researchers, including historians, "first in print" bears considerable weight.

Thus, with multiple names, it is generally up to the appropriate land administration agency to choose an official name. Where this responsibility is not taken seriously, and the Board on Geographic Names is not asked to arbitrate, the decision is made by public usage. Whichever name gets used the most, by the public and by writers, eventually becomes "official" by default, and the other names fade into history. Some examples of this are given later in this chapter.

WHAT DO ARCH HISTORIANS HAVE TO SAY?

It would seem reasonable to assume that arch historians would have clarified the various problems with naming natural spans. Unfortunately, this is not the case. The one historian who has delved into the matter has not served as a model of scholastic objectivity when it comes to naming new spans, to using existing names, or to choosing between multiple names.

In the first volume of his catalogs, arch historian Vreeland discusses how some Colorado Plateau natural spans acquired their names, and notes that the Board on Geographic Names has "a fine set of rules" for naming geographic features, but concludes that:

> "-- being a bureaucracy, the board is subject to pressure from other bureaucracies, politicians, and influential people, so the names approved by it are not always those in the best interest of the public. In view of this, it would be unfair to attach the designation 'official' to those names approved by this board. We will use the word 'accepted' to designate the names of all structures used here."

He does not indicate "accepted" by whom, but concludes that when there is any question, the older name is given preference. From even a casual study of Vreeland's other volumes, however, it quickly becomes apparent that he does not follow his own stated policy. He even ignores or rejects names officially assigned by the Board on Geographic Names, the National Park Service and the Bureau of Land Management, and in many cases deems uncontested names for spans that have been widely circulated in public print as "unacceptable" or "inappropriate," without defining the basis for this judgment. He also describes many significant spans and lists them as "unnamed," even though some have widely used names. The only discernible principle seemingly followed is that of subjectivity -- if he doesn't like a name, or the person or organization that assigned the name, it is not "acceptable." This is unfortunate, because the matter deserves more objective treatment.

In sum, the only arch historian in the field does not recognize the authority of the organization legally responsible for the names of geographic features on federal lands, which means most Colorado Plateau natural spans, and does not follow the widely accepted scholastic principle of "first in print." These failings alone place the results of his historic research in a questionable light.

PRESENT NAMES -- SOME EXAMPLES

Problems with names began almost from the first with Colorado Plateau natural spans. As reported in W. W. Dyer's 1904 account of Horace Long's 1903 trip to the White Canyon bridges, local cattleman Scorup agreed to guide Long to the bridges on "-- the condition that one of the arches should be named the 'Caroline' in honor of his mother." This name was subsequently given to the bridge now called Kachina, the first span the party reached.

Upon reaching the second bridge, now called Sipapu, Dyer reported that Long named it after his own wife:

> "Emulating the example of Mr. Scorup, Long named this bridge the 'Augusta,' in honor of his wife; and it is fortunate that the lady was so appropriately christened."

Later, upon reaching the third natural bridge in that area, now called Owachomo, it seemed so small in comparison to the other two that they named it "Little Bridge." When H. L. A. Culmer reached the White Canyon bridges in 1905, he didn't like the name "Little Bridge" and renamed it "Edwin Bridge" after Colonel Edwin F. Holmes, who helped sponsor Culmer's expedition. This practice of deliberately ignoring established names, or giving them new ones for personal reasons, started early.

In his 1910 report on "The Great Natural Bridges of Utah," Byron Cummings summarized the next step in the naming of the White Canyon bridges:

> "-- President Roosevelt issued a proclamation April 16, 1908, creating the 'Natural Bridges National Monument.' There are three of these large bridges in the monument, generally known as the 'Augusta,' the 'Carolyn' and the 'Edwin' --."

Note that Cummings changed "Caroline" to "Carolyn." He continued:

> "In September 1909, President Taft issued a second proclamation -- by which the former boundaries were changed somewhat -- and the names changed to the Owachomo, the Kachina, and the Sipapu. The earlier names, however, have been so long associated with the structures by the people of the region and through the various magazine articles that have appeared that it will be very difficult to secure a general recognition of the Hopi names now applied by the government."

In his 1922 report, F. J. Pack agreed with Cummings about the new names given the White Canyon bridges, and indicated that they had still not caught on and probably never would:

> "The older names, however, had become so thoroughly

Saddle Arch, Capitol Reef National Park

> entrenched in the minds of the public that these official designations were never widely employed, and, in consequence, were soon lost sight of. It is believed that any effort to bring them into general usage would be of questionable value."

Both Cummings and Pack would doubtless be surprised to find that in the long run the opposite happened. Today, few people would even recognize the original names for the three major spans in Natural Bridges National Monument. The official names eventually dominated.

NAMING

According to a 1947 Desert Magazine article by Kelly and Martin, the Hopi names were given the bridges by W. B. Douglass of the General Land Office:

> "-- bestowed under the mistaken impression that inhabitants of nearby cliff dwellings had been of Hopi origin. Certainly Indian names are appropriate; but they should have been Paiute rather than Hopi, since the latter never occupied San Juan county."

In a June 1952 National Geographic article, Jack Breed reported that to the Board on Geographic Names the original names for the White Canyon spans seemed "a little flat," so it:

> "-- cast around for appropriate Indian names. Unfortunately, the Paiutes of the vicinity weren't in the habit of naming bridges; they referred to them all, indiscriminately and rather inelegantly, as Ma-Vah-Talk-Tump, or 'Under the Horse's Belly.' Undeterred, the Board turned to the Hopi, who were then thought to be descendents of the prehistoric Indians who had inhabited southern Utah."

Subsequently, the Hopi names were approved by the Board on Geographic Names. It should be noted that while the Hopi are not the direct descendents of the Anasazi Indians who once inhabited southeastern Utah, and whose ruined structures were found near the bridges, they are believed to be related to another subgroup of this same prehistoric culture.

Name confusion over the span now called simply "Rainbow Bridge" was even worse, and largely because of numerous attempts to assign Indian names to the span and to the canyon in which it is located. It all started with the April 28, 1910, report of government surveyor W. B. Douglass, who was with the original discovery party:

> "-- the shape of the bridge suggests as a name 'The Rainbow,' or 'Barohoini,' the Paiute equivalent. The Navajos called the canyon 'Sebigehotsow' meaning 'rock with a hole in it.' Their name for (a) bridge is 'Nonnezoshe,' but it is applied only to the artificial (manmade) structure."

In his 1910 report, Cummings used both "Nonnezoshie Bridge" and "Nonnezoshie arch" in his text and photo captions of the span, but said the Indians called the canyon "Nonnezoshieboko" or "Great Arch Canyon." He also noted assignment of an official name for the span:

> "President Taft by proclamation May 13th, 1910, set aside this arch with the land about it as the 'Rainbow Bridge National Monument.'"

In the title of his November 1911 National Geographic article, "The Great Rainbow Natural Bridge," J. E. Pogue used the official name for the span but added the word "natural." Throughout the article's text and photo captions he used the names "Rainbow Arch," "Barohoini" and "Nonnezoshe."

The span's official name was still not universally used a decade later. In his 1922 report on "Natural Bridging in the High Plateaus," Professor F. J. Pack used the name "Nonnezoshie arch." In their 1923 report, U. S. Geological Survey geologists Miser, Trimble and Paige used the official name for Rainbow Bridge and cast doubt upon the Indian names assigned earlier:

> "The rainbow form of the bridge -- has been recognized by the Paiute Indians and suggests to them the name Barohoini, a rainbow. To the Navajo Indians the form suggests the name Nonnezoshe, meaning 'a hole through a rock.' The Indians have, however, not applied either of these names to the bridge."

In his 1925 book, "Mesa, Canyon and Pueblo," C. F. Lummis used Pogue's augmented version of the official name, "Rainbow Natural Bridge," and noted that "The Navajo Indians -- call it Not-se-lid."

In Clyde Kluckhohn's 1933 book, "Beyond the Rainbow," he tried to clarify the Indian name puzzle, but only added to the confusion:

> "The Pahutes called the Bridge, Barohini -- the Rainbow. The Navajos in the Navajo Canyon-Navajo Mountain region call it most often Tsay-nun-na-ah -- Rock-Goes-Across-the-Water. In the Kayenta region and on the Reservation generally they say Nonnezoche -- the Arch. Only ceremonially do the Navajo say Nonnezoche Nadzelid -- Rainbow Arch."

In his 1974 article, "Rainbow in the Rock," outdoor writer Verne Huser noted that "-- the Navajo name for the structure, Nageelid Nonnezoshi, means rainbow of stone," and to this day, many writers find the alleged Indian names for Rainbow Bridge to be irresistibly romantic.

In his March 1934 monthly report on the "Arches National Monument Scientific Expedition," Monument Director Frank Beckwith named and described several of the major spans there -- Fallen Rock Arch, Double Arch, Arch in the Making, Massive Arch, Elephant Arch, Great Oval Arch, Turret Arch, Delicate Arch and Landscape Arch. Not all of these names survived into the present.

The report noted that Double Arch was locally called "The Jug Handles," and described Fallen Rock Arch:

> "This is so named because a part of the rock mass which fell out still lies in the corner of the arch, partly blocking it."

In 1940, the detached rock mass fell out of the opening. The span is now called Skyline Arch. In his 1981 book about Arches National Park, J. F. Hoffman mistakenly noted that Skyline Arch was formerly called "Arch in the Making," but the description of this span in the 1934 report does not fit Skyline Arch. Hoffman's book also provided a chart listing other "variant names" for spans within this third National Park Service area in the Colorado Plateau to be dedicated to natural spans.

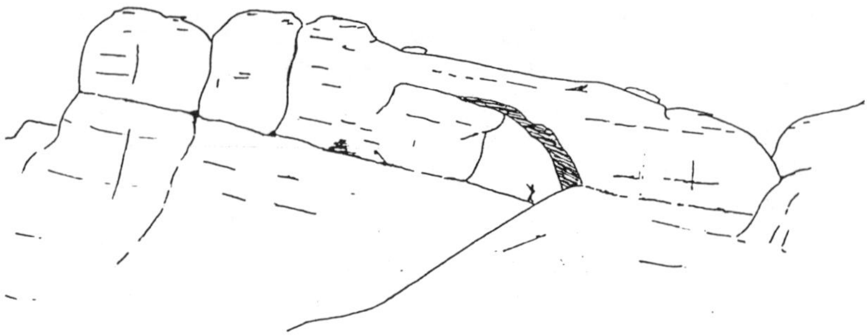

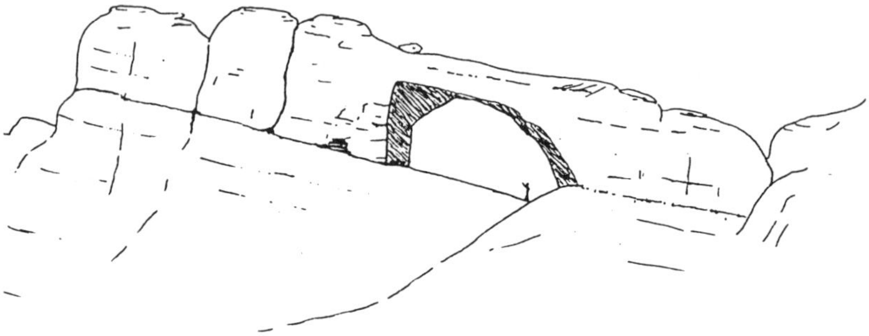

Skyline Arch (Fallen Rock Arch) before 1940 (top) and as it now appears.

In his six-part series on "Stone Rainbows" in 1976 and 1977 issues of Earth Science, R. M. Pearl describes a "Pothole Arch" in the White Rim of Canyonlands National Park. His description fits the natural bridge there that presently bears the name "Musselman Arch." In a 1983 report on the natural spans in the Maze district of the same park, L. Dansie gives the confusing name history of two spans there:

> "The site (described) is Tibbett Arch, one of the larger arches found in the Maze District. The arch is named after the Tibbett family, one of the early settlers in the area. Originally Muffin Arch was named Tibbett Arch. By a decision of the Board of Geographic Names, Muffin Arch was given its present name and Tibbett was given to a previously unnamed arch."

In his 1938 report on the geology of San Juan County in Utah, U.S.G.S. geologist H. E. Gregory noted "three beautiful windows" in Arch Canyon to the west of Blanding, calling them "Cathedral Arch, Angel Arch and Keystone Arch," without indicating whether these were local or newly assigned names. Because these spectacular spans are so remote and seldom seen, the early names have stuck by default. However, for the purposes of this book, and with permission of the appropriate land administration agency, the "Angel Arch" of Arch Canyon has been

Gregory Arch, Arch Canyon

renamed "Gregory Arch" in order to avoid duplication with the present "official" Angel Arch in Salt Creek Canyon of Canyonlands National Park. The name "Gregory Arch" was assigned to honor this early Colorado Plateau geologist, whose original namesake, Gregory Natural Bridge, is now lost beneath the waters of Lake Powell.

A 1939 report by geologists Gregory and Anderson shows and describes a major span "-- known to topographers of early days as the 'Broad Arch'--." The span depicted is within Capitol Reef National Park and is now called Hickman Natural Bridge.

A circa 1920 travel brochure entitled "Wonderlands of Utah" had a picture of a span in Cedar Breaks it called "Arc de Triomphe." F. J. Pack called the same span "Walls of Jerico" in his 1922 report. The present National Park Service name for the span is "Square Arch." Pack's report also depicted an "Ashdown Gorge Bridge" in Ashdown Gorge, the canyon through which the colorful basin in Cedar Breaks National Monument drains. This large span is now called "Flanagan Arch."

NAMING

In an October 2, 1983, article in the Deseret News Magazine, professor of sociology Reed Blake described the history of the naming of Kolob Arch in Zion National Park by noting that early geologists, including Gregory, Williams, and F. J. Pack, had seen the arch but assigned no name. Thus:

> "-- the arch had no official name as of 1953, even though locally it was called the Kolob Arch. Sometime around 1958 the span officially received the local label."

Other reports provide insight into the naming of natural spans outside of National Park Service areas. In his 1910 report, Cummings mentioned naming three spans in the Pritchett Canyon drainage near Moab, Utah. He called them Pikyabo Bridge, Wigwam Cave and Block Bridge. The first span is now called Pritchett Arch, the second span's name stuck, and the third eventually became Window Arch, a typical "generic" name.

In his 1925 book, author C. F. Lummis assigned a name to a famous span in Arizona:

> "For a third of a century I have been calling this the 'Pine Creek Natural Bridge.' That's no name for a wonder of the world! I have never been able to learn what the ancient Apaches called it -- so, since I first made it known to the world, I am going to name it -- 'The Mammoth Natural Bridge.'"

The name didn't stick. The famous span is now called Tonto Natural Bridge. In his 1953 book, R. E. Klinck depicted three Monument Valley arches he called Honeymoon Arch, Ear of the Wind and The Big Eye. The first two names have endured, but the third is now "Eye of Sun Arch."

In a 1955 National Geographic article "Escalante: Utah's River of Arches," its author, Robert Moore, noted that the expedition had given names to most of the spans in the lower Escalante Canyon. Some were named after local usage, some were given new names. Variously, they were named Jacob Hamblin Arch (after an early Mormon scout), Coyote Natural Bridge and Stevens Canyon Arch (after the tributary canyons in or near which they were located), Cliff Arch (because of its location), Broken Bow Arch (after the broken Indian bow found nearby), Gregory Natural Bridge (after the geologist of that name), La Gorce Arch (after the magazine's editor) and Bement Arch (after the Utah official who had explored the area by air earlier).

A July 1964 article in Natural History by Willard Luce depicted a span it called "Looking-glass Arch" to the south of Moab, east of the highway. A 1976 geologic guidebook to the "Northern Colorado Plateau" by J. K. Rigby showed the same span but called it "Window Arch." Today, a large span a couple of miles west of the highway in the same vicinity is called Looking-Glass Rock, and the span east of the highway is called Wilson Arch.

In his unpublished arch log, Dick Wilson noted visiting a small, oddly shaped arch in a spectacular setting high on the west rim of Moab-Spanish Valley on March 1, 1968. Because of the lovely view through the span's opening of Mount Tukuhnikivatz in the nearby La Sal Mountains, he named it "Tukuhnikivista."

R. H. Vreeland (1976) noted that spectacular Royal Arch in northern Arizona has been known by several other names, among them Thin Rock Window, Window Arch, Cove Arch, Ben Wittick Arch, and Paul Gregg Arch.

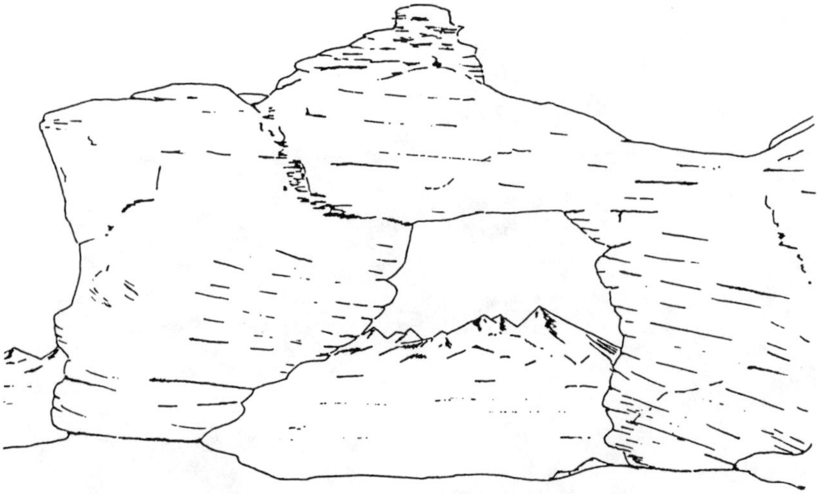

In his article in the Denver Post's Empire Magazine, reprinted in the Winter 1980 issue of Our Public Lands, Zeke Scher noted some outstanding name confusion in Rattlesnake Canyon in western Colorado, an area known for its numerous spectacular arches:

> "Old plats of the General Land Office once showed a Rattlesnake Creek flowing northward through Horsethief Canyon to the Grand River (which was renamed the Colorado River in 1921). U. S. Bureau of Reclamation river surveys in the 1940s dropped the creek name, but still referred to the canyon south of the river as Horsethief. Present topographical maps of the U. S. Geological Survey, dating to 1962, 'created' (the present) Rattlesnake Canyon and transferred the name of Horsethief Canyon to an eight-mile stretch of the Colorado River westward from the end of the Grand Valley."

Scher described several of the spans in the Rattlesnake Canyon area but mentioned the name of only one, Cedar Tree Arch. Confusion over the names of spans in that area came later -- and continues today. Vreeland cataloged fewer than half of the significant spans in the area, and indicated that several were "unnamed," despite the names assigned by the BLM, which administers the area. He also added to the confusion by calling one span "Rainbow Arch," even while admitting that "The local BLM office has called this Cedar Tree Arch," and by calling another major span "East Rim Arch" in spite of its BLM name, Centennial Arch.

In 1979, in some of its wilderness survey literature, the BLM compounded problems with the names of spans in the Rattlesnake Canyon area by issuing a partial list of them. Four were given no names. Some of the names of the spans in Rattlesnake Canyon were those previously assigned, but others were changed or abbreviated. To make matters worse, a second column of Ute names was listed, even though these names were nothing more than a local travel-promotion gimmick. Of these, only one -- Akiti --could easily be pronounced by the average American. And that is where the matter presently stands -- badly in need of some decisions by the Board on Geographic Names.

NAMING

Hole-in-the-Bridge Arch, Rattlesnake Canyon

SUMMARY

To summarize, the Department of the Interior Board on Geographic Names has the legal right to decide on names for geographic features on all federal lands, but does not actively seek out features to name. It has no great interest in naming minor features, but will mediate or give its stamp of approval on request. Federal and state agencies that administer public lands have first priority when it comes to assigning names to geographic features under their jurisdiction, but often take little interest in doing this. Where such agencies do assign names, the public should either accept them, or take the matter to the Board on Geographic Names for arbitration.

Many names of geographic features, including natural spans, simply come from general usage. Such names generally derive from other nearby features, from local or historic individuals, or from some oddity of shape or location. A few are given "generic" names, such as "Window Arch" or "Bridge Arch." With newly reported spans, the reporter is free to assign a name, but should first determine that the span has no earlier name. If it does not, the reporter should not assign the name of a living person and should understand that the assigned name may not be accepted by everyone, even if it is submitted to and ultimately approved by the Board on Geographic Names.

Most scholars honor the rule of "first in public print" when researching name history, but a few seem to give more weight to oral history, preferring the names allegedly given by early cattlemen, sheepmen and prospectors. Land administration agencies are not bound to honor "first in print" names, but are free to choose any name.

The policy followed in this book has been to use the names assigned by the Board on Geographic Names or appropriate land administration agencies whenever they exist, since by far the majority of the natural spans in the Colorado Plateau are on public land. Where a span was within a National Park Service area, but had no widely known name, the NPS was asked to assign a name. Spans within either Forest Service or BLM lands that had no widely known names, or that had not previously appeared in print, were assigned names for the sake of communications. It is the right of anyone to ignore such author-assigned names, or to submit suggested changes to the Board on Geographic Names.

It is this author's contention, however, that no one should have to ask "have you seen unnamed arch?" If a span is worthy of note, it deserves a name, preferably one selected following the general rules established by the Board on Geographic Names.

Funnel Arch, Cane Creek Canyon

THE CASE OF THE COVERT ARCH

We were slowly heading south on the rarely used jeep trail that traveled on top of the immense wedge of sandstone between the colorful depths of Salt Wash and Lost Spring Canyon, just east of Arches National Park. The area should have been included in the park because of its numerous natural spans and miles of beautiful cliffs of Entrada sandstone.

Our Toyota crept along the faint trail over broken rock and occasional patches of sand and reddish soil set with gnarled juniper trees and sparse desert shrubs. The trail closely paralleled the rim of Lost Spring Canyon. About half a mile beyond where Fish Seep Draw joined that canyon from the east, the trail plunged abruptly down a steep eroded slope of red sediments. I stopped and we got out to reconnoiter, take a stretch and admire the spectacular scenery.

Here, the trail dropped to a lower level, one that was predominantly barren slickrock. The trail immediately ahead was very eroded and hazardous, but I could see a faint detour that bypassed the bad stretch. The scene was magnificent in all directions. The canyon rims were whitish sandstone, the Moab member of the Entrada. Below this, and on down to the floor of the immense canyon system, the salmon-hued Slickrock member of the Entrada had been eroded into a three-dimensional maze of rounded, sloping walls and abutments.

The canyon walls to the east and west of us were broken many places by short side-canyons, some of them so narrow they were wedge-shaped. The isolated ridges between some of these canyon spurs formed castles, battlements, minarets and towers of every conceivable design. I took out the binoculars we always carried to study this sandstone labyrinth more closely, in particular one shadowy alcove to the southeast. After one quick look, I let out an excited yip, handed the binoculars to my wife and pointed toward the distant shadow.

"That's an arch -- a big one!"

She confirmed my observation and, as I backed the Toyota to the trail detour and dropped down onto the lower level, we started making plans for trying to reach the span. Beyond the detour I paused to take another look at our "new" arch. I couldn't see it at all, even though we were only yards farther south and perhaps fifteen feet lower. Nor could we see it from anywhere else along the trail as we continued to a point overlooking the Salt Wash-Lost Spring confluence and hiked both rims for some distance.

On our return trip we watched for the arch from the trail to the north of where we had sighted it, but again to no avail. That arch was well hidden! No wonder it did not appear on U.S.G.S maps or in any literature, as we discovered later. But surely we could get to it somehow. Perhaps from the canyon below?

A week or so later we tried this by hiking up Salt Wash from historic Wolfe Ranch in Arches National Park. This was not an easy route to take, but we persisted and eventually managed to reach the complex maze of spur canyons to the east of the Salt Wash-Lost Spring confluence. But spotting our covert arch from below proved to be even more difficult than seeing it from a distance.

After hours of searching the hot arid maze, we reluctantly headed back -- too late in the day! We had to find our way back through the most hazardous part of the route in total darkness, aided only by faint starlight. That was a hike I would not care to repeat!

Several weeks later we tried again. This time we drove into the Yellowcat area to the east of Lost Spring Canyon, then southwest on the Wintercamp Ridge trail. After careful study of the topographic map, we parked and hiked out across the desert wilderness toward where I thought our arch was located. After about a mile, we came to the rim of a deep spur canyon and there, directly across from us, was our elusive arch!

It was a beauty! Its opening was a soaring arc through a narrow flat-topped ridge of Entrada sandstone, but from across the canyon it was difficult to estimate its size. We could see no obvious way down to the opening, so hiked around the canyon head and across the span's top. But we still wanted to get below "our" lovely arch, to see the breathtaking scenery all around framed in its graceful opening. Skilled or foolhardy climbers might have found a way down from its top, but we could not.

Reluctantly, we left. It had taken us a long time to get to the span and the day was almost gone. I didn't want to have to find my way out of the maze of mine trails in the Yellowcat area in the dark. But as we hiked out around the upper end of the canyon that cut below the arch, I spotted a safe route down onto the sloping slickrock of the canyon wall, at an intermediate level that could lead to beneath the arch -- if the slope didn't get too steep for us to walk it.

A week or so later we returned to try that possible route to the opening of the span we were now calling "Covert Arch" -- for very good reasons -- and did get into the arch's opening. There, with a steel tape, we measured its width as 65 feet, estimated its height as 35 feet, and spent a couple of glorious hours enjoying the spectacular setting -- and savoring our ultimate victory over the covert nature of Covert Arch.

I published the first pictures of Covert Arch in 1978, in my trail guide book and map of the area, and described how to find and reach it without all the difficulties we had encountered. Later, several other arches of notable size were reported in the same area, none of them previously documented. Pictures of some of these spans appear in this book, but none found so far exceed Covert Arch for size and beauty.

Although the name for this span has not been submitted to the Board on Geographic Names, it has appeared in print in thousands of published maps and books, and has not been known by any other name, so the name we gave it should stick -- unless the area should some day be incorporated into Arches National Park and be renamed by the National Park Service. We would be happy to relinquish the name we gave the span, in return for protection of the beautiful area in which it is so well hidden.

Covert Arch

Skywalk Arch, near the
Monitor and Merrimac Buttes

Bullwhip Arch, in
Four Arches Canyon

MEASUREMENTS

QUESTIONS AND PROBLEMS

As with the naming of natural spans, at a casual glance their measurement would seem to present few if any problems. A more careful study, however, reveals that measuring things as diverse in shape and location as Colorado Plateau natural spans is anything but easy.

Certainly, it is simple to measure almost any particular span or opening that can be reached, but problems soon arise when the technique used on that span is applied to the measurement of other spans, or when an attempt is made to devise a technique that can be applied to a number of natural spans for the sake of comparison. Difficulties can also be encountered with spans that are so located that they cannot easily be reached for direct physical measurement. There are many such spans.

The first major question that arises about measurement is whether a natural rock opening is represented by its free-standing arc of rock, or its opening. Many natural rock openings simply have no separate arc of rock. They are holes through walls of rock, with their openings not near enough to an edge to produce an apparent arc or length of rock. It immediately becomes apparent, then, that at least two systems of measurement are needed, one for openings that have distinct spans of rock and another for openings that do not. Which of these to apply in marginal cases would remain a problem even then, however.

With openings that have distinct spans of rock, how are these to be measured? By the distance between tangents to the opening that are parallel to each other and at right angles to the axis of the span? By the length of rock that is supported on the span's abutments? This could be considerably greater than the actual size of the opening behind or beneath the span. What if the span has no distinct abutments? Or if it is nearly vertical, as with many jughandle arches? With certain stress exfoliation arches, the span's abutments flare out below the span, sometimes for a hundred feet or more. Measuring the span's length at the bottom of such a flare could give a span with a ten-foot opening a "length" of a hundred feet or more.

If an arch's length is measured by its opening length parallel to the span's axis, this could be misleading with openings that do not have their longest axis parallel to the span. If the opening is to be measured, what counts? Its longest dimension? If so, is an arch with an opening 30 feet by 2 feet "bigger" than one with an opening 20 feet by 20 feet? Should the area of the opening be the criterion? If so, then an arch with an opening 50 feet by 6 feet would be "smaller" than one with a round opening 20 feet in diameter. Further, some openings have such strange shapes that they defy attempts to measure their areas.

What about the height of a span? Is it from the highest place on the span to the first thing solid below it, or from the highest point in the opening? What if the rock or ground below the span slopes steeply? Measurements taken from the two sides of the span would differ. What if the span is an alcove arch high in the rim of a canyon? Is its height the distance to the canyon floor? This could make a small arch with a ten-foot hole

300 feet "high." How high would a jughandle arch be by this standard? Should jughandle and similar arches have their own set of measurement standards, thus creating three distinct systems?

How are the widths of spans to be measured? Some are easy, but what about tunnels? Is the width of the span the tunnel's length? Or its width? Is the width of a span always a horizontal dimension? If so, what about jughandle arches, or the many other angled spans? Horizontal measurements could give false readings of such a span's true "width," yet the semantics of "width" implies a horizontal dimension, as "height" implies vertical.

And how would the standard dimensions of length, width and height fit the natural rock opening formed when two adjacent seep-caves join at the rear, leaving a near-vertical span of rock as a column in a single, wide seep-cave? There are many other natural rock opening types that standard measurement terms fit either poorly or not at all.

The questions about measurement are myriad, as plentiful as natural spans in the Colorado Plateau. The ones noted are just samples, given to illustrate that the measurement of Colorado Plateau natural rock openings is not a simple affair, and to produce in the reader an attitude of skepticism about all published measurements and size comparisons.

Perhaps the biggest problem of all is human subjectivity. A natural span can be dimensionally big, yet not appear impressive to a human viewer. There are many such spans. A span can also be impressive without being particularly big in any dimension. This element of subjectivity is bound to creep into the measuring process, and even into the selection of spans to measure. A 100-foot arch close to a cliff face will simply not get the same attention as a similar arch with a large opening behind and beneath it. A natural bridge with a span of 275 feet that is prominent will be given national monument status, while one with a span of 243 feet that is tucked away in an obscure canyon will receive little attention and no special protection at all. These are real examples. The bridges are Rainbow Bridge, in the national monument of that name, and Morning Glory Arch, in Negro Bill Canyon near Moab, Utah, on unprotected public land.

MEASUREMENT TECHNIQUES USED

The techniques used to measure Colorado Plateau natural rock openings have ranged from rough visual estimates to the use of modern scientific measuring instruments and methods. In between these extremes are: estimates based on people standing beneath a span, or on the height of trees near the span; the use of altimeters to measure height; the use of weighted ropes or measuring tapes; and the use of standard surveying instruments.

Perhaps the most inaccurate methods are visual estimates of all kinds, but especially those not based on some familiar object near the span opening. Estimates based on the size of a nearby tree are worse than those based on the height of a nearby human, because within the semi-arid Colorado Plateau region tree sizes are highly variable and the tree species common there -- pinyon-pine and Utah juniper -- are not widely known. The eroded terrain itself offers nothing familiar for size comparisons.

The use of an aircraft altimeter to measure a span opening is not very accurate, but is sometimes the only practical technique, as when a helicopter was used in the mid-1960s to measure the elevations of the top and bottom of inaccessible Keyhole Arch in Wetherill Canyon, now an arm of Lake Powell. The

Ojo Arch in Little Canyon (left) and Bartlett Bridge in Bartlett Wash (right) are typical examples of the many large, well-defined natural spans that have been ignored because their openings are not conspicuous. The relatively narrow openings behind the spans are not apparent except from directly above or below. Even so, such spans are still significant because of their size and the fact that their weight is entirely supported on their abutments.

differences in elevation gave the span opening a height of 70 feet. This figure was then used to estimate a width of 25 feet.

Unfortunately, while the use of surveying and scientific measuring instruments produces more accurate results, their use will not be entirely suitable until standard definitions and techniques have been devised that can be applied to virtually all natural spans and accepted by all land administration agencies, state and federal geological survey organizations, and interested academic groups. Only after measurement by such a standardized system would it be appropriate to apply such comparative terms as "biggest," "highest" and "widest" to a list of natural spans.

SOME PUBLISHED MEASUREMENTS

In the following pages the dimensions of a number of natural spans are given. In most cases the sources of the figures are also given. Even though the author of this book questions the scientific validity of most measurement techniques used to date, he neither supports nor challenges the validity of specific measurement figures, but presents them as published by those who made them, for whatever they may be worth to the reader. The reader should bear in mind, however, that no two sets of measurements listed for the same span, or for different spans, have been made using identical definitions and techniques.

One of the earliest attempts to measure a Colorado Plateau natural span was noted in Arthur Winslow's report in the April 22, 1898, issue of Science. In this first report of the magnificent span now called Landscape Arch, Winslow described how its size had been gauged:

MEASUREMENTS

> "The dimensions of the bridge, as estimated by the photographer, are about 500 feet in span and about 150 feet in height. A comparison of the bridge with (human) figures shown in the original photograph in the right hand corner and with the tree growth near by indicates that these dimensions are quite possible."

This crude method of measurement, while still used on occasion, is not very accurate, as later proven in the case of Landscape Arch. In his March 1934 report on the first scientific expedition within the newly established Arches National Monument, Director Frank Beckwith noted that:

> "Landscape Arch is -- a very large arch, the span of which is 291 feet -- the greatest span in the monument area. Through this arch the landscape behind it may be seen, hence the name."

This figure for the span of Landscape Arch still appears in most literature, even though in the winter of 1940/1941 a piece of rock fell from beneath one end, making the span "approximately 305 feet in length instead of 291 feet," according to a March 1941 report of the event by the "custodian" of Arches National Monument. More recent remeasurements of Landscape Arch are discussed later in this chapter.

Attempts to measure the immense White Canyon bridges began early, soon after their official discovery. In W. W. Dyer's report in a 1904 edition of Century Magazine about the 1903 Scorup-Long expedition to the White Canyon bridges, Dyer noted that:

> "The travelers had with them no scientific instruments for making accurate measurements, but by a series of rough triangulations Long obtained results which are doubtless correct within narrow limits."

These dimensions are shown in the following chart, together with measurements of the same spans reported later by others for comparison. Dyer also couldn't resist making some size comparisons between Augusta (Sipapu) Natural Bridge and some other objects familiar to many Americans:

> "-- its height is more than twice and its span more than three times as great as those of the famous natural bridge of Virginia. This bridge would overspan the Capitol at Washington and clear the top of the dome by fifty-one feet. And if the loftiest tree in the Calaveras Grove of giant sequoia in California stood in the bottom of the canyon, its topmost bough would lack thirty-two feet of reaching the underside of the arch."

In his report in the winter 1972 edition of the Utah Historical Quarterly, C. R. Steen gave the "recent measurements" of the three famous natural bridges and, in the accompanying diary that H. L. A. Culmer kept on his 1905 trip to the bridges, the dimensions recorded at that time. These dimensions are also listed in the following chart.

In his 1910 report on The Great Natural Bridges of Utah, Cummings compared the size of Augusta (Sipapu) with familiar buildings in Salt Lake City, but went on to note that the sizes of the White Canyon bridges reported by the Scorup-Long

Angel Arch, Canyonlands National Park, viewed from behind. For scale, note the silhouetted figures beneath the span.

expedition were incorrect:

"The White Canyon bridges -- were unknown to the outside world until Mr. Scorup of Bluff led Mr. Long to them in 1903. The result of that trip was an article in the Century Magazine of August, 1904, which did a great deal to awaken interest in these natural bridges and direct attention to the scenery of Utah; but no careful measurements were made, their size was overestimated, and the comparisons with the United States Capitol and other buildings (were) erroneous and misleading."

MEASUREMENTS

In the November 1911 issue of National Geographic, author J. E. Pogue compared the sizes of the White Canyon bridges with Rainbow Bridge, noting that all the figures came from "surveys of W. B. Douglass," the official government surveyor at the time. These figures are also listed in the size chart for the White Canyon bridges. The similarity between Douglass' figures and those listed by Steen in 1972 indicates that Steen's figures may not have been as "recent" as purported. Douglass measured the White Canyon bridges in 1908.

It is interesting to note that Pogue's chart of large spans listed Rainbow Bridge, the White Canyon bridges, Virginia Natural Bridge and a span in France, but did not mention other large Colorado Plateau spans that were certainly known at the time, such as Landscape Arch, Kolob Arch and Morning Glory Arch near Moab. The article also repeated the earlier doubtful size comparison of Sipapu Natural Bridge with the Capitol building.

DIMENSIONS OF WHITE CANYON NATURAL BRIDGES

	AUGUSTA (SIPAPU)				CAROLINE (KACHINA)				LITTLE BRIDGE (OWACHOMO)			
REPORTER	HIGH	SPAN	WIDE	THICK	H	S	W	T	H	S	W	T
LONG, 1903	357	335	40	60	197	208	127	125	142	211	33	18
CULMER, 1905	265	320	35	83	182	350	60	60	111	205	30	10
DOUGLASS, 1908	220	268	31	53	210	277*	44	93	106	180	28	9
CUMMINGS, 1910	222	261	28	65	205	186	49	107	108	194	--	10
PACK, 1922	222	261	28	65	205	186	49	107	108	194	25	10
NPS, 1970	220	268	31	53	210	206	44	93	106	180	27	9
STEEN, 1972	220	268	31	53	210	206	44	93	106	180	27	9
VREELAND, 1976	220	268	31	53	210	210	44	93	106	180	27	9
NPS, 1984	220	268	31	53	210	204	44	93	106	180	27	9

* Average of two measurements taken.

NOTE: From the diversity of dimensions in the above chart among the first four reporters for the same natural spans, it is easy to conclude that uniformity of technique is vital if measurements are to be meaningful and usable for size comparisons. For the most part, those taking the measurements did not define exactly what they were measuring, or were ambiguous. Height figures were from the "stream beds" to the tops of the spans, not the tops of the openings.

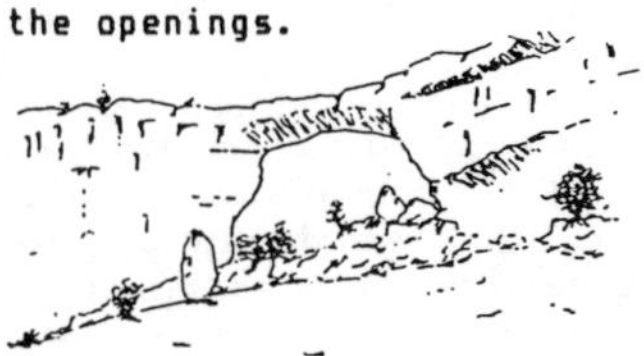

Douglass made the first measurements of Rainbow Bridge during the 1909 discovery trip, using the men and equipment in his surveying party:

"Mr. Wetherill had succeeded in scaling the south walls of the cliff and returned with the information that the top of the bridge could be reached by means of a long rope. I dispatched my two chainmen and flagman with the necessary ropes. They lowered themselves over a 50-foot ledge to a point on the south arch of the bridge where sufficient footing was afforded to enable them to clamber to the top.

"Two steel tapes having a combined length of 333 feet were lowered over the edge to the bed of the stream below. Later I was lowered to the bridge and made measurements as to width."

Douglass did not define what the figures given in his 1910 report represented. His measurements are shown in the following chart, together with those given for Rainbow Bridge by other writers.

In his report in the November 1911 National Geographic, Pogue noted that the height of the opening beneath Rainbow Bridge was only "267 feet," but that this figure was "augmented by a gorge cut by the stream to a depth of 80 feet below the level of the supporting bench."

Pogue's more objective method of measuring height from the base of a natural bridge's abutments was echoed by U.S.G.S. geologist H. E. Gregory in his 1917 report, in which he measured the height of Rainbow Bridge:

"The arch rests on a bench of bedded sandstone below which the inner gorge of Bridge Creek is sunk to a depth of 80 feet. Between the chord uniting the springers at the level of this supporting bench and the lower side of the keystone point the vertical distance is 187 feet, and the arch at its summit is 42 feet thick. The total height of the crown above water is therefore 309 feet -- exceeding that of the second greatest natural bridge in the United States, Sipapu, in White Canyon, Utah, by 89 feet."

Why is it more objective to measure a natural bridge's height from "a chord uniting the springers" than from the stream bed below the span? Because if figures from the stream bed are used, then the span would seem to grow "higher" as the stream bed cut deeper, a subjective concept that is false to objective fact. Natural rock spans do not grow "higher," even though the gorges below them may be cut deeper. The same concept should also be applied to the measurement of many other types of natural spans.

In his 1922 report on the natural bridges of Utah, Pack rejected Gregory's size comparison between Rainbow Bridge and Sipapu:

"So far as is known (Sipapu) is the largest natural bridge in the world. The Nonnezoshie (Rainbow Bridge) is an arch."

Pack thus added the confusion about the definitions of natural spans to the confusion surrounding their measurement. In the late 1970s, the National Park Service announced that measurements of Rainbow Bridge, taken in connection with monitoring the span for damage from water beneath it, had shown it to be 19 feet shorter and 3 feet narrower than previously thought. Thus, while a natural bridge cannot grow higher, it can grow shorter -- either from erosion or from taking another measurement.

1974 Bureau of Reclamation photo by V. Jetley of Rainbow Bridge. Note the measuring crew on top of the span.

DIMENSIONS OF RAINBOW BRIDGE

REPORTER	HEIGHT	SPAN	WIDTH	THICKNESS
DOUGLASS, 1909	309	278	33	42
CUMMINGS, 1910	308	274	--	--
CLELAND, 1910	308	273	--	--
POGUE, 1911	309	278	33	42
GREGORY, 1917	309	278	--	42
VREELAND, 1976	309	278	33	42
NPS, LATE 1970s	290	275	--	--
ROYLANCE, 1982	309	278	33	--
NPS, 1984	290	275	33	42

Tonto Natural Bridge in Arizona provides another example of a difficult span to measure. Authors Cleland (1910), Miller (1946) and Winstead (1980) gave its height as "128 to 150 feet," "183 feet," and "183 feet," respectively, and its width as 140, 140, and 150 feet, respectively. They all agreed on a length of 400 feet, probably an example of copying earlier figures. Considering the irregular shapes of the tunnel's openings at each end, it is unlikely that independently-made measurements would be identical.

In an article in the September 1949 issue of National Geographic, author Jack Breed tells of finding and measuring "-- a new arch, uncharted and unnamed," which the expedition promptly named Grosvenor Arch and measured as "152 feet high, 99 feet wide, and only four feet thick at the top of the span." As with so many other spans, these height and width figures are misleading. Actually, the span is one of many that are simply openings high in a cliff rim. In the case of Grosvenor Arch, the height figure is meaningless, and the larger of its two openings is fairly round and considerably less than "99 feet" in diameter.

In an expedition reported in the September 1955 issue of National Geographic, writer W. Robert Moore cataloged the measurement of the eight major natural spans in the lower Escalante River canyon system. Other spans in the upper reaches of the same canyon system were measured later by others. The dimensions recorded by Moore for the 1955 expedition, and by arch historian Vreeland for the same spans and the ones not reached by Moore, are listed in the following chart:

ESCALANTE CANYON SYSTEM NATURAL SPAN DIMENSIONS (1)

LOCATION	SPAN NAME	MOORE HEIGHT	MOORE WIDTH	VREELAND HEIGHT	VREELAND WIDTH
MAIN CANYON	ESCALANTE N.B.	--	--	130	100
	SAND CREEK ARCH	--	--	2 (2)	15 (2)
	STEVENS ARCH	160	225	160	225
COYOTE GULCH	COYOTE N.B.	38	68	38	68
	JACOB HAMBLIN ARCH	90	170	80	102
	CLIFF ARCH	45	40	45	40
DAVIS GULCH	LA GORCE ARCH	75	100	75	100
	GREGORY N.B. (3)	75	177	75	127
	BEMENT ARCH	80	100	80	118
WILLOW GULCH	BROKEN BOW	100	94	100	94
PHIPPS WASH	PHIPPS ARCH	--	--	26	85
	MAVERICK N.B.	--	--	16	20
FENCE CANYON	ZANE GREY ARCH	--	--	45	70
UNNAMED SPUR	BOWINGTON ARCH	--	--	20 (2)	40 (2)

NOTES: 1. Figures are for opening dimensions.
2. Figures are dimensions of a horizontal opening.
3. Span now inundated by Lake Powell.

In late 1983, a research group from Brigham Young University in Utah, led by Professor Reed Blake, measured Kolob Arch in Zion National Park using "electronic distance measuring equipment." As reported in the October 2, 1983, edition of the Salt Lake City Deseret News Magazine, this remeasurement led to a public proclamation that Kolob Arch was "the world's largest natural span." Previously, Landscape Arch at 291 feet wide, and Rainbow

Distant view of Kolob Arch, Zion National Park, to the left of the pouroff. Harvey Halpern photo.

Bridge at 287 feet wide, had been thought to be the "largest," but Kolob's new measured width of 310 feet placed it in "first" position.

In the spring of 1984, another group from Brigham Young University, led by Professor Dale Stevens, measured Kolob Arch using techniques somewhat different from those used by the first group. As might be expected, the resulting measurements also differed. The "span" figure obtained by the second group was 18 feet less than that of the first, but still left Kolob Arch the "largest" by one foot -- if the 1941 rockfall from one abutment of Landscape is ignored.

As noted by Blake in a 1977 article in Mountain West magazine, and described further in his 1983 Deseret News Magazine article, Kolob Arch had been measured first in 1953 by two Oregon climbers, using "a compass, a measuring tape, and a rope with a rock tied to the end." Even though their methods were crude in comparison, the 1953 figures matched the 1983 and 1984 figures rather closely, except for width. The 1984 measurement for height was made from the back of the span. It would probably have matched the earlier figures more closely had it been made from the front of the span, or the two figures averaged.

KOLOB ARCH, ZION NATIONAL PARK

MEASUREMENT	HEIGHT	SPAN	WIDTH	THICKNESS
1953	240	290-310	65	80
1983	230	310	--	76
1984	177	292	41	82

Soon after the new figures for Kolob were published, National Park Service officials at Arches National Park requested that Landscape Arch be remeasured, applying the modern techniques used to remeasure Kolob Arch. This was a reasonable request, since comparisons should be made between spans only if they have been measured by the same techniques. Landscape Arch was measured by Blake's team in the summer of 1984, and by Stevens' team in the autumn of the same year. Blake reported the span of Landscape Arch to be "287 feet; 291 feet on the slope," while Stevens found it to be 306 feet or 325 feet, depending on the definition of "span" used. Stevens concluded that "If one were to use the two most frequently used dimensions of light opening and span length, Landscape Arch would be the largest --," thus leaving the matter of which arch is the "largest" still unresolved.

At best, however, with spans as different in shape as Kolob, Rainbow, Landscape, Sipapu, Morning Glory and Stevens, size comparisons are still questionable and subjective. These limitations should be borne in mind while examining the following chart of major Utah spans, compiled from Blake's 1977 and 1983 articles:

MEASUREMENTS

MAJOR UTAH NATURAL SPANS

SPAN NAME	LOCATION	SPAN IN FEET
Kolob Arch	Zion National Park	310 (1)
Landscape Arch	Arches National Park	291
Rainbow Bridge	Rainbow Bridge N. M.	287
Sipapu N. B.	Natural Bridges N. M.	261
Stevens Arch (2)	Glen Canyon N. R. A.	225
Kachina N. B.	Natural Bridges N. M.	206
Owachomo N. B.	Natural Bridges N. M.	200
Double Arches	Arches National Park	163
Natural Arch (3)	Zion National Park	156
Broad Arch (4)	Capitol Reef National Park	133
South Window	Arches National Park	130
North Window	Arches National Park	98

NOTES:
1. The 1953 figures were 290-310.
2. Not listed in Blake's 1983 chart.
3. Not listed in Blake's 1977 chart.
4. Now Hickman Natural Bridge.
5. Note that 243-foot Morning Glory Arch is not listed.
6. Wrather Arch, over 200 feet, is in Arizona.

The BYU remeasurement of Kolob Arch established an excellent precedent for the scientific measurement of Colorado Plateau natural spans. Perhaps the techniques and definitions used by one of the two teams will be standardized and become widely used, thus introducing an element of scientific objectivity into the heretofore highly subjective and variable subject of span measurement.

What are the "largest" natural spans in the Colorado Plateau? This question cannot be answered with any accuracy because of the lack of a standardized measurement system to date, but a list of those having openings generally accepted to be greater than 200 feet in one dimension can be given. These are, in alphabetical order: Kachina Natural Bridge, Kolob Arch, Landscape Arch, Morning Glory Arch, Rainbow Bridge, Sipapu Natural Bridge, Stevens Arch, and Wrather Arch. There are literally dozens of spans with openings between 100 and 200 feet in one dimension.

SUMMARY AND CONCLUSIONS

1. Of the hundreds of natural rock openings within the Colorado Plateau, only a few have been measured. For the most part, these have been measured because of a general interest in trying to determine which is the "largest" or "highest" or "widest," terms that so far have not been objectively defined.

2. To date, there has been no standardization in the techniques and definitions applicable to the measurement of natural spans and none have been proposed, let alone published or generally accepted. A special committee of the Utah Academy of Science has been established to devise span measurement standards, but it may be years before these become generally known and accepted throughout the Colorado Plateau region. None of the measuring techniques used to date could be universally applied to all shapes and locations of Colorado Plateau natural spans.

3. All size comparisons between natural spans made to date are objectively meaningless, because the measurements upon which they are based have not been made by standardized procedures. All claims that certain natural spans are the "largest" in any dimension are thus questionable.

4. While there is general interest in determining which spans are the "largest," human subjectivity and aesthetics, as well as dimensions, play parts in the establishment of such superlatives. Interest in a graceful and conspicuous span is always greater than in an obscure and less lovely span.

5. The diversity of shape and location of Colorado Plateau natural spans is so great that it would be extremely difficult, if not impossible, to devise a single standard measuring system that would be universally applicable. Readers who question this are challenged to study just the spans shown in this book, then devise a set of standard definitions and techniques that could be applied to every one, and that would produce the same results on a particular span no matter who did the measuring.

Morning Glory Arch, Negro Bill Canyon

MEASURING THE GEMINI BRIDGES

The year was 1968, and we were on our way into the sand and slickrock wilds to the west of Moab, in the heart of southeastern Utah's canyon country, seeking to find, photograph and measure a pair of twin natural bridges that local tour guide Lin Ottinger had discovered eleven years earlier, then "mislaid" in the maze of canyons about ten miles to the northeast of Canyonlands National Park. He thought that he now knew where they were, so we were on our way, anxious to document the elusive spans.

After an hour or so of tortuous travel over jeep trails often almost too faint to see, Lin stopped and reconnoitered on foot, descending a series of wooded sandstone terraces toward a barely visible canyon rim. The rest of our small party followed. After a hundred yards or so, Lin paused and pointed dramatically -- toward two wide strips of rock with a narrow gap between them and a large oval hole to their right. The bridges! But they were not very impressive from this angle, so we hurried on down the slope for a closer view -- then gasped as we reached them!

The huge spans bridged a hanging-canyon spur to a sheer-walled gorge almost 300 feet deep. The floor of the hanging canyon spanned by the flat-topped bridges was about 100 feet below their upper surfaces, and they seemed massive enough to support a railroad.

After taking several photographs, we set about measuring the twin spans, using the crude but fairly accurate tools Lin had brought along for this purpose -- a long wire, a board, and a steel measuring tape. Measuring their width in the center was easy, as was measuring their length between the sheer and parallel canyon walls. However, measuring the opening widths required a little ingenuity. The wire, weighted with the board, was dropped along the wall of each opening, a sighted mark was made on the bridge top, and the distances between these marks were measured.

The height of each span was measured by dropping the board-weighted wire from its center until it hit the canyon bottom, marking the wire, hauling up the board, then measuring the length of wire to the mark. To get some idea as to the thickness of each span and the height of each opening, we lowered the board until someone sighting from a distance indicated that it was about even with the top of an opening, then marked and measured the wire. Our efforts produced the following figures:

ORIGINAL MEASUREMENTS OF THE TWIN GEMINI BRIDGES

DIMENSION	WEST BRIDGE	EAST BRIDGE
Length across the top	89	70.5
Width in the center	26	26
Thickness in the center	40	18
Depth below bridge top-center	100	103
Opening height	60	85
Opening width	67	44

Maximum distance between bridges, 10 feet.

These figures were published in the Moab newspaper with this note:

> "The bridge measurements were made with a long piece of wire, a board and a steel tape and are obviously subject to refinement. If someone would like to volunteer to provide scientific measurements for the record, the TIMES-INDEPENDENT will be happy to publish them."

Getting into the canyon beneath the twin bridges proved to be tricky. On the first attempt we entered the wrong branch of the complex Bull Canyon system in which the spans were located, but on the second try we found the correct canyon branch and the half-mile spur whose upper end the bridges spanned. They were impossible to see until we were almost directly below them. We found no indication that other men had ever been there.

Shortly after these trips to the Gemini Bridges, I wrote to the Board on Geographic Names about the twin spans, outlining their known history, giving their location on BLM land, proposing a name, and attaching a photograph of them. Six months later, in a decision dated March 7, 1969, the Board accepted our proposed name, "Gemini Bridges," for all federal use.

Following our exploratory trips to the bridges and their official naming, I wrote two articles about them. One appeared in the March 1969 issue of Western Gateways Magazine, and the other was published in the November/December 1970 edition of Outdoor World Magazine. Once the twin bridges were publicized, several local people claimed to have had prior knowledge of them, but when asked for specifics failed to come even close to their actual location. I found that at least one local bush-pilot knew of the spans, but as the Ottingers noted in a letter-to-the-editor responding to local claims of prior knowledge of the twin spans:

> "There is no doubt that men have, indeed, seen these magnificent twin bridges before, but there is also little doubt in our minds that most of these men were Indians and cattlemen, long since dead. We believe that the probability is very low that anyone now living in this area saw these bridges before 1957, except perhaps fleetingly from the air, where their unique nature would not be apparent."

Later, I took leaders of the Moab Annual Jeep Safari to the spans, so they could serve as a destination for a new route on this popular annual affair. Even with this guidance, however, the bridges were so difficult to find that for the next two Safaris the leaders of that trail route became lost trying to reach them, to the utter disgust of those following.

Today, with the roads and trails to the bridges much improved and marked by the BLM -- albeit with "Gemini Arch" on their signs -- hundreds of people see these outstanding natural spans each year. To date, however, no one has provided any better measurement figures. Not that it really matters. From the tops of these magnificent spans, or from the canyon below, the awesome beauty of these rare natural wonders speaks for itself!

Metate Arch, Devils Garden Outstanding Natural Area

PROTECTION

PROTECTED NATURAL SPANS

Many of the better known natural spans on federal public land in the Colorado Plateau are protected within special areas, three of them dedicated to the natural spans for which they were established. All the protected areas are listed in the following chart. There are many significant spans in less well-known areas that are not protected. Some of these are listed and discussed later in this chapter.

PRESENT FEDERALLY PROTECTED AREAS WITH NATURAL ROCK OPENINGS

NATIONAL PARKS	STATE
Arches N. P.	Utah
Bryce Canyon N.P.	Utah
Canyonlands N.P.	Utah
Capitol Reef N.P.	Utah
Petrified Forest N.P.	Arizona
Zion N.P.	Utah
NATIONAL MONUMENTS	
Canyon de Chelly N.M.	Arizona
Cedar Breaks N.M.	Utah
Colorado N.M.	Colorado
Dinosaur N.M.	Utah/Colorado
Natural Bridges N.M.	Utah
Rainbow Bridge N.M.	Utah
NATIONAL RECREATION AREAS	
Glen Canyon N.R.A.	Utah/Arizona
PRIMITIVE AREAS, BLM	
Dark Canyon P.A.	Utah
Grand Gulch P.A.	Utah
OUTSTANDING NATURAL AREAS, BLM	
Devils Garden O.N.A.	Utah
El Malpais O.N.A.	New Mexico
Escalante Canyons O.N.A.	Utah
Phipps-Death Hollow O.N.A.	Utah
The Gulch O.N.A.	Utah
NATIONAL HISTORIC LANDMARK, BLM	
Desolation Canyon N.H.L.	Utah
WILDERNESS AREAS	
Paria Canyon W.A.	Utah/Arizona

LIMITS TO PROTECTION

National Park or Monument status does not give full legal protection to natural features and values because some parks and monuments still have mineral claims and leases, and even private land holdings, within their borders. Even so, the probability that significant physical damage would be done within their borders by mineral development is low. The chance of damage is reduced still further by the establishment of Congressionally approved wilderness areas within such parks and monuments.

National Recreation Area status provides considerably less protection from the development of existing mineral claims and leases within such areas, but since most natural spans are well above ground level the chance of damage to them from development is quite small. Again, the establishment of wilderness areas within a National Recreation Area will add protection.

Primitive Areas and Outstanding Natural Areas are established administratively by the BLM to protect special natural, recreational or cultural resource values within certain areas. Primitive Areas and Outstanding Natural Areas that are not eventually converted to wilderness by Congress, may lose their special protection. This conversion is quite probable, however, since these areas have been managed for protection of widely recognized values since the 1970s. Some Outstanding Natural Areas may be converted to wilderness, but those that are not may still be protected. Such limited protection could be removed, however, by an administrative decision.

In the case of Desolation Canyon, National Historic Landmark status provides only limited protection from mineral development. Even so, the physical locations of most natural spans within Desolation Canyon and its tributaries are such that the chance of any destruction from mineral development activities is quite small.

In addition to the vicissitudes of politics and the existence of mineral claims and leases, there are also other dangers to natural spans within federally-protected areas. Some are natural, such as climatic changes, weather extremes and earthquakes. Others are manmade hazards.

As noted in earlier chapters, weather modification programs can and do create changes in precipitation patterns, types and totals, all of which affect natural span life expectancies. No one knows what part of the heavy runoff and flooding of various Colorado Plateau rivers and streams in the 1980s was caused by weather modification activities, but scientists who have studied the subject are certain they did contribute to the problem.

The construction of reservoirs can destroy natural spans by flooding and by collapse from continual wetting, as with Gregory Natural Bridge, Padre Arch and other spans within Glen Canyon National Recreation Area. In his July 1967 article in National Geographic, author W. M. Edwards noted that while La Gorce Arch in the Escalante canyon system would remain above Lake Powell's rising waters, the canyons below it would be flooded. The article also noted that the National Geographic party had barely been able to get a boat under Gregory Natural Bridge. The following summer, the water rose to within inches of the underside of the span, but fell enough by late autumn to permit this author to boat under it one last time.

Lake Powell's rising water has provided easy boat access to some of the spans in the lower Escalante that remain above water, but it has adversely affected land access to some of the other spans farther up the canyon system, despite earlier claims to the contrary by authorities.

Direct vandalism to Colorado Plateau natural spans began with their discovery. Early explorers often carved their names, dates of visitation and other things on the abutments of arches and natural bridges, and early visitors added their own petroglyphic efforts. In the 1941 edition of the WPA guide to the state of Utah, one writer reported how Zeke Johnson, custodian of Natural Bridges N.M., told of handling one such flagrant act of vandalism in 1926:

> "A (local) group of boys was going out to the bridges from Blanding, and I told them not to scratch their names on rocks inside the Monument. One man who went along with the boys told his boy that that man Zeke Johnson had no police authority over those rocks, and he and his son wrote their names on top of Sipapu Bridge and dug it deep. Well of course I explained the law and told him to get back out and rub it off, and he told me to go 'places.' The invitation to go some place was repeated several times, and I found it necessary to write to the director of national parks in Washington D. C. The defendant got a letter telling him that unless he immediately complied with Custodian Johnson's instructions he would have a free ride at his own expense from Blanding to Washington D. C., to be arraigned before the U. S. Government. He soon hunted me up and asked what he should do. I said just do as I told you, track back out there and take off those initials. He offered me twenty bucks if I would do it, but I told him nobody could do it but him, and it was a five-day horseback trip in those days."

Needless to say, human nature being what it is, such vandalism still goes on, despite all efforts by park authorities to prevent it, especially within parks that receive heavy visitation.

Another quaint practice pursued by some "arch collectors" also detracts from the naturalness of many natural spans. In order to gain some kind of personal esteem, such collectors leave their names in cans or bottles in cairns or on ledges beneath remote natural spans, apparently as a means for telling the world "I was here." This practice was adopted from mountain climbers, who leave such records on the tops of mountains they have ascended, but is essentially meaningless with natural spans. Such name-and-date records certainly have no bearing on the discovery or documentation of a span, and they disturb a span's natural beauty with unnatural rock cairns and human artifacts -- another form of vandalism.

Developments adjacent to a protected area can also adversely affect natural features within that area, including natural spans. Records show a considerable increase in vandalism, including the scratching of names on rocks, in the vicinity of large construction projects. With such proposed large-scale projects as the establishment of a nuclear waste dump on the borders of Canyonlands National Park, archaeological sites and such nearby natural features as arches and bridges would be lost, or at least suffer a drastic increase in vandalism. There are a number of natural spans, some of them quite delicate and unusual, within Davis, Lavender and Indian Creek canyons and Harts Draw, the canyon systems that would be disturbed by this project.

Another kind of hazard that natural spans face even within protected areas is blasts of sound that could cause their collapse. Some of these come from mineral development work adjacent to protected areas. Others come from aircraft "sonic booms." Explosive blasts will continue to be a danger, especially to fragile spans such as Delicate Arch and Landscape Arch, as long as permitted by the mineral development laws that govern federal land.

The sonic booms of military aircraft flying over or near

national parks was also a serious and continuous hazard during past decades. As reported in the April 20, 1972, edition of the Moab Times-Independent, after growing public criticism of this practice, the National Park Service announced that an Executive Order had been issued:

> "-- which states that the U. S. Air Force will no longer permit routine supersonic flights over or near national parks. There are some instances in which supersonic operations will be allowed over the parks but for the most part, these are in case of emergency. Supersonic flights must not only avoid passing over national parks, they also may not fly near them, according to the new regulation. For each 1,000 feet of altitude, the pilot must allow one-half mile between the flight path and the park boundary. The regulation also prohibits supersonic flights below 30,000 feet (over land) so the high speed planes must allow fifteen miles between the nearest park boundary and the flight path."

For more than a decade, this order brought quiet to the national parks in southeastern Utah, but by the mid-1980s sonic booms were heard again. For example, in the Park Service daily report for February 10, 1984, sonic booms were heard in both Arches and Canyonlands national parks, and booms farther south had become so common that Natural Bridges reported there were "no sonic booms" that day. On March 2, three sonic booms were reported by Arches and two by Canyonlands. Others were heard by Moab residents during this period, although no "emergencies" were announced by authorities. The problem continues, and with each boom the thought occurs to many concerned Moab residents -- "did that one get Landscape?"

Landscape Arch, Arches National Park

HISTORY OF PROTECTION

The U. S. Forest Service was established in 1905. The first federal Antiquities Act was passed in 1906. Although Yellowstone National Park was created in 1872 and Yosemite National Park in 1890, the National Park Service was not established until 1916. From that time on, this organization has administered the National Parks, Monuments and Recreation Areas within the Colorado Plateau, except for Flaming Gorge National Recreation Area which is administered by the U. S. Forest Service.

According to Park Service archaeologist C. R. Steen in his article in the Utah Historical Quarterly of Winter 1972, one of the purposes behind the rush to pass protective legislation in the early 1900s was to promote tourism to the West:

> "With the wave of enthusiasm for preserving both natural and historic features of the country there went a growing concern to attract tourists. The lone traveler -- hunter, artist, or recorder of places and events -- was long a part of the western scene. With easier and-faster methods of travel the trickle of tourists became an ever increasing stream, and sporadic efforts were made to open up the wilder sections of the West and to make their features known to the general public."

The manner in which one National Monument was established probably contributed to Steen's conclusion. Although some of its features such as Landscape Arch were known to local residents in the late 1800s, the first effort to protect two areas that are now within Arches National Park were made in 1923 by F. A. Wadleigh, an agent of The Denver and Rio Grande Western Railroad System, for the purpose of promoting tourism on its line that ran to the north of the area through Thompson, Utah. This commercial inspiration was recognized as valid at the time in a General Land Office survey report on the area dated June 29, 1928:

> "-- both of the areas described (The Windows and Devils Garden areas) contain enough interesting formations to well warrant their withdrawal as National Monuments. Several areas more remotely located from the Union Pacific Railroad System have been exploited with marked success that do not compare with the scenic features of either of these proposed monuments."

The part played by the railroad in getting recognition for the area was noted in a July 17, 1924, article in the Moab newspaper:

> "The unique scenic attractions in the section locally known as 'The Windows,' situated some ten miles north of Moab, will probably be set aside by the department of the interior as a national monument, as a result of a movement initiated by the Denver & Rio Grande Railroad. It is possible that the monument might be extended to include Castle Valley, noted for its scenery, where castles, huge spires and other unusual formations have been carved out of the red sandstone by the action of erosion."

Another less than altruistic attitude of the times concerning the protection of natural features was also expressed in Wadleigh's 1923 letter:

PROTECTION

> "This man Ringhoffer (who first reported the Devils Garden area to Wadleigh) has a mine prospect about four miles from the 'Devil's Garden,' and he is of the opinion that there are ore deposits in that section, although not in the 'garden.' With the exception of this one prospect, there is nothing in the region. So cattle and sheep range there during the winter, but there are no settlers thereabouts and no one could possibly be injured, as I see it, by making a National Monument out of the region."

This attitude of creating a protected area only if that area were useless to anyone for any other purpose was further illustrated in a General Land Office report on the same area dated September 26, 1924:

> "-- there are no reservoir sites involved and no rights of way are on record. There is no timber in the region of more than nominal value for fuel purposes. There are no forage grasses of any importance. The area has a nominal value for grazing. There are no water holes --. The area -- will include no land that has value for any other purposes whatever. In consideration of the facts set forth, I recommend the withdrawal of the above area as a National Monument."

This unwritten general federal policy of withholding and protecting public land only if it "has no value for any other purposes" continues today, and is the principal limiting force behind the establishment of Colorado Plateau federal park and wilderness areas. For example, the first area considered in the establishment of a Canyonlands National Park was some one million acres in size. Because of pressures from mining and grazing interests, the park eventually established was a quarter that size. Similar pressures have kept Arches National Park smaller than it should have been. Many BLM and Forest Service areas, some of them containing outstanding natural spans and topography, have also been dropped from wilderness consideration because of opposition from mineral, grazing and other development interests.

Another practice by mineral interests that has long inhibited the establishment of protected areas within the Colorado Plateau was illustrated in a June 26, 1925, General Land Office report concerning the area later to become Arches National Monument. The author of the report noted that while there were:

> "-- numerous applications for oil and gas and potassium prospecting permits -- no wells have been drilled. Some slight evidence of mineral was observed -- but not sufficient to warrant the expenditure of money in development -- no mining claims were found --"

This speculative filing for mineral rights, with little or no evidence of actual mineral value, still prevails today and has been responsible for scuttling many protective efforts by various federal land administration agencies. Today, major energy corporations sometimes blanket entire regions with speculative claims, ignoring existing claims, in blatant violation of both the letter and spirit of federal and state laws governing the establishment of mineral claims.

Further, many speculative mineral claims have been kept "active" by the filing of falsified "development" statements, to avoid spending the effort and money required by law for the actual "proving" of claims. As shown by the rate of actual development, the percentage of valid mineral claims within the Colorado Plateau is extremely low in proportion to the number of claims filed. This has inhibited the withdrawal of many areas deserving of protection, including some that contain outstanding natural spans.

Fortunately, the three park areas dedicated to natural spans were established well before the energy industry discovered the region, otherwise they would probably not exist today.

Natural Bridges National Monument was established April 16, 1908. Its boundary was adjusted in 1909, 1916 and 1962.

Rainbow Bridge National Monument was established May 30, 1910. It consists of only 160 acres and its boundary has never been changed.

Arches National Monument was established April 12, 1929. Its boundary was changed in 1938, 1960 and 1969. It became a National Park on November 12, 1971.

Canyonlands National Park, established September 12, 1964 and slightly enlarged in 1971, probably contains as many significant natural spans as Arches National Park, but no inventory of them has been taken to date.

PROTECTION

RAINBOW BRIDGE ENDANGERED

A national monument was established May 30, 1910, to protect Rainbow Bridge. In an article about the bridge in the November 1911 edition of National Geographic, author J. E. Pogue of the United States National Museum noted this:

> "The government has already made of this natural wonder a national monument, thus preserving it for all time against vandalism and commercialism and conserving it for the enjoyment of all."

Unhappily, Pogue's dream of the span's immunity from commercialism proved to be illusion. National monument status did not give Rainbow Bridge perpetual protection. Once the federal Bureau of Reclamation and its development-minded supporters conceived of the idea of a series of stair-step reservoirs on the Colorado River, Rainbow Bridge's solitude, sanctity, and possibly its very existence, were lost.

In his article "Three Roads to Rainbow" in the April 1957 issue of National Geographic, staff writer Ralph Gray noted that until then Rainbow Bridge remained so remote and inaccessible that:

> "-- since its recorded discovery by white men -- in almost half a century fewer people have seen it than view Grand Canyon in two average summer days. -- Soon Rainbow Bridge will lose some of its age-old isolation. Glen Canyon Dam, now under construction by the Bureau of Reclamation, will back water from the Colorado into Forbidden Canyon. A new highway running north from U.S. 89 will lead to the dam. Thus motorists in the future will be able to drive to a boat landing and skim over the man-made lake to within shouting distance of Rainbow."

Professor of geography Steven C. Jett, in his introduction to the 1980 reprint of Clyde Kluckhohn's 1927 book about his trip to Rainbow Bridge, summed up the problem that Glen Canyon Dam created:

> "The Colorado River Storage Act of 1956 which authorized Glen Canyon Dam also called for protection of Rainbow Bridge National Monument, into which water would otherwise penetrate as Lake Powell filled. Envisioned was a barrier dam, across Rainbow Bridge Canyon, to prevent backup of water into the Monument. An upstream diversion dam plus a tunnel to Forbidden Canyon would evacuate Bridge Creek's flow. Predictably, when appropriation time came, these protection-providing dams were judged impractical to build, and no funds were forthcoming."

The subject of how, why and whether to protect Rainbow Bridge was explored in great detail in articles in the August 26, 1960, and May 19, 1961, editions of Science, with one side of the public argument represented by writer W. R. Halliday, and the other by A. M. Woodbury. Another excellent article on the subject was written by author Verne Huser for the February 1974 issue of American Forests Magazine. A summary of the entire controversy to date is hereby presented.

Rainbow Bridge as it is. For scale, note figure on top of span.

THE RAINBOW BRIDGE CONTROVERSY -- WATER UNDER THE BRIDGE

The groundwork for the controversy was laid back in the mid-1800s, when commercial paddlewheel steamers were still plying the lower Colorado River as far up as Callville, Nevada, which is now beneath Lake Mead. This was long before the railroads had penetrated to western Arizona, back when the famous advice to "go West, young man" was still new. The cry for western development was loud and clear and not to be opposed. Indeed, opposition to the unlimited development of America was unheard of then.

Plans for the development of the lower Colorado were being made, with dams for flood control and water for agriculture having priority. Within the next few decades, the lower Colorado was changed from a wild, unpredictable and destructive torrent of silt-laden water into a stairstep series of placid reservoirs, from which the cement arteries of a network of irrigation canals fed off into the desert in all directions. The federal agencies responsible for this modern "water conservation" magic then turned to thoughts of how to best develop and use the vast reaches of the upper Colorado River, where it flowed through Colorado, Utah and northern Arizona.

Fortunately, by that time there were people in this nation who were asking if all this rampant and unrestricted development was really necessary. Thus, when Congress was asked to fund the next batch of great concrete dams on the mighty Colorado and its major tributaries, they heard not just one side of the matter, but two -- that of the developers and that of the conservationists.

PROTECTION

To pacify the conservationists, Congress passed a law governing the development of the upper Colorado that contained a clause prohibiting any development that would allow reservoir waters to intrude into national parks and monuments. This limit allowed the developers to continue with most of their projects, including the highly controversial Glen Canyon Dam, but put a stop to planning for dams that would have affected parts of Grand Canyon National Park that were then designated as Grand Canyon National Monument, plus several other National Park areas.

Resigned to the loss of the magnificent Glen Canyon system, but secure in their knowledge that no reservoir waters would be allowed to enter any established park or monument, the conservationists relaxed and stopped their strong opposition to Glen Canyon Dam. In 1963, with the dam nearly completed, water began to fill the reservoir's 180-mile long maze of unspoiled canyons that extended from just below the Arizona-Utah border to deep within the virtually unknown "canyon country" of southeastern Utah. It took many years for the huge reservoir to reach any significant depth, to begin to intrude deeply into Bridge Canyon, where Rainbow Bridge stood serenely within its small and remote National Monument, supposedly protected.

That's when the Rainbow Bridge controversy first came to public attention. Belatedly, it became obvious that when the reservoir water reached a certain level, it would be allowed to invade Rainbow Bridge National Monument in spite of the law against such an intrusion. Those who had planned and built the dam had known this all along, and were not concerned with its illegality. Conservationists tried to get Congress to fund the protection for the monument that had been planned, but to no avail. The waters crept ever higher, and the conservationists grew more frantic.

As the controversy heated up, Congress cast doubt on whether or not it had really meant its provision against water in parks and monuments, by its firm refusal to provide funds for keeping the water out of Rainbow Bridge National Monument. Conservationists attempted to arouse the voting public against the outrage of "water under the bridge" by loudly proclaiming that the magnificent beauty of Rainbow Bridge would be marred by an ugly puddle of stagnant water under its soaring arc of stone. They claimed that such water would be both illegal and unsightly.

But for some time no one dared to challenge the contention of the U.S. Geological Survey that standing water under Rainbow Bridge would not physically endanger the unique natural span. This official position was stated in a U.S. Geological Survey report that was so ridiculously unscientific that it was not released to the public.

Finally this writer, who did manage to obtain a copy of the U.S.G.S. report and who was familiar with the region's geology, grew impatient with waiting for established conservation groups to contest the U.S.G.S. claim that Rainbow Bridge was in no danger. He wrote a detailed report describing how standing water under the great stone span would physically endanger it, would contribute to its premature collapse.

The complete report was published in a magazine, a shorter version appeared in a conservation journal, and summaries were printed in several magazines and regional newspapers. One version of the report was sent to various major universities, for study by their geologists who were familiar with the unusual geology of the upper Colorado River basin. In their review comments, all agreed with the report, and a few even contributed data that added emphasis to its arguments and conclusions.

Despite this support from the scientific community, the report fell on deaf ears. The reservoir waters continued to rise, fluctuating with the seasonal flows, but making a net gain almost every year. Finally, the water began to enter Rainbow Bridge National Monument. Every other effort having failed, a coalition of conservation organizations then took the matter to the federal court in Salt Lake City. The judge studied the applicable law, which clearly prohibited reservoir waters in national parks and monuments, then ordered the immediate release of enough water to drain the encroaching water from Rainbow Bridge National Monument.

The Bureau of Reclamation, the federal agency responsible for control of the dam, with plenty of urging from frantic development interests, quickly took the matter on to the federal appeals court. To keep the reservoir water level permanently below the elevation of the boundary of Rainbow Bridge National Monument would mean that a significant part of the reservoir's storage capacity would be lost, with many unwanted economic and political consequences.

The federal appeals court promptly made a political decision. It reversed the lower court's decision, on the highly questionable premise that since Congress had funded Glen Canyon Dam, and had consistently refused to fund any protection measures for Rainbow Bridge National Monument, that Congress had not really meant what it said in that federal law about no reservoir waters in national parks and monuments.

The appeals court's decision was a classic case of a court creating law, so conservationists promptly took the matter on to the U.S. Supreme Court. But that august body knew very well when not to meddle with what was by then a political hot potato. It refused to review the appeals court's decision, thus letting it stand.

The water under Rainbow Bridge continued to rise, and eventually reached its upper limit. It still stands there today, with some seasonal variation -- a monument to the cupidity of our elected representatives and our federal "water reclamation" agencies, and to the blind momentum of a national development policy that has outlived its usefulness. The only thing left now is to wait and see what happens.

This author's report describing the danger to Rainbow Bridge claimed that the glorious stone span's life expectancy would be drastically shortened by water standing beneath it. The final paragraph in his letter to the publisher of the magazine that printed the full report read:

> "I sincerely hope that I am wrong, or that the proof of my correctness waits until I am decently dead. My personal world would be achingly incomplete without Rainbow Bridge standing in lonely, aloof majesty astride a bone-dry canyon, far from the threatening waters of Lake Powell."

Rainbow Bridge as it was, but will never be again -- "- standing in lonely, aloof majesty astride a bone-dry canyon." For scale, note the tiny figures standing in the bridge's shadow.

SOME UNPROTECTED DESERVING SPANS AND AREAS

As noted earlier, there are many deserving individual natural spans, and areas that contain numerous spans, that are not protected. Most are on multiple-use federal public land, but a few are on state or private land. Some of these are listed and discussed in the following pages.

Wilderness status may eventually protect a few of these spans and areas. Designation of an area as a Wilderness Study Area gives it some temporary protection, but that protection is not permanent unless the WSA is later given full wilderness status by Congress, and a WSA has no chance of becoming wilderness unless it is presented to Congress. Often, WSAs are dropped from any further consideration for political reasons, or because mineral, grazing or timber interests have protested the withdrawal.

While knowledge of the existence of a large natural span within a WSA might incline Congress favorably toward granting wilderness status, all too often such information is not even included in wilderness inventories. For example, in the Forest Service study of the area that took in upper Arch Canyon, there was no specific information about the three magnificent arches there. Thus, on the practical side, neither the BLM nor Forest Service wilderness programs offer much chance of protection for the following spans and areas:

TONTO NATURAL BRIDGE. This outstanding natural tunnel is on private land. The large area on top of the span is even under cultivation, and the owner charges visitors a fee to see it. Things might have been different, however, had someone in the federal government seized the opportunity that was once offered. Author C. F. Lummis reported in his 1925 book, "Mesa, Canyon and Pueblo," that at the time of his visit to the span:

> "The hermit who owns it -- wished to dedicate his homestead to the Government, and to have it accepted, made accessible, and cared for as a national monument -- as it is worthy to be."

MORNING GLORY ARCH. Within the Colorado Plateau proper, this is the only known natural span with an opening wider than 200 feet that does not have some kind of special protection. It is located in a short spur of a canyon system that has wilderness qualities, but the span itself is on a corner of state land, which means that it is subject to whatever activity the state may approve, including mineral development. The only chance to protect the span would involve a land exchange between the state and BLM, then special action by the BLM. If the main canyon does become a part of the wilderness system, this will at least afford some small protection to Morning Glory Arch by limiting access to it. Meantime, a mineral claim holder recently bulldozed up the main canyon to within a short distance of the huge span.

ARCH CANYON. This is an area that was proposed for national monument status at the same time as Arches National Monument. It contains not only three magnificent arches and many spectacular sandstone spires, but countless prehistoric Indian cliff dwellings and matchless canyon beauty. Its lower end is set with cottonwoods, pinyon and juniper, and its upper several miles are populated by mountain evergreens, including Ponderosa pine and Douglas fir. A stream of clear water flows its length, continuous in wet seasons, intermittent in others, and wildflowers color its tiny meadows for much of the year. Raptors nest in its towering cliffs.

Although the first plea to protect the two areas now within Arches National Park was made by a railroad agent for commercial purposes, a Dr. Frank R. Oastler of New York City also took a hand in the matter at the request of Stephen Mather, then Director of the National Park Service. In a letter to Mather dated March 13, 1926, Oastler noted that:

> "I am sending you the description of the Devil's Garden and Window Castles and also a description of Arch Canyon, all with photographs. Arch Canyon is within one days travel of the White Canyon Bridges and could be added to them as a national monument. Arch Canyon is passed on the way to the White Canyon Bridges. It is really a very remarkable canyon

> and I am fearful that somebody may obtain a lease of this area and use it for commercial purposes."

Mather acknowledged this letter, but evidently did not keep Oastler informed of what action was being taken to protect the areas. In a May 7, 1929, letter to Horace Albright, then the Park Service Director, Dr. Oastler expressed surprise and pleasure at the establishment of Arches National Monument, then asked about Arch Canyon:

> "There is also a canyon that we named Arch Canyon, not far from the White Canyon Bridges in Southern Utah, which certainly should be made a National Monument and could easily be attached to the White Canyon Bridge Monument. This canyon is about 9 miles long, and is one of the most beautiful that I have ever visited. It would be a shame to have it get into the hands of exploiters."

On May 11, Oastler wrote a similar letter to A. W. Cammerer, Assistant Director of the National Park Service, again mentioning Arch Canyon:

> "Is it not possible to have Arch Canyon made a Monument also, and included in the White Canyon National Monument? Arch Canyon is probably the most interesting canyon in South Eastern Utah, and it certainly should come under the jurisdiction of the Park Service."

The Park Service responded to Oastler in a letter dated May 14, 1929:

> "Arch Canyon was first brought to the attention of the Service in 1923 when Mr. Demaray accompanied Congressmen Cramton and Anthony on a pack trip across Southern Utah. This party saw Arch Canyon and remarked on its beauty. Later on their return the location of the area was looked up and it was found to be within the La Sal National Forest. As the area if further withdrawn as a national monument would have been under the administration of the Forest Service, no further steps were taken by the Service. Arch Canyon still remains in the La Sal Forest."

The upper four miles of Arch Canyon is, indeed, within the Manti-La Sal National Forest, together with its three great arches and most of its two major tributaries, but the lower seven miles of spectacular beauty and abundant prehistoric ruins is under BLM jurisdiction. At this date, neither part of Arch Canyon is being considered for wilderness status. It is open to exploitation, just as Oastler feared. Thus, because of split jurisdiction, a magnificent canyon system lies unprotected.

Following are some more individual spans and concentrations of spans that deserve some kind of special protection from damaging development:

> Grosvenor Arch, in Kane County, Utah, is so uniquely beautiful that it deserves a wide protective zone around it.

> Pool Arch, in the Behind-the-Rocks area of Grand County, Utah, is exceptional in size, beauty and location and should be protected by wilderness status.

Cathedral Arch, Arch Canyon. For scale, note figure in wash.

The Lost Spring Canyon complex, east of Arches National Park, contains numerous beautiful arches and should be added to the park.

The Gold Bar Bend area of the Colorado River gorge near Moab, Utah, has three outstanding arches and should be protected from further development activities.

The Rattlesnake-Mee-Knowles canyon complex near Colorado National Monument, Colorado, contains more than a dozen spectacular natural spans, plus several smaller ones, and should be added to the monument.

PROTECTION

The San Rafael Reef in Emery County, Utah, contains several reported natural spans and further exploration may disclose others.

The Bull Canyon complex, in Grand County, Utah, has eleven reported spans, and may have more.

The Mineral Canyon complex, in Grand County, Utah, contains one major span, plus another recently collapsed and many others in various stages of formation.

The Pritchett-Hunter Canyon complex, in Grand County, Utah, has ten unusual natural spans, including two of outstanding beauty.

The Cockscomb and vicinity, in Kane County, Utah, and Coconino County, Arizona, contains a number of known natural spans, including several within one short canyon, and may have more because it has never been thoroughly inventoried.

The Harts Draw canyon complex, in San Juan County, Utah, has two natural bridges and two arches, all quite unusual, and may have more.

There are numerous outstanding natural spans within the Navajo, Hopi and Uintah/Ouray Indian Reservations. Some are protected already, such as those in Monument Valley, but others are subject to mineral development damage.

Hunter Arch, Hunter Canyon. Rodney Greeno photo.

SUMMARY

Many of the better known natural spans within the Colorado Plateau are protected within special federal areas. The protection within such areas is not total, however. The few natural spans on private or state land are subject to commercial exploitation or damage from development activities. One of the eight largest known spans in the United States is on state land that must, by law, be exploited for any possible economic gain.

The prospect for the further protection of natural spans by the establishment of wilderness is very limited. Too many have not even been reported in wilderness inventories, and too many areas that contain outstanding natural spans have been dropped from wilderness consideration under pressure from development interests.

In an article in the May 9, 1926, edition of the New York Times Magazine, National Park Service Director Stephen Mather predicted national monument status for the Windows and Devils Garden areas to the north of Moab, Utah. He then went on to say:

> "The National Park System contains many marvelous prodigies, yet none of them exceeds in interest the extraordinary specimens of natural sculpture and architecture found in these two small areas in Utah."

Those familiar with the wild backcountry of southern Utah and adjoining regions of Arizona, Colorado and New Mexico know full well that Mather's words apply equally to vast areas of the Colorado Plateau that are unknown to the general public and still not protected. Most of these areas are known to stockmen and the mineral industry, however, and are hence open to the kind of exploitation and development these commercial activities bring.

Maybe an alternative to the present patchy, incomplete system for protecting the region's outstanding natural values should be considered. Perhaps instead of protecting isolated enclaves of land within the Plateau, the private and state lands within the region should instead be isolated, and all the rest protected as a unit, say as "Colorado Plateau National Park." Doubtless this is an idea whose time has not yet come, and may never come, but the concept is hereby presented by someone who knows and loves the region.

THE CLAIMING OF CORONA

You had to have been there at that time -- to know what a 2 by 2 inch stick of wood jammed into a pile of rocks could mean.

--- * --- * --- * --- * --- * --- * --- * ---

It happened during the second big uranium boom to hit the Colorado Plateau, the one triggered by the Great Arab Oil Embargo and the crudely contrived "shortages" of petroleum products in this country that doubled and tripled retail prices. Our government's belated efforts to make America "energy-independent" -- as long as it was nuclear or fossil fuel energy -- set off a frantic search for more uranium, the "clean" fuel.

Hundreds of individual prospectors headed for Colorado Plateau backcountry, radiation detectors in hand, but the big energy corporations used another approach. They had thousands of mineral claim forms neatly typed in their eastern corporate offices, all properly numbered to cover every square foot of huge areas of land their geologists said were promising. Each batch of forms was then sent west.

There, a couple of fellows with a pickup truck were hired, given a bundle of forms, a map, and several hundred 2 by 2 inch wooden stakes several feet long, then told to get busy. After a few months of this mockery of the 1872 Mining Act, a drive into many previously unspoiled regions of the Colorado Plateau revealed vast patterns of 2 by 2 stakes, each with a foil-wrapped claim form tacked to it, all in neat rows marching off toward the horizon.

Sometimes such forests of sticks meant nothing, just mad shouts in the wilderness by mindless corporations -- "This is mine, all mine!" Other times the sticks were followed by roaring platoons of machinery -- huge bulldozers ripping roads into wild and trackless terrain, followed by screaming drill-rigs making patterns of holes in the virgin land.

To avoid such depressing sights my wife and I curtailed our backcountry explorations almost entirely for a couple of years. We couldn't stand the sight of all the senseless devastation being wrought under the guise of patriotism. We settled for short hikes into nearby slickrock mazes, far from the crazy world of roaring machinery and vast claim-stake orchards.

One of our favorite hikes was in the Gold Bar Bend area of the Colorado River gorge, a few miles downriver from Moab Valley. There, a highly scenic paved road followed the big looping meanders of the river, at the base of high cliffs and domes of colorful Navajo Sandstone.

At Gold Bar Bend, where overly optimistic prospectors once panned the sandbars for placer gold, several side canyons and a great amphitheater broke one sheer wall of the river gorge. There, high above the road level in a distant slickrock rim, a lovely pothole arch was visible -- Gold Bar Arch, sometimes called Pinto Arch. And out of sight about a mile away, in an even more spectacular setting, were Bowtie Arch and Corona Arch. Corona was often called Little Rainbow for its resemblance to famous Rainbow Bridge, but it was "little" only by comparison with that enormous span. A local pilot had flown his plane through Corona Arch many times.

We climbed the steep fill of the railroad track that came out of a nearby tunnel that bypassed several miles of river gorge, then scrambled up a narrow and very eroded cut that had once been a jeep trail. At the top of this ascent we paused to admire the broad sweep of the river gorge below us, and to muse over the countless tons of water-rounded rock and gravel that the river had long ago left behind on an immense dome of sandstone that was now more than a hundred feet above the river level.

After a breather, we continued toward the still higher cliff of sandstone ahead, pausing to admire a blaze-red clump of Indian Paintbrush here, a patch of white Desert Primrose there. Beyond a fairly level stretch of sand and slickrock set with blossoming cacti and low desert shrubs, we reached the foot of the cliff. From there on, the route was on solid rock. After skirting around the base of the cliff for some distance, we climbed up some notched steps and a steep slope to a still higher level, then entered an immense, echoing alcove several hundred yards across.

A broad rock terrace looped around the wall of this alcove at mid-level, below a hundred feet or more of sheer cliff, and above an immense ledged hole that dropped hundreds of feet into a distant canyon. On the opposite side of this gigantic amphitheater we could see Bowtie Arch and, not far to its right, the graceful curves of Corona arching down from the cliff to the terrace -- a breathtakingly beautiful sight, altogether!

We hurried on around the wide terrace toward the arches, anxious to admire the symmetry of Bowtie with its sky-window opening at the top of an elongated spring-seep cave set with horizontal bands of wet mosses. We were anxious to stand beneath the soaring arc of Corona, to feel its impersonal immensity, to gain the sense of perspective that comes from being a mere human in a setting of gigantic sandstone shapes, to be reminded once again of how insignificant and transient mankind really is in the flow of eons-long geologic processes and the life of a planet.

After spending a pleasant half hour with Bowtie, we went on toward Corona. As I walked the sloping sandstone terrace I kept my eye on the striated arc of the huge span, studying its every lovely line and curve. Then my wife let out a choked cry --

"What's THAT!?"

"That" was a rock cairn supporting a 2 by 2 stick with a shiny pad of aluminum foil tacked to its upper end! Quickly I glanced around. Back the way we had come, but closer to the cliff than we had walked, was another cairn and stake -- and ahead, directly beneath Corona Arch was still another! We were shocked and silent as we hurried on to Corona and stood beneath its majestic arc looking at the mineral claim marker there. Some distant corporation had claimed Little Rainbow for its own, to exploit as it saw fit -- to leave untouched, or to devastate if the whims of commerce and chance traces of uranium left beneath it ages ago so dictated.

I stood there under Corona Arch and raged -- that such beauty should be subject to the rape, ruin and run tactics of someone seeking a few pounds of yellow mineral without which mankind would be much better off.

I stood beneath a Little Rainbow and wept bitter tears.

White Craig Arch, White Mesa. Photo courtesy Bill Crawley.

AMERIND ATTITUDES

THE QUESTIONS

During the controversy over whether to permit Lake Powell water to intrude into Rainbow Bridge National Monument, conservationists and Native Americans contended that Rainbow Bridge was sacred to certain tribes, and hence should not be desecrated by the intrusion of reservoir waters. Evidently, the federal courts and agencies did not think the historic sacredness of the natural bridge was enough reason to forgo the modern economic benefits of a full reservoir, because Lake Powell's water was allowed to enter the monument. Still, the matter deserves more consideration than it was given.

Was the huge span actually sacred to the Hopi or their Anasazi predecessors? To the Navajos or Paiutes who inhabited the region during historic times? Were other natural spans also a part of the religious beliefs of Colorado Plateau Native Americans?

A number of early writers mentioned observed Indian attitudes toward Rainbow Bridge, but apparently no serious effort was made, even by the archaeologists who were in the original "discovery" party, to find objective evidence that might settle the questions scientifically before any possible evidence was destroyed. Following are some of the comments and observations made by early writers and visitors to Rainbow Bridge, before the one bit of objective evidence found there was permanently lost to vandalism.

Professor Byron Cummings, whose archaeological team was a part of the 1909 "discovery" expedition to Rainbow Bridge, mentioned in his 1910 report, "The Great Natural Bridges of Utah," only that:

> "-- in the shadow of Nonnezoshie stand the remains of what appears to be an ancient stone altar."

In his February 1910 National Geographic article of the same title, Cummings made an interesting observation, then added a bit of speculation and a piece of hearsay evidence:

> "Few even of the Indians are well acquainted with (the region to the north of Navajo Mountain). It is celebrated as the place where Hoskinimi, one of the most revered leaders among the Navajo, successfully evaded Kit Carson, in 1866, when the latter taught the Navajo such a terrible lesson; but not even Hoskinimi seems to have penetrated as far as the Nonnezoshie.
>
> "In its shadow on the bench at one side are the remains of what was probably an ancient fire shrine. One can easily imagine a group of cliff-dwellers gathered around the sacred fire with offerings to the Sun Father and the Earth Mother. The Pahutes look upon it with awe, and Mr. C. A. Colville, who took a party there in November (1909), tells us that their Pahute guide, Whitehorsebiga, would not pass beneath the arch because he had forgotten the prayer that must be said before doing so."

AMERIND ATTITUDES

Mr. W. B. Douglass of the General Land Office, whose survey team made up the other half of the party that reached Rainbow Bridge in August, 1909, noted in his April 1910 report that:

> "On my second visit to the bridge (in October, 1909), I had with me a Navajo known as White Horse who after passing under the arch would not retrace his steps but invariably, at considerable personal inconvenience, would clamber around its north end. On being pressed for an explanation of his peculiar conduct he would only shake his head mysteriously and point to the sun.
>
> "Later, Mrs. John Wetherill, an accomplished Navajo scholar, obtained from an old Navajo the information that the arch represented the rainbow or sun path, and that he who passed under the arch could not retrace his steps without uttering a certain prayer under penalty of death. White Horse's trouble was he did not know the prayer, and was taking no chances. Nothing could induce the old Indian to repeat the prayer for fear the 'lightning snake' would bite him."

Neil M. Judd, the young archaeologist who accompanied the Cummings team to Rainbow Bridge in 1909, said nothing about an altar beneath the span in his 1919 report to the National Park Service about the trip, but he had this to say in his article, "Beyond the Clay Hills," in the March 1924 issue of National Geographic:

> "According to Navajo mythology, a rainbow was here turned to stone, thus permitting certain hero gods to escape flood waters in the canyon. An ancient shrine still stands beneath the left-hand buttress."

Decades later, in an August 1967 article for Arizona Highways, Judd recalled another bit of evidence:

> "Present-day Indians have, or formerly had, a deep veneration for the unusual, the sublime, and the awesome. In 1909, I clearly recall, gentle old Dogeye-begay, the horse wrangler, would not ride beneath Rainbow Bridge because he did not know the prayer that would insure his safe return."

In his article in the November 1911 issue of National Geographic, J. E. Pogue quoted Douglass' story, but added his own observation from a trip to Rainbow Bridge earlier that year, plus an unsupported conjecture:

> "Nearly beneath the arch are the remains of an ancient altar, built doubtless by the cliff-dwellers, indicating that the bridge was probably an object of superstitious worship even to this ancient people."

Theodore Roosevelt visited Rainbow Bridge in 1912. In his report of that trip in the October 11, 1913, issue of The Outlook, he observed:

> "Next morning early we started on our toilsome return trip. The pony trail led under the arch. Along this the Ute drove our pack-mules, and as I followed him I noticed that the Navajo rode around outside. His creed bade him never pass

> under an arch, for the arch is the sign of the rainbow, the sign of the sun's course over the earth, and to the Navajo it is sacred. This great natural bridge, so recently 'discovered' by white men, has for ages been known to the Indians. Near it, against the rock walls of the canyon, we saw the crumbling remains of some cliff dwellings, and almost under it there is what appears to be the ruin of a very ancient shrine."

In his 1922 book, "Tales of Lonely Trails," writer Zane Grey described an episode observed during his 1913 trip to Rainbow Bridge:

> "Just before we reached it Nas ta Bega halted with one of his singular motions. He was saying his prayer to this great stone god. Then he began to climb straight up the steep slope. Wetherill told me the Indian would not pass under the arch."

In the November/December 1920 issue of Natural History, C. L. Bernheimer described his trip to Rainbow Bridge earlier that year and provided a photograph of the span with this caption:

> "Midway between the figure of the man at the left and the arch is an Indian prayer shrine -- upright stones around a flat stone in the center -- and a similar shrine is on the opposite side of the creek just beyond the dark rock in the central part of the picture."

In his report of Bernheimer's 1921 trip to Rainbow Bridge, reported in the November/December 1922 issue of Natural History, E. H. Morris noted that there were "two ancient altars in the shadow of the eastern base," not just one as reported by earlier writers, but placed them in different locations than indicated in Bernheimer's photograph.

AMERIND ATTITUDES

In his 1932 book, "Trails Through the Golden West," Robert Frothingham repeated information about Rainbow Bridge that John Wetherill had obtained from a local Navajo called Sharkie:

> "He told John of an altar that had been built beneath the great arch by the Navajo's ancestors, which was so ancient as to be beyond the memory of the oldest men; the medicine men of earlier days had made pilgrimages to this place and offered prayers to the Great Spirit on behalf of the tribe."

In his 1927 book, "To the Foot of the Rainbow," Clyde Kluckhohn noted:

> "Navajos do not like to visit the region of the Bridge; they believe that it is the abode of great spirits, and if the Navajo does go he will not pass under the Bridge without repeating a certain prayer --"

In his 1933 book, "Beyond the Rainbow," Kluckhohn noted that:

> "-- Navajo songs of the Bridge have existed for many centuries, and -- generations of cliff-dwelling men had built their propitiatory smokes beneath it long before the Navajos came. Here, according to the Navajo Rain Chant, dwelt the Rain God --"

In her 1934 book, "Traders to the Navahos," Mrs. John Wetherill claimed to have recovered from an Indian child, and translated, the long-lost prayer that Navajos were supposed to say before going under Rainbow Bridge:

Mountain where the head of the War-God rests,
Mountain where the head of the War-God rests,
Dark Wind, beautiful chief, from the tips of your fingers a rainbow send out,
By which let me walk and have life.
Black clouds, black clouds, make for me shoes,
With which to walk and have life.
Black clouds are my leggings;
A black cloud is my robe;
A black cloud is my headband.
Black clouds go before me; make it dark; let it rain peacefully before me.
Before me come much rain to make the white corn grow and ripen,
That it may be peaceful before me, that it may be peaceful before me.
All is peace, all is peace.

Iridescent Wind, beautiful chieftainess, from the palm of your hand a rainbow send out
By which let me walk and have life.
Iridescent clouds, iridescent clouds, make for me shoes,
With which to walk and have life.
Iridescent clouds are my leggings;
An iridescent cloud is my robe;
An iridescent cloud is my headband.
Iridescent clouds, come behind me; make it dark; Earth Mother give me much rain,
To make the iridescent corn grow and ripen,
That it may be peaceful behind me, that it may be peaceful behind me.
All is peace, all is peace, all is peace, all is peace.

Karl Luckert, in his 1977 book, "Navajo Mountain and Rainbow Bridge Religion," quoted a very similar prayer, which had been found in the Richardson Collection in the Arizona State University library. It had been recorded in 1927 from a Navajo woman who lived in Kayenta. The minor differences between the two versions could be attributed to translation.

In the Spring 1973 issue of Plateau, Professor Stephen C. Jett summarized the situation in an article titled "Testimony of the Sacredness of Rainbow Natural Bridge to Puebloans, Navajos, and Paiutes." He concluded:

1. There was a "masonry structure, possibly a shrine," at the site in earlier years, but it was gone by 1930.

2. While it was of "Puebloan" origin, "The putative shrine certainly does not prove that the Hopi consider Rainbow Bridge to be sacred. Nevertheless, prehistoric inhabitants of the Bridge region contributed to the historic Hopi population, and Hopi migration legends place certain clan ancestors near Navajo Mountain. I know of no evidence of contemporary Hopi worship at the Bridge --"

3. Douglass' guide to Rainbow Bridge in 1909 "denied that the Paiute had any special feeling about the Bridge -- (this) suggests that veneration of Rainbow Bridge is not characteristic of Paiute culture, although some Paiutes seem to have adopted from the Navajo the idea of the Bridge's having supernatural associations."

4. Thus, "-- there appears to be adequate evidence to conclude that twentieth century Navajos, and some San Juan Paiutes influenced by the Navajo considered, and probably still consider, Rainbow Bridge to be a sacred place."

5. "Most, if not all, of the information regarding the sacredness of Rainbow Bridge is related in some fashion to John Wetherill and his associates. The only alternative to concluding that the Bridge was considered sacred would be to believe that these people perpetuated an elaborate hoax for twenty years and deliberately gulled everyone from scientists to a President of the United States. I do not find this particularly plausible."

In his introduction to the Rio Grande Press 1980 reprint of Kluckhohn's 1927 book, Jett described what may have been the first time any Navajo saw Rainbow Bridge:

> "During 1863-64, when Kit Carson's troops were harassing the Navajo -- about a score of Navajos -- fled from their exposed homeland -- and hid themselves in the baldrock country behind Navajo Mountain. It was probably at this time that (a Navajo called 'Blind Salt Clansman') became the first Navajo to see Rainbow Bridge. He was tracking horses, keeping his eye on the trail. It wasn't until he was actually under the Bridge that he looked up and saw it. As a Singer or medicineman, he came to see an association between the prismatic rainbow, its petrified representative, and life-giving rain. From this association plus the cloud-gathering abilities of Navajo Mountain, a new set of rain-calling rites evolved which included pilgrimages to the Bridge, where prayers and offerings were made."

AMERIND ATTITUDES

If this tale is true, then Navajo involvement with Rainbow Bridge is fairly recent, only since the 1860s. To living Navajos, however, even that would be "ancient -- beyond the memory of the oldest men." Luckert's 1977 book provides an excellent study of Navajo traditions and beliefs concerning Rainbow Bridge and the canyon in which it stands, based on background research and personal interviews of several knowledgeable Navajos, including three "singers." He also concluded that Navajo relationships with Rainbow Bridge probably began with "Hashkeniinii," who escaped from Kit Carson's soldiers by fleeing into the area to the north of Navajo Mountain in 1863.

Granary Arch, Davis Canyon, Canyonlands National Park.

OTHER SPANS

What about Amerind attitudes toward other natural spans? An earlier chapter summarized what is known about the beliefs of Indian tribes in the vicinity of the Columbia River concerning a legendary natural bridge of lava that spanned that great river, then collapsed sometime in the late 1700s or early 1800s. Further details about that span and the part it played in Indian legends can be obtained from the references quoted.

There are also a few references in Colorado Plateau literature that indicate Indians in this region might have supernatural feelings about natural spans other than Rainbow Bridge.

Two observations can be made about the relationship between the prehistoric Indians that occupied the region and its natural spans:

1. Few if any natural spans have rock graphics near them. Although Arch-of-Hands in Mystery Valley, Arizona, may be one exception to this, there is no way to determine whether the pictographs there are prehistoric Anasazi or historic Navajo. Rock art images are often associated with both Anasazi and Fremont culture ceremonial sites. If these cultures considered natural spans to be sacred, why are there almost never any petroglyphs or pictographs near them? Surely, the Anasazi and Fremont Indians who occupied much of the Colorado Plateau region for more than a thousand years knew about most or all of its major natural spans.

2. There are a number of natural spans with prehistoric Anasazi or Fremont dwelling or granary ruins beneath or very near them. Three examples of such arch-ruin associations can be found in Canyonlands National Park. There are arch-dwelling combinations on Aztec Butte and in Horse Canyon, and there is a large granary under an arch in Davis Canyon. There is also a dwelling ruin under an arch in the upper end of Fable Valley west of the Abajo Mountains, and there are several more in the Monument-Mystery Valley area of the Navajo Indian Reservation. Theodore Roosevelt reported seeing "cliff dwelling" ruins "near" Rainbow Bridge on his 1913 trip to that span. There are also dwelling ruins quite close to Kachina Natural Bridge in White Canyon. Would there have been such casual daily associations with natural spans if they had significant religious meaning?

Prehistoric ruins beneath Fable Arch, in Fable Canyon.

AMERIND ATTITUDES

In his 1973 article in Plateau, Jett reported finding a few petroglyphs on the southern buttress of an arch called The Window in Canyon de Chelly, but again the 'glyphs could have been Navajo. There is considerable Navajo rock art in Canyon de Chelly. Jett also noted asking a Navajo whether there was a universal proscription against passing under natural spans. The reply was "yes," it was "a religious thing." Jett then went on to conclude that:

> "There is some evidence that veneration of Rainbow Bridge is an elaboration of a traditional Navajo attitude toward all natural arches and bridges. Testimony indicates that a prayer is required if one is safely to pass under Rainbow Bridge or, probably, under any natural bridge."

SUMMARY

There is very little objective evidence upon which to base a firm scientific conclusion as to whether any Native American culture -- prehistoric, historic or contemporary -- held or holds Rainbow Bridge or any other natural span to be sacred. There has been, however, ample personal testimony from early explorers and Amerind individuals, supported by general ethnological evidence, to make some kind of special relationship with the region's larger spans highly probable. Smaller spans are far less likely to have been involved.

Although most early writers about Rainbow Bridge noted some kind of evidence of a "shrine" or "altar," or an Indian reaction to the bridge, they differed considerably in the details, once again illustrating that history is an inexact discipline that cannot serve as a firm basis for objective scientific studies. Even so, the sheer quantity of historic testimony makes it quite probable that Rainbow Bridge had religious significance to some late historic Navajos, and still does to a few contemporary Navajos.

A great majority of the natural spans in the Colorado Plateau are north of the San Juan-Colorado rivers. Historically, there has been very little Navajo or Paiute occupation of this region, especially in Utah. Further, there is little evidence to indicate how the Anasazi and Fremont cultures that did occupy this vast region for so long felt about natural spans, other than the inferences that could be drawn from a few arch-ruin associations, and the general lack of rock graphics associated with natural spans.

Thus, the probability of finding any concrete evidence of prehistoric or historic Amerind attitudes toward the majority of the natural spans in the Colorado Plateau is very low. The subject will probably remain a mystery forever.

There are prehistoric ruins on the highest ledges beneath huge Paul Bunyans Potty in Horse Canyon, Canyonlands National Park.

LION COUNTRY

There was quite a bit of snow on the ground in the higher ground around Moab Valley, but the young hiking enthusiast who was visiting us on his mid-winter break from university studies didn't mind. He went hiking anyway, in the maze of slickrock fins and domes that borders nearby Mill Creek Canyon. He dressed warmly, took a light lunch and was off for the day.

That evening he returned, with cold-rosy cheeks but also an excited gleam in his eyes. It was easy to see that he could barely wait to tell us about something. After warming up a bit, he described finding an arch in a snow-capped expanse of slickrock, and of meeting a hunting cougar -- an adult mountain lion -- of catching the eye of that sinuously beautiful cat for a timeless instant! Of standing frozen -- by a mixture of cold, fear and enchantment -- wondering what the great tawny feline would do.

I wanted a picture of his arch, so our lucky friend put a dot on a topo map where his arch was located, in an area so vertical that the map's contour lines were almost too close together to read.

The following spring, my wife and I drove up into the wild country, parked, and started hiking the route we had recommended that past winter to our young friend, the one he had taken to his arch and lion.

There was no trail, just a way to go with a bit of persistence. We would work our way between the close walls of two high sandstone fins, climb up onto another, walk its length, then descend and climb onto still another, gradually making progress toward where the arch should be. It was slow hiking, of a type found only in certain areas of the Colorado Plateau.

Eventually, we ran out of fins. Ahead of us was a steep slope of almost pure white sandstone that ended in an appalling drop into a narrow, sheer-walled canyon that seemed bottomless from our vantage point -- the North Fork of Mill Creek Canyon. Cautiously, we inched down the slope toward the cliff, then parallel to it and around another giant slickrock slope that appeared as we descended.

Soon, there seemed nowhere else to go. The deep gorge lay before us, with its inner rim just a few yards away, and an unscalable sandstone cliff barred further progress upstream, toward where the arch was supposed to be. Between us and that cliff was a narrow cleft, cut into the solid rock along an ancient joint line by ages of rain runoff, a gap too wide to jump across and too deep to climb across.

As I squatted on the rock slope to enjoy the fantastic view and calm my frustration, my wife descended the last few yards toward the rim of the gorge. As she approached that end-of-the-world, I shouted my usual precautions and, also as usual, she blithly ignored me. To take my mind off of the risk she was taking, I studied the narrow cut that barred our approach to the upstream cliff, then followed the cliff-line with my eyes.

In the bright spring sun the scene was a study of white-on-white, with depth perception hampered by the brilliant reflected light. But it looked like there was a gap between the cliff and the next fin toward me, a gap just barely discernible to the sun-blinded eye. I got up and headed toward the upper end of the cut that lay between me and the cliff. As I neared it, I saw that it was possible to inch around the cut on a narrow ledge, then continue into the inconspicuous gap on a slope of slickrock so steep it just barely afforded traction for the rubber-soled shoes I was wearing.

I was so intent upon watching my footing, and trying to

avoid looking at the drop below me, that I was well into the gap before I paused to see what was ahead. I let out an echoing whoop, then continued on into the narrow cleft between immense white sandstone fins. Not far ahead, a seeping spring sent a tiny trickle of water under a broad arc of rock. Although the water soon disappeared, its length was set with the healthy green of thriving vegetation, including one highly optimistic young tree.

As I approached the arch, a couple of lizards and a half-grown toad scurried out of sight and the smell of moisture filled the arid desert air. The well-hidden arch harbored its own little ecosystem there in the midst of a wilderness of almost unrelieved barren sandstone. Winter snow-melt, slowly seeping down through a hundred feet of solid rock, had met less permeable rock that forced it to travel laterally, thus creating a trickle of moisture between two sandstone fins. In time, from that unpromising start, the seeping water had created a miniature swamp and shaped a graceful arch -- an arch of unusual beauty -- in an unusual setting, a hidden rock-pocket.

After spending a few moments enjoying the lovely scene, I retraced my steps until I could see my wife, who by then had grown a bit concerned at my apparent disappearance, and showed her the way around the cleft and to the arch. Back in the little hidden Eden, we sat on a rock ledge and admired the arch -- or natural bridge -- for half an hour, then looked around for an alternate way back toward our car. The way we had come had not been easy.

There was a chance. After skirting around the base of the fin that hid the arch, we found we could climb up the steep edge of that fin by carefully picking our footing, but were not sure we could get all the way to the top. At intervals as we ascended we looked back into the hidden pocket with its arch and tiny oasis, to admire its strange beauty from above, and to orient ourselves for an easier descent that way the next time.

That was not to be. At each later visit to Pocket Arch we tried to find our backdoor access to the span, but to no avail. Pocket Arch was determined to stay hidden, to remain inaccessible to all but the most determined. We have yet to find the easy way down to that lovely span in its tiny, secretive hanging-cleft between massive sandstone fins -- the one that is so easy to see when it is time to climb back out.

And we have never spotted our friend's cougar. No doubt that tawny fellow spends the warmer months in some happier hunting ground.

Double Arch, Arches National Park

AESTHETICS

AESTHETICS

According to Webster's Unabridged Dictionary, "aesthetics" is "that branch of philosophy which deals with the beautiful." This chapter, then, deals with the beauty of natural rock openings, that one aspect of nature so difficult to define, yet so inseparable a part of human appreciation of its endless wonders.

Some of the earliest writers about North America's natural spans failed to recognize that beauty was an important aspect of public interest in natural arches and bridges. Geologist H. F. Cleland, who wrote a long and scholarly paper on America's natural spans soon after the widely publicized discoveries of Utah's White Canyon bridges and Rainbow Bridge, had this to say about public interest in these phenomena in his 1910 report to the Geological Society of America:

> "There are few topographical features that excite greater interest than natural bridges. This is due, no doubt, to some extent to their rarity, and to some extent to the questions which arise as to the force or forces which have been at work to produce such structures."

Thus, according to Cleland, public interest in natural spans lies in curiosity about their rarity, their novelty, and perhaps about how they form -- not in the appreciation of their beauty. Cleland's opinion may have been the product of his time. Certainly, there were far fewer known natural spans in 1910 than now. Only a handful of the hundreds of major natural spans now known to exist within the Colorado Plateau were known at that time, and perhaps most of the better known spans east of the Rockies were not particularly beautiful. Yet most of the early Colorado Plateau explorers waxed eloquent about the beauty of the arches and bridges they encountered in that vast and mysterious region.

As with a sense of humor and other aspects of human nature, a sense of and appreciation for natural beauty varies drastically among individuals. While this variability adds spice to human affairs, it also leads to conflicts in which those who do not see beauty think nothing of destroying things others consider beautiful, while those who do see and appreciate beauty are often willing to fight to preserve it. In essence, this is the story of the Colorado Plateau -- those who are not capable of seeing and cherishing its unique beauty, versus those who are -- those who would develop its natural resources at the expense of its beauty, versus those who would preserve it from destructive forms of development.

Another professor of geology, R. E. Janssen, in his September 1943 report in The Scientific Monthly, associated human interest in natural spans with a similar interest in manmade spans:

> "Alluring, entrancing, mystifying -- the sight of a beautifully executed bridge unfailingly captures the imagination of the observer. Whether fashioned of stone by

AESTHETICS

Nature or designed of architectural steel by man, a sturdy bridge seldom fails to arrest one's attention. This may be true, perhaps, because our fancies are at once quickened to an appreciation of the labors which brought the bridge into being."

A more recent writer attributed the almost universal human interest in natural spans to other subtle aspects of human nature. R. M. Pearl introduced his series of articles on "Stone Rainbows" that appeared in several 1976 and 1977 issues of Earth Science with this conjecture:

"Natural bridges and their kin -- natural arches, natural windows -- stand among the more conspicuous of the earth's scenic features. The large ones, the bridges, are spectacularly so. But even the small windows, usually high on a rock wall, tend to be obvious out of proportion to their size, for they interrupt whatever monotony may be present in an otherwise unbroken panorama of cliff. They may admit the blue sky, especially in the arid lands of the American West. Possibly, the clouds float through them. Perhaps, even, a waterfall drops from the opening. They are thus noticed beyond their relatively limited extent. As indication of the popular appeal of this kind of scenery is the creation of national and state parks and monuments and other public reserves for the preservation, protection, and presentation of natural arches all over the world."

While there may ultimately be as many reasons why humans like to stand gazing raptly at a natural span as there are humans, it might still be possible to divide humanity's interest in spans into five major categories: curiosity about such unusual phenomena; fascination with their size; puzzlement over how they form; appreciation of their beauty of shape and setting; and, little or no interest in such things. Doubtless, many of us experience more than one of these forms of interest, and a few of us experience all but the last one.

Readers who are primarily interested in the largest, longest, widest, and highest spans are referred to the chapter on measurements. Readers preoccupied with how natural spans form are referred to the chapter on formation modes. Those whose primary interest is in rarity should see the following chapter on cataloging. This chapter is for readers who appreciate the beauty of natural spans, perhaps associated with the other curious aspects of these natural phenomena, perhaps not.

If pressed to explain why they hunt or fish, many men will sheepishly admit that they pursue their favorite sport largely because it gives them an "excuse" to enjoy the beauty and silence of unspoiled nature. Many of us, however, need no excuses. We have learned to enjoy beauty for its own sake, wherever we can find it, with no need to apologize to ourselves or others for this. Individuals who have reached this stage of human development will find much to enjoy within the Colorado Plateau -- including its numerous arches, bridges, windows and tunnels. While not all natural spans are aesthetically pleasing to the eye, a great many of them exhibit outstanding beauty in both form and setting.

Kirk Arch, Salt Creek Canyon, Canyonlands National Park

SOME HISTORIC IMPRESSIONS

Beginning with the first major natural spans reported in the Colorado Plateau, the reporters -- whether scientists, authors, artists, or practical-minded government surveyors -- were impressed with their beauty, their splendor, their magnificence, as well as their sheer size.

W. W. Dyer, in his 1904 Century Magazine report of the 1903 Scorup-Long expedition to the White Canyon natural bridges, noted how, as Scorup and Long approached the spans, Scorup was concerned over whether his earlier impressions of their size and beauty were real, and not just exaggerations:

> "On the morning of March 15, Long and Scorup were early in the saddle. They were in the immediate vicinity of the bridges, and Long has confessed to a rising excitement as they turned their horses' heads up the canyon. Scorup himself showed signs of nervousness, as if apprehensive that his memory had magnified the size and grandeur of what he had seen nearly eight years before, and had thus prepared a disappointment for them both."

Scorup need not have worried, as anyone knows who has ever stood beneath any of these three awesome, magnificent natural bridges, and as Dyer noted once they reached the first bridge, now called Kachina:

> "Extravagant indeed must have been their expectations to experience any disappointment at sight of the collosal natural bridge before them. Yet, from the scenic point of view, this bridge was the least satisfactory of the three which they visited."

After taking the span's rough measurements, they named it Caroline, as agreed earlier, and then headed up White Canyon toward Sipapu:

> "Remounting their horses, Long and Scorup passed under the mighty mass of the Caroline and pushed on up the canyon. At a distance of three and a half miles they found themselves in the presence of what is doubtless the most wonderful natural bridge in the world -- a structure so lofty and magnificent, so symmetrical and beautiful in its proportions, as to suggest that nature, after completing the mighty structure of the Caroline, had trained herself for a finer and nobler form of architecture."

It should be remembered that Rainbow Bridge was formally discovered and reported in 1909, so Augusta was able to retain its undisputed title as the "most wonderful natural bridge in the world" for only five years.

Dyer's descriptions of the White Canyon bridges excited a number of people, among them Edwin F. Holmes of Salt Lake City. Holmes proposed and outfitted an expedition to the bridges, to be led by well-known local artist, H. L. A. Culmer. Who better than an artist would be qualified to comment about the aesthetic qualities of natural bridges? In his diary of the 1905 trip, Culmer noted that after two days of hard travel the party reached Little Bridge (Owachomo) first:

> "We take a few photographs, but I am too tired to sketch or measure and we leave that for tomorrow. We are camped right under it, and the impression I have is that it is wonderfully lofty, graceful in style and very symmetrical."

The following day, the party measured Little Bridge, renamed it "The Edwin Bridge" after the expedition's sponsor, explored some nearby prehistoric ruins and painted a sign on the span's base "in good New Era black paint." The sign noted the names Culmer, Whitaker, and Holmes, the name of the sponsoring Salt Lake Commercial Club, the span's measurements and the date, April 14, 1905. This kind of defacement was normal to those times, but is strictly prohibited today. The group then moved on to Caroline (Kachina). Culmer reported sketching this bridge, but had nothing to say about it beyond a detailed description. In his diary, he had more to say about Augusta (Sipapu):

> "Apr 16 --05. A notable day. I saw and sketched the great Augusta bridge, though a week could be spent in painting it from various points of view. It is a most magnificent and shapely structure --"

In his writings for publication, Culmer described the White Canyon bridges in more eloquent terms, but still retained his original impressions of the three bridges. In his article in the May 14, 1905, edition of the Salt Lake Tribune, Culmer called Caroline "huge and clumsy," and Augusta a "magnificent and shapely structure that must even rank among the most splendid achievements in nature."

In his article in the September 1908 edition of The Technical World Magazine, Culmer again described Caroline in uninspired terms, calling it "somewhat clumsy and ungraceful in structure, but still picturesque." His impression of Little Bridge/Edwin Bridge/Owachomo was somewhat better:

> "It is paler in hue than the others, and while it does not express equal strength and power, it is far more graceful than they."

Culmer saved his best superlatives for Sipapu:

> "-- one of the most magnificent and shapely structures ever achieved by nature -- Augusta Bridge must henceforth rank with the greatest of natural wonders, and take its place with Niagara, the Yellowstone geysers and the Grand Canyon in Arizona, as one of the masterpieces of American landscape. A sense of enormous strength pervades it, a sense that it has endured for ages, and will endure for ages yet to come."

Sipapu Natural Bridge

AESTHETICS

Early descriptions of Rainbow Bridge revealed that even the most case-hardened and practical of men were impressed with its beauty. Government surveyor Douglass, who was a member of the 1909 discovery party, said in his 1910 report:

> "Not only is this bridge the greatest known as to size, but it is more delicately proportioned than any of its rivals. A graceful, rainbow-like arch of buff sandstone --"

In his 1910 report, archaeologist Byron Cummings, who was also a member of the 1909 expedition to Rainbow Bridge, called Augusta/Sipapu natural bridge "the crowning glory of the three bridges," and called Rainbow Bridge "a graceful arch of magnificent proportions."

After his discovery trip to Rainbow Bridge, Cummings visited several natural spans in Grand County, Utah. He was struck by the beauty of one in particular, now called Pritchett Arch, which he said "deserves to be classed with (the White Canyon bridges) and counted among the great natural wonders of our continent," even though it had a span of only "122 feet." This arch, and Kachina, are fine examples of how beauty and size do not necessarily go together. Kachina Natural Bridge is immense, but not generally known for its aesthetic appeal, while much smaller Pritchett Arch is considered by almost everyone who has seen it to be outstandingly graceful, a span of rare beauty.

In his 1910 report, Cleland noted Tonto Natural Bridge's rarity, beauty and size, all in one sentence:

> "Not only is the bridge unique in respect to its origin, but (it) is, moreover, one of the most beautiful natural bridges in the United States, and, with the exception of those in San Juan County, Utah, one of the largest."

In J. E. Pogue's article about Rainbow Bridge in the November 1911 edition of National Geographic, he noted that even with that immense span, size was not the only factor in creating personal impressions:

> "A mere recitation of figures must fail to convey an adequate idea of the imposing nature of the bridge. It is not the size alone, though this far exceeds the greatest masonry arches constructed by engineering skill; nor is it solely the graceful lines or curvature of maximum stability, but rather all of these, that combine to make this the most remarkable single arch now known."

Theodore Roosevelt chronicled his 1912 trip to Rainbow Bridge in a 1913 edition of The Outlook. He described the forbidding country through which he had to travel to get there in grim terms:

> "The sullen rock walls towered hundreds of feet aloft, with something about their grim savagery that suggested both the terrible and grotesque. All life was absent, both from them and from the fantastic barrenness of the boulder-strewn land at their bases. The ground was burned out or washed bare. In one place a little stream trickled forth at the bottom of a ravine, but even here no grass grew -- only little clusters of a course weed with flaring white flowers that looked as if it throve on poisoned soil. The cliffs were channeled into myriad forms -- battlements, spires, pillars,

buttressed towers, flying arches; they looked like the ruined castles and temples of the monstrous devil-deities of some vanished race."

But when he reached the bridge, Roosevelt's mood changed drastically:

"At last we turned a corner, and the huge arch of the Bridge rose in front of us. It is surely one of the wonders of the world. It is a triumphal arch rather than a bridge, and spans the torrent bed in a majesty never shared by any arch ever reared by the mightiest conquerors among the nations of mankind. At this point there were deep pools in the rock-bed of the canyon, with overhanging shelves under which grew beautiful ferns and hanging plants. Hot and tired, we greeted the chance for a bath, and as I floated on my back in the water the Bridge towered above me."

That night, Roosevelt saw an entirely different aspect of Rainbow Bridge, one not seen by many even now:

"The night was cloudless. The round moon rose under the arch and flooded the cliffs behind us with her radiance. After she passed behind the mountains the heavens were still brilliant with starlight, and whenever I waked I turned and gazed long at the loom of the mighty arch against the clear night sky."

Covert Arch, Lost Spring Canyon

Zane Grey, a man who had a way with words, made a trip to Rainbow Bridge in 1913. His impression of the great span appeared in his 1922 book, "Tales of Lonely Trails":

"This rainbow bridge was the one great natural phenomenon,

> the one grand spectacle which I have ever seen that did not at first give vague disappointment, a confounding of reality, a disenchantment of contrast with what the mind had conceived.
>
> "But this thing was glorious. It absolutely silenced me. My body and brain, weary and dull from the toil of travel, received a singular and revivifying freshness. I had a strange, mystic perception that this rosy-hued, tremendous arch of stone was a goal I had failed to reach in some former life, but had now found. Here, was a rainbow magnified even beyond dreams, a thing not transparent and ethereal, but solidified, a work of ages, sweeping up majestically from the red walls, its iris-hued arch against the blue sky."

Zane Grey also spent the night beneath Rainbow Bridge, and described that experience even more vividly than Roosevelt had:

> "We were to camp all night under the bridge. Soon the rosy and golden lights had faded. All the walls turned pale and steely and the bridge loomed dark. Presently I moved away into the strange dark shadow cast by the bridge. It was a weird black belt, where I imagined I was invisible, but out of which I could see. There was a slab of rock upon which I composed myself, to watch, to feel.
>
> "White stars hung along the dark curved line. The rim of the arch appeared to shine. The moon was up there somewhere. The far side of the canyon was now a blank black wall. Over its towering rim showed a pale glow. It brightened. The shades in the canyon lightened, then a white disk of moon peeped over the dark line. The bridge turned to silver.
>
> "I did sleep a few hours under Nonnezoshe, and when I awoke the tip of the arch was losing its cold darkness and beginning to shine. The sun had just risen high enough over some low break in the wall to reach the bridge. I watched. Slowly, in wondrous transformation, the gold and blue and rose and pink and purple blended their hues, softly, mistily, cloudily, until once more the arch was a rainbow."

In contrast to Roosevelt's grim impressions of the approaches to Rainbow Bridge, E. H. Morris, in his 1922 description of the 1921 Bernheimer expedition to the span, saw the same land quite differently:

> "Along the boundary line between Arizona and Utah, within the mighty northward curving bow formed by the Colorado River and its largest eastern tributary, the San Juan, there lies a country of unusual charm to the lover of nature and to the archaeologist. The region is a vast plateau furrowed by hundreds of miles of labyrinthian canyons at every turn of which the traveler may expect new and pleasing vistas to unfold before his eyes and will seldom be disappointed. The magnificent desolation which is the dominant tone of the scenery weaves around the wayfarer an enchantment which ever calls him back --"

Hickman Natural Bridge, Capitol Reef National Park

Morris went on to describe Rainbow Bridge by saying that:

> "-- in the estimation of the writer there exists nowhere a more majestic and impressive natural object -- so true are the proportions, so perfect the symmetry of the gigantic bow of stone --"

He also described the canyon between the bridge and the Colorado River, a stretch of beauty that will never be viewed by modern man:

> "Few of those who visit the bridge go on down the canyon to the Colorado River, thereby missing some of the most beautiful scenery the country has to offer. In many places the reddish cliffs, streaked and mottled with darker stains, are so high that they seem almost to meet overhead. There are deep, clear pools in the streambed, and cool springs well up through crevices in the rock. Some four miles below the bridge, Bridge Canyon empties into West Canyon through a slot so narrow that it resembles a gigantic doorway, whence even at midday one stares from twilight shadow into a

> dazzling glare of sunlight. The remaining two miles to the river consist of a succession of pools larger than those in Bridge Canyon. In many of these pools beaver dwell, and some of them the crafty animals have considerably enlarged by the construction of dams. In one of the ponds thus empounded there were catfish which in the clear water looked as black as lumps of coal.
>
> "There are several small cliff houses in West Canyon. In one, situated beneath an overhang where the clear streamlet loses itself among the turbid waters of the Colorado, there was a crude vessel fashioned from a piece of driftwood, and several names were written in charcoal upon the cliff. Some of the prospectors who are wont to ply along the river in the wintertime, washing gold from the gravel bars, had taken shelter in the ancient home of a people whose hearts knew not the lure of the yellow metal."

These were but the first traces of what was to be the almost total intrusion of human development into the pristine beauty of Rainbow Bridge's remote canyon Eden.

In his 1929 book about the same expedition, Bernheimer pointed out that it was ridiculous to compare Rainbow Bridge with anything else, that it was in a class by itself:

> "Objects to be compared must be like in kind. To line up and compare a blue-ribbon horse with a blue-ribbon cat, the

Fortress Arch, Horse Canyon, Canyonlands National Park

> Grand Canyon with the Yosemite, the ocean with Niagara Falls, all wonderful units, is as futile as a comparison of the Rainbow Bridge with any other natural marvel. There is no other arch yet found in nature to compare with it. There are other natural bridges -- but they are distinctly bridges, and not a rainbow-shaped arch, a delicate semi-circle, free-standing on one side, on the other partly embedded in the rock from which it was born. The dimensions of the Rainbow Bridge, its symmetry, its graceful sweep, its delicate balance, its daintiness, despite its bulk, its picturesque setting and colouring, make it an unique and stupendous monument. It is beautiful from whatever angle it is seen. It is one of the most inspiring marvels of the ages."

Bernheimer went on to note that since the area around Rainbow Bridge was so remote it had not been grazed by Indian sheep and goats, and hence plant life was profuse there. He described several dozen species of flowering plants, shrubs and trees there, including one plant unique to certain semi-arid western regions:

> "Brigham tea, a wiry weed growing in the Bad Lands, is named after Brigham Young, who taught his people to use it in place of China tea. I have tasted this Brigham tea and found it as good a substitute for tea as Postum is for coffee."

Charles Bernheimer was so entranced by the beauty of Rainbow Bridge country that he made three trips to it in the early 1920s. The published chronicles and unpublished field notes of those trips make fascinating reading.

Even geologist Frederick J. Pack was forced by the beauty of several of Utah's natural spans to lapse into vividly descriptive language. In his 1922 report on natural bridging he described Sipapu Natural Bridge as the "sublime Augusta -- a most impressive sight, and one never to be forgotten." He was also impressed by an arch in Cedar Breaks, and once again illustrated that size is not the only factor in the human perception of natural spans:

> "Right in the midst of this glorious spectacle of painted fantasies is the 'Wall of Jerico' -- neither its height nor its span greatly exceeds 50 feet, yet its coloring, its irregularity of form and particularly its environment give it a peculiar distinction not present in the other bridges."

Cedar Breaks is now a national monument, and the National Park Service has given the name "Square Arch" to the opening in the "Wall of Jerico."

In his 1932 book, writer Robert Frothingham went into raptures over the beauty of Rainbow Bridge, as he first saw it:

> "-- we finished on foot the last few hundred yards of the day's journey, following the serpentine, trickling stream that empties into the Colorado River six miles farther on. There it stood in its towering majesty: a 'rainbow turned to stone' in very truth.
>
> "Never in all my life have I seen anything so superb, so overwhelming. Mere words could not describe it then -- they

cannot now. As this supernal beauty burst on my sight, the resounding call of the Psalmist passed through my thoughts:

'Lift up your heads, O ye gates, and be ye lift up, ye everlasting doors!'

"And color! Color! A dozen shades of red, pink, brown, yellow, and mauve blending into each other, with an exquisite rosiness of hue so inexpressibly soft and tender as to exert a tremendous emotional appeal. It was suggestive of the skin of a beautiful woman. One wanted to smooth it, to lay his cheek against it. And it seemed to reach upward to the very arch of heaven itself. And such proportions! 'Graceful as a bow just bent.' An epic in sandstone!"

Even geologist Richard M. Pearl was briefly carried away while describing Rainbow Bridge and the three White Canyon bridges in his 1976 report "Stone Rainbows":

"Symphonies they are -- of color, of symmetry, of grandeur -- among the major wonders of the world --"

What remains to be said about these splendid spans that has not already been said? Yet there is still room to talk about the beauty of Colorado Plateau natural spans that are less well known, some of which are just as lovely in their own way.

Turret Arch, from the North Window, Arches National Park

SOME OUTSTANDING SPANS

To compare natural spans for "beauty" or "magnificence" or any other such expressions of aesthetic impact is even more subjective than to compare them for size. While it should not even be considered in judging aesthetic qualities, size nonetheless does have a subjective impact on most viewers. A span can be impressive for its size alone, even though it is not particularly beautiful. Kachina Natural Bridge and Morning Glory Arch are examples.

Similarly, many exquisitely shaped natural spans fail to elicit general public interest simply because they are not very large. In preparing the list of "outstanding natural spans" on the inside back cover of this book, the author has avoided comparing beauty by using the term "outstanding," yet has met the problem squarely by noting that the list is subjective, a list of personal preferences based on one or more of several factors.

While there are doubtless many other spans within the Colorado Plateau that should be on that list, they are not there simply because the author has not personally visited and photographed them. Many small but beautifully shaped spans were also left off of the list for lack of space.

MULTIPLE SPANS

If natural spans in general are rare, multiple spans are even less common, and thus of particular interest. If the forces of nature that form natural openings and free spans of rock produce such phenomena only occasionally, then the chances of their forming double and triple spans must be considerably smaller -- but not zero. Multiple spans do exist within the Colorado Plateau, and doubtless many more than are listed here. Again, as with outstanding spans, the limits to the list are those of the author who, even in his twenty years of exploring the region, has not visited all the natural spans presently known to exist -- although he has visited hundreds, some of which he has found himself, and continues to search for others.

In listing the following multiple spans it is noted that there is, as always with natural spans, room for argument about definitions. The basic problem arises from whether "arch" is defined as an opening in rock, or as the free-standing span of rock partly surrounding the opening. In general, if the rock has two holes through it within close proximity, the spans can be said to be "double." If there are three holes near each other, then the span can be called "triple."

There are exceptions to these general rules, however, such as when there are two holes in a small mass of rock, leaving three spans, or where a long span is split lengthwise, making it double. An example of this is Cleft Arch. If the holes interconnect, this can also confuse things. It also becomes a matter of judgment as to when multiple holes are close enough to make the resulting spans multiple, or just two close single spans. In the following list, where there is a basis for varying judgment, this is noted. This list is certainly not complete, but will give readers a start on "collecting" multiple spans.

DOUBLE NATURAL SPANS

SPAN NAME	STATE	GENERAL LOCATION	
Anniversary Arch	Utah	Klondike Bluffs, Arches N.P.	
Aztec Butte Arches	Utah	Aztec Butte, Canyonlands N.P.	(1)
Brimhall Arch	Utah	Waterpocket Fold, Capitol Reef N.P.	
Bullwhip Arch	Utah	Four Arches Canyon, west of Moab.	
Cleft Arch	Utah	Lavender Canyon, Canyonlands N.P.	
Cly Arch	Arizona	Monument Valley area	
Dry Fork Arch	Utah	Lavender Canyon, Canyonlands N.P.	
Double Arch	Utah	Upper Muley Twist, Capitol Reef N.P.	
Double Arch	Utah	Windows Section, Arches N.P.	
Double-O Arch	Utah	Devils Garden Section, Arches N.P.	
Druid Arch	Utah	Elephant Canyon, Canyonlands N.P.	
Gemini Bridges	Utah	Bull Canyon, northwest of Moab	
Grosvenor Arch	Utah	Southeast of Kodachrome Basin	
Hole-in-the-Bridge Arch	Colorado	Rattlesnake Canyon, west of Fruita	(2)
Inchworm Arch	Utah	Lost Spring Canyon	(4)
Mask Arch	Utah	Devils Garden Section, Arches N.P.	
Mastodon Arch	Utah	Salt Creek Canyon, Canyonlands N. P.	
Mystery Canyon Bridges	Utah	Mystery Canyon, Glen Canyon N.R.A.	(3)
Nels Johnson Bridge	Utah	Near Hickman N.B., Capitol Reef. N.P.	(2)
Partition Arch	Utah	Devils Garden Section, Arches N.P.	(4)
Smith Arch	Utah	San Rafael Reef	
Spiderweb Arch	Arizona	Hunts Mesa, Monument Valley	
Square Kiva Arch	Arizona	Canyon deChelly N.M., north rim drive	(1)
The Spectacles	Arizona	Monument Valley area	(5)
Twin Arch	Utah	Fiery Furnace, Arches N.P.	
Turret Arch	Utah	Windows Section, Arches N.P.	(4)
Triple Arch	Utah	Cottonwood Canyon, Glen Canyon N.R.A.	(6)
Undine Bridges	Utah	Labyrinth Canyon, Green River gorge	

TRIPLE NATURAL SPANS

SPAN NAME	STATE	GENERAL LOCATION	
Davis Arch	Utah	Davis Canyon, Canyonlands N.P.	(7)
Fraggle Rock	Utah	Sevenmile Canyon, northwest of Moab	(8)
Miser Arch	Utah	San Juan arm, Glen Canyon N.R.A.	(9)
Triple Arch	Utah	Little Canyon, northwest of Moab	(10)

NOTES:

(1) This is a case of three openings making two rock columns in the mouth of a cave, resulting in double spans.

(2) This broad arch has a relatively small hole in its roof, making it double in its center.

(3) These two spans are very close together, making them double in one sense of the word. They are now beneath Lake Powell.

(4) This is a case of a small opening close to a large one. Some might say the second opening is too small to qualify.

(5) These two separate arches are close enough to each other to have earned a single name.

(6) This multiple span was named Triple Arch, but its two openings produced only two rock spans in the rock mass in which they formed.

(7) This multiple span is a toss-up -- it could be a double or triple, depending on personal viewpoint.

(8) Similar to Hole-in-the-Bridge Arch but much smaller, this wide span has two holes in its roof, making it triple, even though one of the holes is partially blocked by an enormous boulder perched above it.

Twin Arch, Fiery Furnace, Arches National Park

(9) This multiple span is like Fraggle Rock but larger and without the boulder blocking one of its two extra openings.

(10) Although this arch high on a rim was known for decades, its triple nature was not discovered until it was reached during an attempt in 1971 to climb an adjacent spire called The Bride.

AESTHETICS

AESTHETICS VERSUS PROGRESS

Were it possible to send one of our modern land survey satellites back through the centuries to the time when the first Pilgrim settlers reached the shores of North America, then compare its infinitely detailed photographs with an equivalent set taken now, few people would fail to be shocked at the comparison, at the changes wrought by mankind's technology and booming population. America the nation has become progressive, rich and powerful, but at a terrible cost to America the continent, America the diverse and balanced ecosystem found by the Pilgrims, America the Beautiful.

This progression toward a burgeoning, thriving human population and away from an ecosystem balanced to serve countless life forms -- including primitive humans -- continues today at a rampant pace. While the nature of natural spans has tended to protect most of them from human progress, that protection has not been absolute.

Despite efforts to protect the more obvious and massive spans on public land by the establishment of parks and monuments, development activities have spoiled the beauty of the settings surrounding some spans on private land. Uncontrolled mineral search and mining and other developments on public land have spoiled the settings of still other lovely spans.

Big industrial development projects on public land within the Colorado Plateau, such as the construction of reservoirs and utility corridors, have already completely destroyed many arches and bridges and threaten others. Massive Gregory Natural Bridge and dozens of lesser spans have been inundated and forever lost beneath the deep waters of Lake Powell. Padre Arch collapsed when the rising water reached its base. Famous Rainbow Bridge has become a symbol of the continuing clash between those who appreciate the aesthetics of natural spans, and the kind of human progress that ignores the higher values of human civilization.

The battle between aesthetics and progress is quite one-sided -- most of the power lies on the side of progress, with elected officials serving as its front-line troops. At the peak of the controversy over whether or not to flood Rainbow Bridge National Monument, Utah's Senator Moss distributed thousands of copies of a report in which he spelled out the heart of the controversy:

> "The issue boils down to a value judgment. I am compelled to this irrefutable conclusion: we must continue to protect our water and power development in the interest of Utah. This we can do with our magnificent Lake Powell at full storage without harming Rainbow Bridge structurally, or in my view, scenically."

A lot of people disagreed with the senator's aesthetic judgment. While water under Rainbow Bridge was to provide lovely reflections of the span, to the knowledgeable that new beauty was badly marred by inner knowledge. There is a parallel with the architectural beauty of humanity's numerous ancient castles, pyramids, temples and cathedrals. To many, that beauty is spoiled by the knowledge that most of them were built at enormous costs in human suffering, deprivation, indentured labor, slavery and death.

Similarly, the beauty of Rainbow Bridge reflected in the tongue of stagnant, algae-laden water beneath it is spoiled for many by the certain knowledge that the lovely reflection will be paid for by reduced life expectancy of the unique span. In effect, we now enjoy that augmented beauty at the expense of future generations. The only question remaining is -- how long will it be until visitors to Rainbow Bridge National Monument hear something like this from their guides:

> "--and here, gentle persons, is where magnificent Rainbow Bridge once stood, awesome in its size and beauty, revered by the Native Americans who lived near it, proclaimed by scientists, writers, historians and all who ever saw it standing as the epitome of grace and poise, a towering bridge that spanned this remote canyon in a single mighty leap -- but condemned by those who placed other things above its incomparable majesty. Unbelievable as it may seem, the rubble and boulders you now see in the canyon, plus others beneath the green water, once all fit together to make a single, stupendous arc of stone.
>
> "To see how it looked for ages before the arrival of modern man, and for a short time even after that, please step over into the air-conditioned museum and look at the historic 'photographs' preserved in inert argon gas -- study the solar-powered, compu-generated holo-display, created from the computer analysis of hundreds of such archaic 'photos.' Note that we have dedicated this museum to all who declared the mighty stone rainbow to be safe from 'water under the bridge' -- Utah's Senator Moss, the Bureau of Reclamation and the Geological Survey -- to the countless, faceless, nameless bureaucrats who had not personally experienced the beauty of Rainbow Bridge and thus thought it no great loss, and especially to the industrial interests who knew of the danger but simply didn't care"

THE ENGULFED CATHEDRAL

NOTE: A slightly longer version of the following personal story was published in a southeastern Utah newspaper during the peak of the controversy over the intrusion of Lake Powell water into Rainbow Bridge National Monument. It thus seems appropriate to share that story here, with readers from a different time, in hope that by reading it some of them will gain an empathy for the "little" beauties of the Colorado Plateau.

Why such a loud public outcry over hypothetical damage to one natural wonder, but virtually no interest at all in other very real losses of ancient and irreplaceable beauty?!

True, thousands have viewed and become familiar with Rainbow Bridge since the rising waters of Lake Powell have made it accessible. But thousands have also seen and appreciated the beauty that abounds on a smaller scale in the countless branching sidecanyons of the lake.

True, for sheer size the soaring span of the "frozen rainbow" has no match, anywhere. Yet does relative size provide a true measure for beauty? Is a snow-capped mountain, for all its ponderous mass, more beautiful than a full-blown rose, or an exquisitely cut diamond? Is the tiny iridescent blur of a ruby-throated hummingbird less lovely, or less worthy of preservation, than the soaring, dark majesty of a condor or osprey or eagle?

By the scale of public outcry it would seem so! Yet there are those who also mourn the losses of "little" beauty. Here, then, is the story of one such loss. The story may be unique, but the kind of loss it describes is not. The flooding waters of Lake Powell assure the regular drowning of countless vignettes of loveliness and beauty, and it would seem only fair, just once, to tell the story of one such "little" loss.

From the very first, we were impressed with the strange mood of Mystery Canyon. The canyon started out like numerous others, with the water twisting and turning between high, sheer walls of mineral- and lichen-stained sandstone. Soon, however, it began to grow narrower and darker, with the walls only ten or fifteen feet apart, with frequent branchings and with each fork looking alike in size. We followed several wrong turnings to dead ends as we idled up, and still farther up this truly maze-like canyon.

The canyon walls closed in even more as we continued on, grew higher still, and overhung the water threateningly. Only occasionally did we glimpse the sky or a shaft of friendly sunlight. A mood of ominous foreboding grew, but we persisted, hoping either to reach the end, or at least a place too narrow for our boat. What seemed like miles drifted by as we idled our quiet craft around bend after dark bend. We grew more silent and depressed with each passing minute.

In hushed tones we discussed the oppressive mood of this place. Was it just us? Were the lack of light in midday, the dank cellar-like humidity, the slime-covered, close-confining, looming rock walls simply arousing within us some primordial, instinctive fear? Or was there something more than that? Was this dark, gloomy gorge perhaps haunted by the lost and lonely spirits of the ancient Indians who once lived nearby? Were these spirits being made miserable and restless by the new waters encroaching upon their sacred domain? Were they telling us, in their own limited way, of their unhappiness at this violation, this desecration?

Several times we impulsively started to turn around and

retrace our path before reaching an ending, something we had never done before. But each time we went on. Finally, to our immense relief, we were stopped -- by a huge mass of floating driftwood which choked the canyon from wall to wall. The passage went on, far out of sight, but now we had a good excuse for retreating. And retreat we did! Almost in a panic, we turned the boat around, with only inches to spare, and as rapidly as we could, left that foreboding and forbidding gorge!

That, our first visit to Mystery Canyon, took place several years ago, when the water level of Lake Powell was much lower than it is at present. The dark, narrow depths that so oppressed us then have since been submerged in a fitfully rising tide of green water.

Nor did we discover on our original trip into Mystery the enchanting beauty of the lovely twin natural bridges that grace one of the several branchings of this maze-like canyon system. That came two summers ago when, for a second time, we set out to penetrate Mystery Canyon, this time to see if the fearful, ominous mood that we had first encountered was still there.

Once again our quiet boat idled up the narrow, rock-bound canyon, but this time the dark vertical walls were covered with rippling, changing, kaleidoscopic patterns, created by sunlight reflecting from the breeze-riffled water. It was close to high noon on a mid-summer day, and the sun straight overhead penetrated all but the most overhung turnings of the winding canyon. Our mood was cheerful -- and stayed that way. Gone were the unhappy forebodings of our first venture into this rat's-maze of deep water and vertical rock.

We had resolved to explore Mystery to the fullest. This meant following each of its many branchings as far as our boat could go. To do this, we followed a system of "keep to the left." This meant that at each fork we would go left, continue until stopped by shallow water or walls too close for the width of our boat, then retrace our path and turn left again at the next fork encountered. Frequently, we would have to back our boat carefully out of narrow places, sometimes by pushing on the rock walls with our hands until we reached a place wide enough to turn the boat around.

So we kept to the left. The first left branching was as wide and sunlit as the main channel. The second ended abruptly after a half mile of narrower, unusually straight travel.

The third left turning was the charm! Here the channel was perhaps twenty feet wide, with sheer walls reaching out of sight above, but with sunlight sparkling brightly on the algae-green water. That third left branching was short, just a few yards long, and there at its end was a place of sheer loveliness!

Picture a rounded, cylindrical amphitheater formed by soaring rock walls and floored by sun-mottled water, with a knife-slit alcove above the floor set off by two graceful, rounded bridges of age-stained sandstone -- the first bridge massive and angled up toward the right, the second smaller and more delicate, but the very image of water-sculptured beauty! A spotlight of sunshine lit both lovely spans, sparkled on the water and decorated the dark, shadowed walls with shimmering patterns of reflected light.

We quieted the murmuring burble of our boat's idling engine and stood enthralled by the beauty of our enchanted natural cathedral with its twin flying buttresses. All of our earlier unpleasant memories of Mystery Canyon were blotted out forever by the almost supernatural charm of what we were now viewing.

At first glance, because of the sheer scale of the canyon walls, our sisterly bridges seemed to be quite small, but a closer examination proved otherwise. They were only "small" when compared to such giant spans as Rainbow or Landscape or the Gemini twins. The relative size of our "little" bridges was soon established in the most direct manner possible.

As I stood in our boat photographing the setting and savoring the bright mood that prevailed, I heard a quiet splash, the boat rocked gently, and I then saw a blonde head drifting slowly toward the alcove which contained the bridges. As she reached the steep cleft below the larger span, my wife's voice echoed back --

"There's a steel peg in the rock just above the waterline!"

Evidently, someone had scaled the sheer drop below the bridges long before the water had made them easily accessible. By using the peg, my wife was able to climb above the water and stand precariously below the first bridge, with one foot on a tiny projection of rock on the steep slope.

My cameras clicked, and recorded a scene never to be viewed again -- or at least never so long as Glen Canyon Dam endures. And the height and breadth of the sisterly bridges were recorded, by simple comparison with the five feet, two inches of my wife's height!

By this measure, the larger span stood about 40 feet high and 20 feet wide. The smaller span was less than half this, but had been more delicately carved by eons of fitful water flow. The measurements, however, were insignificent in comparison with the complex, three-dimensional loveliness of the two spans and the setting which they graced!

A year later we returned to our enchanted alcove -- and came away with heavy hearts! Our visit was later in the season, in September, and no bright sunlight lit our beautiful twin bridges. Nor did they soar as gracefully as before. The inexorably rising water was drowning them, inch by torturous inch.

The underside of the first span, the one my wife had climbed up under, was just ten feet above the water's surface, and most of the intricate convolutions of the eroded rock below the more delicate bridge were also now under dark water.

Sadly, we studied our half-drowned cathedral-in-rock, telling ourselves that this was probably the last time we would ever see it. We inched the boat slowly under the sloping span of the first bridge, a sacrilege perhaps, and once again my wife slipped over the boat's side and swam slowly, almost reverently, into the narrow opening. She paused for a time under the second bridge, then returned and stood between them on a narrow underwater spur of rock, knee-deep in the dark green waters that were inevitably drowning our personal place of "little" beauty.

Reluctantly, I took a few photographs for the record, and reluctantly we left, with somewhat the same feeling as one bears away from the funeral of a dear friend.

We returned, just one more time, on a sad pilgrimage to the desolate site of our engulfed cathedral. It was not a happy trip. We questioned whether our unhappiness, our sense of loss, was any less than that which might be suffered by viewers of a fallen Rainbow Bridge. We thought not.

Yet each month, each week, for many years, such losses have occurred within the immense expanse of Lake Powell's 1800 mile shoreline. Multitudes of such places of "little" beauty have been drowned forever by the flooding waters of the lake. All but a few such losses have occurred unnoticed and unmourned -- but perhaps this story will add to the numbers of those who will truly miss and mourn the loss of Mystery Canyon's Engulfed Cathedral, and myriad other lost "little" beauties of the Colorado Plateau.

Hand Holt Arch, Lavender Canyon, Canyonlands National Park

CATALOGING KNOWN SPANS

HOW MANY ARE THERE?

How many natural spans are there in the Colorado Plateau? As discussed in earlier chapters, any attempt to answer such a question quickly becomes entangled in definitions, semantics, measurements and other complications. Even ignoring such technicalities doesn't help, because no matter how loosely or precisely the phrase "natural spans" is defined, there is no factual answer to the question. At this time, no one actually knows how many natural rock openings there are within this vast, little-known region.

Once it became known in the late nineteenth and early twentieth centuries that the Colorado Plateau region had a number of unusual natural spans, public interest in these phenomena grew, and efforts to catalog the spans began. These first efforts were made by literature search, however. No one made a personal effort to count spans in the field. Even today, with access into the region's rugged backcountry vastly improved over what it was earlier, the task of cataloging known natural spans is difficult and time-consuming, and demands a level of energy and persistence that has yet to be exhibited by anyone.

What about a ball-park figure? In his 1969 book, "Scenes of the Plateau Lands," geology professor W. Lee Stokes of the University of Utah noted that "the number of arches, bridges and windows that decorate and enliven the plateau lands runs into the hundreds," but he made no attempt to catalog these hundreds.

There are, indeed, "hundreds" of known "sizable" or "significant" natural spans within the Colorado Plateau region of the Four Corners States, but more exact numbers depend upon how these terms are defined. There are probably at least 300 whose significance no one would question, plus 400 to 1000 more that could meet less rigorous definitions, and tens of thousands more if a 2-foot size minimum is used for openings.

These figures are, of course, just estimates based on the author's two decades of exploring Colorado Plateau backcountry, but to date there simply are no better figures based on fact. At this writing, no one has made a truly comprehensive on-the-ground count of the natural spans within this region, although limited surveys have been made of specific small areas, such as Arches National Park.

One thing can be said with certainty, however. In any complete catalog of natural spans on the North American continent, those within the Colorado Plateau region of the Four Corners States would easily outnumber those of all the other states put together, perhaps including Canada and Mexico too, and the spans within Utah would probably outnumber those of Arizona, Colorado and New Mexico combined.

SOME WHO HAVE TRIED

Not too surprisingly, very few people have made serious attempts to document or catalog the natural spans of the Colorado Plateau. Among those who have are several earth scientists -- geologists or geographers -- a museum curator, an aerospace engineer and a couple of amateurs -- those who tackled the job out of personal pleasure and curiosity.

The earliest effort at listing all known "natural bridges," the general term used to describe all natural spans at that time, was made in 1910 by a geologist, Herdman F. Cleland. Unfortunately, that was too early in the exploration history of the Colorado Plateau to produce more than just a few for that immense wilderness region, and apparently Cleland made his list without ever having been there. His cataloging of the spans in that region was done from literature sources rather than field research. This approach was also used by most of those who later attempted the task.

Cleland's 1910 catalog of American "natural bridges" listed only seven spans in the entire Four Corners States -- 3 in Arizona, 4 in Utah and none in Colorado and New Mexico -- and one of the Arizona spans is not within the Colorado Plateau. For Arizona he listed Tonto Natural Bridge, Black Rock Natural Bridge and a petrified log bridge in what is now Petrified Forest National Park. For Utah he listed Rainbow Bridge and the three White Canyon spans, Kachina, Sipapu and Owachomo. He did not even note the many Utah spans reported by the Powell survey party and other explorers of the 1800s.

In a report dated March, 1933, Junius Henderson, museum curator for the University of Colorado, made an attempt to bring Cleland's catalog of spans up to date by means of a literature search, but again no field work. He added only one specific span for Arizona -- Hope Window -- but mentioned another unnamed opening reportedly visible from "twenty miles away." For Colorado he added Ela Natural Bridge, in Glade Park, and a few other "insignificant" openings, but made no mention of Juanita Arch near the Dolores River or the several spectacular arches in the Rattlesnake-Mee-Knowles canyon complex to the north of Glade Park.

For New Mexico, Henderson quoted a 1907 report of a "curious natural bridge -- carved not by water, but by sand-laden winds," thus helping perpetuate the myth of wind formation, but did not mention La Ventana, a huge arch on the western edge of the Zuni Indian Reservation near Grants.

Despite creating and perpetuating several errors, Henderson did somewhat better for Utah. He noted Utah's three national monuments -- Rainbow Bridge, Natural Bridges and "The Windows - near Moab" -- and added that there were other areas in the state that "should be set aside, to prevent them from falling into private ownership." He also:

-- noted "a small bridge and a number of small windows" in Bryce Canyon National Park;

-- mentioned "a strange and beautiful arch" in Cedar Breaks, which became a national monument later that year;

-- described an arch beside the highway to the south of Moab (Wilson Arch);

-- mentioned in passing the several spans reported by Powell and Dellenbaugh in Desolation Canyon of the Green River;

-- listed one span, "Owl Bridge," reported by geologist Gregory near Navajo Mountain;

-- quoted Winslow's 1898 report of a "500-foot" span that was a "monstrous product of wind erosion -- not far from

Ela Natural Bridge, Glade Park, Colorado

Moab" (Landscape Arch, in Arches National Park);

-- mentioned Cummings' report of "Pikyabo Natural Bridge (Pritchett Arch) and Wigwam Cave," both also near Moab;

-- listed as being in Utah the Monument Valley spans referred to in Judd's 1924 National Geographic report, although most if not all of these are actually in Arizona;

-- quoted a 1907 report that there were "dozens of natural bridges -- on the southwestern slope of the Blue Mountains" (Abajos) but named only the White Canyon spans there;

-- noted the report of a "Maple Canyon Bridge" in the Salt Lake Tribune, but gave no location for it; and

-- noted another report of two spans to the north of Navajo Mountain -- "Owl and Crag."

CATALOGING KNOWN SPANS

In 1968, Moab freelance writer Dick Wilson started to catalog the natural spans of southeastern Utah, starting with Arches and Canyonlands national parks. By 1971, Wilson's formal log reached 143 spans, and he had unrecorded notes for several dozen more. At that time he discontinued his project for personal reasons, before publishing any of his results. A July 1971 National Geographic article on "Canyonlands" by Rowe Findley noted that "a young journalist from Moab -- has tallied 150" arches in the region.

In 1973, professor of geography Dale J. Stevens of Brigham Young University in Utah started a survey of the natural spans within a quite limited area of the Colorado Plateau, and is continuing his efforts as time permits. He was the first earth scientist to even attempt such a count in the field. All earlier reports by scientists were based largely or entirely on literature research.

In a series of reports published in 1976 and 1977, another geologist, Richard M. Pearl, defined several types of "stone rainbows" and specifically noted the major spans within National Park Service areas and a few other places. His mislocation of two spans, "Little Rainbow Bridge" (Corona Arch) and "Pinto Arch" (Gold Bar Arch), within Canyonlands National Park indicates that at least some of his research on Colorado Plateau spans was not made in the field. While Pearl's reports were not an attempt at cataloging individual spans, they did list quite a few as examples of his general types of spans.

In 1976, Robert H. Vreeland, an aerospace engineer, started cataloging all the "significant" natural spans in the United States, on a state-by-state basis, but dividing the Colorado Plateau into smaller segments.

Vreeland found that most states had no spans or just a few, but the Colorado Plateau, the heartland of the Four Corners States, was another matter. His cataloging of that region is continuing at this time, but is far from comprehensive even within areas supposedly complete. Vreeland will probably never finish his inventory of Colorado Plateau spans, because no one man could live long enough to perform such a task in the field. The Plateau is simply too vast and rugged, and much of it is inaccessible except to expert backpackers or skilled climbers, or by helicopter.

In 1981, Park Service officials started a comprehensive inventory of the various types of natural spans within Arches National Park, based on the system and definitions established in 1973 by Dale Stevens. The study was organized and performed by park ranger Edward McCarrick, starting with approximately 135 known openings. An interim report on this effort was issued at the end of 1983. Using Stevens' minimum of a 3-foot opening, the report listed these results:

> "As of March, 1983, the number of documented openings that meet the criteria of an arch stood at 235, excluding two which had fallen since the current recording procedure had started.
>
> "The number of recorded arches based on currently established criteria is now 288, which includes three that have fallen. This number is further analyzed as follows:"

Arch openings over 3 feet	288
Arch openings over 10 feet	177
Arch openings over 30 feet	81
Arch openings over 50 feet	51
Arch openings over 75 feet	18
Non-arch openings over 3 feet	66
Non-arch openings under 3 feet	32

McCarrick noted in a personal communication that while 32 non-arch openings under three feet were logged, "hundreds were seen." The report broke down the 288 figure into the following types, as defined by Stevens:

Free standing arches	137
Pothole arches	13
Jughandle arches	17
Alcove pillar arches	4
Alcove beam arches	2
Perforated alcove arches	1
Cliff wall arches	51
Cliff side arches	49
Natural bridges	13
Natural tunnels	1

The report indicated that the inventory was not complete. McCarrick mentioned in a conversation that he knew of at least a dozen openings of significant size that had not been recorded in the interim report, and that there were doubtless still others because the survey had been limited to documenting known and reported openings, plus any others found during the survey. The park had not yet been systematically searched on an area-by-area basis.

The report concluded by recommending that the survey be completed "as expeditiously as possible" because "visitors expect, and have a right to -- know how many arches" there are in a national park dedicated to arches. This would seem to be a reasonable expectation, but with this task seemingly never-ending even within a relatively small area of the immense Colorado Plateau, it is apparent that the task of inventorying the entire region would be virtually impossible -- a dream to pursue, but one with no ending.

Ring Arch, Courthouse Wash, Arches National Park

ARE THERE ANY UNDISCOVERED SPANS?

In his 1922 report, "Natural Bridging in the High Plateaus," Frederick J. Pack answered that question this way:

> "Despite the fact that so many of these bridges are already under observation in this most inaccessible Plateau country others doubtless remain yet to be revealed. Further exploration may not possibly result in the discovery of structures equal in stately magnitude and imposing grandeur to those now known but it is sure to disclose many more of close secondary importance."

Pack went on to explain why it would be so difficult to do a systematic survey of natural spans in the Colorado Plateau by using the areas around White Canyon and Cedar Breaks as examples:

> "The country in which the great (White Canyon) bridges are situated is so inaccessible and irregular that the explorer easily passes within a few rods of them without suspecting their existence. Cattlemen ranged their stock in the vicinity of the bridges -- for nearly twenty-five years before they discovered them. None of (the) spans were probably seen by white men until 1883; and even then Indian guides led the way. Then again, both of the bridges situated within less than fifteen miles of Cedar City were first seen only about five years ago."

CATALOGING KNOWN SPANS

In reporting his 1973 inventory of natural spans in the general vicinity of Arches National Park, Dale Stevens noted that:

> "The most intensive work was done within Arches National Park, but in spite of this effort more arches were found outside the park boundaries than within."

As noted earlier, even while this book was being written, several new discoveries were reported in regional news media. Others were found by this writer and by friends who were especially alert for new spans to document for the book.

"Are there any undiscovered arches"? This question, or one very similar, has been asked by innumerable visitors who have just become aware of the beauty and mystery of canyon country natural spans. Although the question is usually poorly phrased, and semantically ridiculous, it still deserves a serious answer, after the usual pause for clarification of terms.

What does the average new arch enthusiast mean by "undiscovered?" Never before seen by human eyes, including prehistoric and historic American Indians? Seen by Indians and early Hispanic or American explorers, but not documented? Seen by cattlemen, sheepmen or prospectors, but not documented? Seen but not documented by contemporary amateur explorers?

The most reasonable and meaningful definition of the term "undiscovered" is, of course, "undocumented." That is, not described or depicted, not identified as to location, not reported and written up in some manner within a publication readily available to the public. As noted in earlier chapters, a "discovery" is meaningless if it exists only within the heads of the discoverer and a few of his acquaintances, or if the discoverer simply scratches his name in the rock near the span, or on a paper register left under the span in a container. Such "proofs" of the date of discovery or visitation are not reliable. This author has personally noted numerous names-in-rock with obviously faked dates. Unscrupulous "discoverers" can easily discard earlier paper registers and substitute their own, and containers under spans can be destroyed or moved by the elements.

A discovery is also meaningless if reported, but the report does not graphically pin down the span's physical shape and at least its approximate location, or if the report is not available to the public, as in a diary, personal journal or a private record. For example, arch historian R.H. Vreeland noted in one of his reports that while a span he dubbed "Uranium Arch" had no doubt been seen earlier by uranium prospectors, he was the first to bring it to public attention, after spotting it from the air in 1971. Subsequent research has shown that photos of this span existed within the files of local photographer and arch collector Dick Wilson some time before Vreeland made his find public. Even so, since Wilson did not publish his log, credit properly goes to Vreeland, who did.

With "undiscovered" defined as "not described, located and published in a document readily available to the public," are there still any undiscovered natural spans within canyon country?

The answer to this is -- "certainly, there are many." How many? Who could answer that question, except perhaps as an order of magnitude, and only then after some kind of lower limit on size of opening has been established.

But with some kind of reasonable lower limit on size -- say with openings of more than 75 square feet -- what kind of numbers of undiscovered natural spans might there yet be in canyon country?

CATALOGING KNOWN SPANS

Extrapolating from the number of previously undocumented spans that this writer has personally found, or had reported to him by others over the past two decades, and using the "not publicly documented" definition, there are quite probably still many hundreds of significant undiscovered natural rock openings -- arches, bridges, windows and tunnels -- within the Colorado Plateau. Reduce the size requirement to, say, 50 square feet and the probable numbers would jump an order of magnitude, to the thousands.

Most, but certainly not all, of these undiscovered spans are probably within regions not readily accessible to the public, such as the several Indian reservations within the Colorado Plateau region, or in regions so remote and difficult to penetrate with wheeled vehicles or on foot that they have simply not been explored thoroughly by the kinds of people who care enough about such geologic curiosities to investigate and report them.

There are many such little-known regions within the Colorado Plateau, even in this next-to-last decade in the Twentieth Century, when mankind is supposed to know and dominate the entire planet. Following are some regions within the Plateau that are likely to contain numerous unreported natural spans:

The unnumerable tributary canyons of Lake Powell, in Utah and Arizona.

The region between the Colorado-Green rivers and the San Rafael Desert, in southeastern Utah.

The sandstone canyons and reefs of the San Rafael Swell, in central Utah.

The vast region to the south and west of the Kaiparowits Plateau, in south-central Utah.

The immense, maze-like drainage system of the Escalante River in southern Utah.

The canyons within the Dolores Triangle in Utah, and on east into Colorado.

The canyon system of the Dolores River in western Colorado.

The San Juan River system in New Mexico and Utah.

Numerous large areas within the Navajo, Hopi, Zuni and other Indian reservations within the Four Corners States where massive sandstone formations are exposed.

The countless canyons of the Green River, from Flaming Gorge south through the Book Cliffs.

The sandstone wilderness that completely surrounds the La Sal Mountains of southeastern Utah.

The canyon systems to the south and west of the Abajo Mountains in southeastern Utah.

The immense wilderness between the Escalante River canyon system and the Waterpocket Fold in southern Utah.

The maze of foothill canyons that surround the Henry Mountains in southern Utah, and Navajo Mountain in Arizona.

The Dirty Devil River canyon system in southern Utah.

The Grand Canyon's maze of canyons in Arizona.

The Cockscomb in south-central Utah and northern Arizona.

In fact, other than a few tiny settlements, a state or federal park here and there, and a sparse webbing of roads and backcountry trails, the entire region roughly defined by U.S. 89 in the west, the Grand Canyon and U.S.160 in the south, the Rocky Mountains in the east and the Uinta Mountains in the north -- a region some 70,000 square miles in size -- has barely been touched when it comes to cataloging natural rock openings. Add to this the vast reaches of the Colorado Plateau within New Mexico, and in Arizona south of the Grand Canyon and U.S.160, and there are another several thousand square miles to explore for unreported natural spans.

Some dauntless amateur arch collectors have already spent decades chasing down and photographing known spans, but they have barely started if they want to be comprehensive. There is room for everyone in the search for uncataloged natural spans within the Colorado Plateau. In that vast wild region, everyone has a chance to discover and report a "new" arch or bridge or window or tunnel, or some other exotic kind of opening eroded over the eons through native rock.

Macomb Arch, East Canyon Wash

CATALOGING KNOWN SPANS

HOW TO HUNT FOR UNREPORTED SPANS

A detailed discussion of this subject could easily fill another book, but here are a few useful basics for "arch hunters":

> Choose one of the prime regions listed earlier. Within that region select areas for closer study on the basis of the exposed geologic formations and combinations most likely to form natural openings. Geologic maps that graphically depict such information are available from the U. S. Geological Survey, from some regional Park Service, Forest Service and BLM visitor centers, and from some state geological bureaus.
>
> Within the selected areas, look for the kinds of combinations of rock formations and water drainage that create openings by flowing or seeping water. If an area is found that has any openings at all in it, the entire area should be searched carefully because the probability of more openings is high.
>
> Low-level aerial surveys can be helpful, but can be dangerous in this region of extreme vertical relief and tricky air currents. Further, lighting is a variable. A large span may be invisible by morning light but quite obvious in the afternoon. Other spans may not be visible at all from the air. Despite such problems, a number of spans have been discovered first from the air.
>
> For a detailed search of an area, the most thorough method is on foot, even though this method is not perfect and has its own problems. As noted by Pack, in broken canyon country it is possible to be within a few feet of a sizable span and miss it. To compensate for this, comprehensive on-the-ground searches must be very systematic and thorough.
>
> Any kind of systematic search of an area is going to take considerable time. While a casual effort may produce a discovery by sheer chance, an intensive effort in a favorable area is certain to produce results. But a "new" discovery should not be announced before a complete search of existing literature has been made. Many "new" discoveries made by enthusiastic amateurs have turned out to be rediscoveries, with the original discovery well documented and made years or even decades earlier. Several examples of this are noted in earlier chapters. The bibliography in this book can serve as a working list of existing literature.

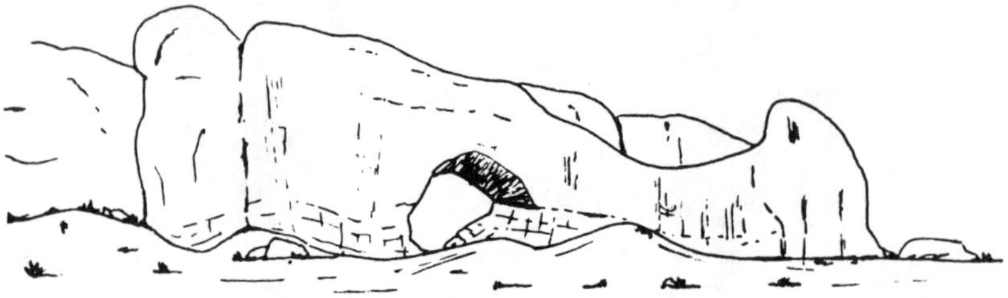

CATALOGING KNOWN SPANS

SUMMARY

No matter what definitions or size minimums are used, no one knows exactly, from direct field research, how many natural spans there are within the Colorado Plateau. Educated guesses are all that is available at this time, even within such well-defined areas as Arches National Park, although a comprehensive inventory of this park has been started.

Despite several historic and current attempts to inventory Colorado Plateau natural spans, there is at present no comprehensive list or catalog covering this region, nor is there likely to be in the near future. The task is simply too big.

This book is not an attempt to catalog Colorado Plateau natural spans, although some effort has been made to be representative, to provide examples of the many distinct types of spans found in the region.

There are still plenty of undiscovered spans in the region, especially if "discovery" is defined as "publicly reported." This makes the region a happy hunting ground for "arch hunters."

Balcony Arch, Behind the Rocks

FINDING FRED'S ARCH

I ran into Fred, a long-time friend, in one of Moab's grocery stores. Fred was a Hoosier who had fallen in love with Utah's canyon country a couple of decades earlier. The tourist-season jet boat tours he ran down the Colorado River gorge from Moab Valley were his hobby. His serious business was trying to explore every nook and cranny of southeastern Utah, a business he energetically pursued twelve months a year, using every means of transport available, by land, air and water.

After a brief chat, I asked Fred if he had any pictures of arches to contribute to the book I was writing -- a good possibility considering how much he got around, but chancy considering his luck with cameras. Canyon country is rough on cameras, and Fred's really took a beating.

Fred had a new arch for me -- maybe. He had spotted and photographed one from the air, but even after several tries had been unable to find it in the broken canyon wilderness where it was located, to the west of Moab. Did I know of an arch in that canyon system?

Yes, I knew of one and had photographed it recently. With Fred's aerial photos, I could see if his arch was the one I knew about, or help him find it on the ground if it was new.

Later that day, Fred came by with his three aerial slides. After carefully studying them and a topographic map of the area, I showed him where his arch should be. I felt that it might be the known span on that canyon rim, but could not be certain without a trip to the span, or pictures taken from the same angle for comparison. Fred had been searching too far downcanyon, but he should be able to go right to it now -- assuming my map and photo reading were accurate. He couldn't check this out right away, however, because he was leaving in a couple of days for a late-winter jaunt down the Baja California peninsula.

I couldn't wait. My curiousity piqued, I set out the next weekend to find Fred's arch, in early March. With me were my wife, Terby, and Linda, her co-worker at the local National Park Service office.

The day was sunny and perfect for hiking, but as we drove out of Moab Valley winter snow still lingered on north-facing canyon walls and slopes, and when we later left the pavement to travel a dirt road toward where the hiking would begin, melting snow patches made the rough road muddy in places.

I drove the miles I had measured on the topo map, turned off for a few yards into a vast wilderness of sandstone, sand, and pinyon-juniper forest, then parked on a flat expanse of slickrock. I walked a few more yards to an overlook and studied the terrain below with binoculars, comparing it to my topo map. We were in the right place, but needed to angle a bit eastward while descending a series of wooded Kayenta sandstone terraces to the rim of the deep inner gorge where we should be able to see Fred's arch on the opposite rim.

Fifteen minutes later I recognized the contours of the gorge Fred's slides had shown. In another five minutes we stood on the edge of a sheer drop, excitedly studying the big span on the other side of a narrow, vertical-walled canyon some three hundred feet deep. We had found Fred's arch, exactly where I had earlier placed a dot on the topo map, a bright arc of sandstone casting its curved shadow on the floor of the immense pour-off alcove behind it. Linda was ecstatic! This was her first "new discovery." We shared her excitement, even after twenty years of canyon country exploring and many such "firsts."

I took several pictures, but the lighting was too flat. The top of the huge span was not well defined against the cliff wall behind it. From Fred's pictures, I though it might be possible

to get below the arch by descending into the canyon at a point upcanyon. We had brought along 100 feet of rope. We were not skilled "aid climbers," but thought the rope might provide a margin of safety in getting down short awkward stretches.

We hiked along the rim of the sheer-walled gorge, looking as we went, occasionally crunching through crusty snow on the sheltered, steeply terraced slopes above the inner gorge. As we paused now and then to study the canyon below, the bright silence was broken by only two tiny sounds -- the musical dripping of water from the melting snow, and the lilting, echoing cries of a canyon wren somewhere below us. After an enchanting quarter hour and quarter mile, we spotted a possible descent route and sat on a flat boulder just inches from the rim, to study the route with binoculars.

It was possible -- but impossible! About two thirds of the way down, the only way on to the bottom crossed a steep slickrock slope -- but that slope was coated with snow and ice, with a forty foot drop onto jagged rock the penalty for slipping! The only alternate route down that last stretch reached the canyon floor just above an undercut drop of some thirty feet, with no way around it. Such sudden and impassable drops are common in canyons cut into Wingate sandstone, as so many are in canyon country.

After a discussion, we decided to give up trying to get below Fred's arch, for then at least, and try instead to get to the canyon's opposite rim above and behind the arch. Maybe photography would be better there.

This took some doing. We returned to our vehicle, backtracked for miles then, guided by the general topography of the terrain, drove slowly down an old seismograph trail, one of the many such trails that crisscross a land otherwise wild and natural, trails bulldozed through virgin pinyon-juniper forest in search of the minerals that are rarely there.

After a couple of miles, the trail reached some antiquated mineral claim markers and then dwindled to nothing. Reluctant to make new trail in this remote place, except on solid rock or in sandy washes, I soon parked and we continued on foot. After reaching the rim of the canyon we wanted, and getting oriented to the topo map I carried, we headed downcanyon toward Fred's arch.

It was still there -- as it doubtless had been for hundreds, perhaps thousands of years. The walls of the immense alcove behind the span were sheer or undercut, and the arch itself was below the alcove rim -- too close and too big to photograph from that rim. I scrambled a few feet down a steep slope beyond the arch, looking for a good vantage point, but that was also too close. I finally ascended several big terraces behind the alcove and stood on top of a huge boulder for my pictures, while Terby stood on the alcove rim for perspective. She looked tiny.

How big was the arch? There was no way to measure it without descending into the pour-off alcove with sophisticated rappelling equipment and skill in its use, but the arch's opening was at least 50 feet wide, with an even greater height -- no small span by any scale of measurement.

We got home about dusk that evening, leg-weary from our steep, trailless hiking but exhilarated at our success in finding Fred's arch -- and still enchanted by the wild, unspoiled beauty of the canyon whose rim the arch graced. Next weekend, we planned to investigate a strange canyon-rim tunnel that Fred had also mentioned.

Was Fred's arch a new one? A subsequent comparison between my pictures of Fred's arch and the known span in the same canyon rim revealed that while Fred's arch was new to him, it had long since been found, photographed and reported by others. But to our little "discovery party" of three, that lack of newness did not diminish the challenge of finding its location from an aerial photograph, or the pleasure of getting to it on the ground.

And it did not change the indisputable fact that there are many significant natural spans in canyon country yet to be found and reported, a fact that my wife and I have seen proven again and again, both before and since the "discovery" of Fred's arch.

ACCESS TO KNOWN SPANS

AFTER DISCOVERY

Once a new natural wonder is discovered and properly publicized, there are going to be people who want to see it. Some will want to be among the first to visit the wonder, to be secondary discoverers. Some will want to study or photograph or write about it. Most will merely want to see what all the publicity is about -- curiosity -- or to admire its novel beauty if it has unusual aesthetic qualities.

For whatever their reasons, people want to visit newly announced discoveries, including arches, bridges and windows. This often presents a problem -- access -- within wild and rugged Colorado Plateau country. This chapter presents a number of examples contrasting past and present access to the region's natural spans, discusses the various access methods now available, then provides a general summary of literature that can be useful in finding known natural spans within the region. It also provides two special lists, one of spans that are readily accessible by ordinary highway vehicle, and another of the few "natural bridges" within the Colorado Plateau across which it is possible to drive a four-wheeled vehicle.

ACCESS THEN AND NOW

Tourism to Colorado Plateau natural spans began almost immediately after the discoveries of the first really big ones were announced in public news media around the turn of the century.

Despite the publicity given the two Powell expeditions in 1869 and 1871, and the fact that both Powell and Dellenbaugh chronicled a number of natural rock openings, there was no rush to visit these first Colorado Plateau spans to be reported. This reluctance was probably due to the casual way in which the spans were reported, and because of the extreme difficulty, danger and expense involved in getting to them, hidden as they were within the wild gorges of the Green and Colorado rivers. In those days, public access to these gorges was simply impossible. Only a few intrepid fur trappers and well-equipped explorers dared to try.

Black Rock Natural Bridge in Arizona, which was reported in the late 1800s, was also in rather wild backcountry, but was not nearly as difficult to reach, so it was visited occasionally. But the well-covered discovery of the White Canyon bridges really started significant public visitation. Although Cass Hite and James Scorup reportedly visited the bridges in 1883 and 1895 respectively, the first published account was Dyer's description of the Long-Scorup expedition of 1903.

Perhaps the first genuine "tourists" to the bridges were H. L. A. Culmer and party in 1905. They traveled from Salt Lake City to Thompson by train (one day), from Thompson to Bluff by horse-drawn wagon (seven days), and from there to the bridges by pack train. It took three days to reach the first bridge, Caroline (Augusta). On the return trip, they went from Bluff to Cortez, Colorado, by "light outfit" then by stage to Dolores. From there they traveled by narrow-gage railroad via Telluride and Montrose to Grand Junction, and then back west to Salt Lake

City by the main rail line. The entire trip took thirty days, and Culmer's diary of the trip is an adventure story in itself.

In his February 1910 National Geographic article about Utah natural bridges, Byron Cummings gave prospective tourists details for reaching Natural Bridges National Monument, and enthused about the area like a modern tourist bureau:

> "To reach this interesting region people from the north and west should leave the Denver and Rio Grande Railroad at Thompsons and take the stage to Moab, a ride of 35 miles. From Moab one must travel by private conveyance to Monticello, 60 miles farther. At Monticello saddle horses and pack animals can be secured for the trip of fifty miles over Elk Ridge to the bridges. Parties coming from the east or south should leave the railroad at Mancos, Colorado, taking the stage to Bluff, a distance of 75 miles. From Bluff one must travel with saddle and pack animals 55 miles northwest to the bridges. Both these routes seem long hard trips by stage and on horseback, but if one enjoys outdoor life every hour of the trip will be a delight and the atmosphere will prove a veritable 'fountain of youth'."

In his November 1910 report, "The Great Natural Bridges of Utah," Cummings noted the establishment of Rainbow Bridge National Monument on May 13, 1910, then added:

> "In the following September, Mr. Wetherill guided Mr. A. R. Townsend of Telluride, Colorado, and his sister to this bridge and to Miss Townsend undoubtedly belongs the distinction of having been the first white woman to penetrate this wild region."

In his November 1911 National Geographic article about Rainbow Bridge, J. E. Pogue cautioned eager tourists about trying to reach the monument on their own:

> "The most exact directions for reaching the bridge would be inadequate, so obscure and devious is the trail leading thereto; hence the services of a guide are indispensible. A minimum of 18 days should be allowed for the round trip. The trip is an extremely arduous and toilsome one, and would be fraught with danger to an inexperienced traveler, but under competent guidance may be accomplished with no special hazard, though hardships and inconveniences, and many of them, must be expected."

When Theodore Roosevelt visited Rainbow Bridge in 1912, three years after its official discovery, he was guided there from Kayenta, Arizona, by John Wetherill. They traveled "light," that is with "no tentage." In his 1913 account of the trip, Roosevelt described efforts to reduce the load still further in the roughest country:

> "On the two days spent actually getting into and out of the very difficult country around the Bridge itself we cut down our luggage still further, taking the necessary food in the most portable form, and, as regards bedding, trusting, in cowboy fashion, to our slickers and horse blankets."

While describing the most difficult part of the trip, Roosevelt mentioned seeing a grim reminder of its hazards:

"We worked up and down steep hill slopes, and along the tilted masses of sheet-rock ending in cliffs. At the foot of one of these lay the bleached skeleton of a horse. It was one which Wetherill had ridden on one of his trips to the Bridge. The horse lost his footing on the slippery slide rock, and went to his death over the cliff; Wetherill threw himself out of the saddle and just managed to escape."

In a letter describing his 1920 trip to Rainbow Bridge, Charles Bernheimer also mentioned seeing the bones of horses or mules and noted the prevalence of "slick rocks," and that such rock:

"-- offers to the horses an insecure footing. The faithful brutes know this and it is pathetic to see how they watch their steps."

While describing his second trip to Rainbow Bridge in 1923 in the March 1924 issue of National Geographic, Neil Judd noted one such near-disaster:

"-- flat-faced, lop-eared Macduff, slowest and most obstinate of all our pack mules -- struck the corner of his pack, fell backward and rolled over. Stub caught him on the very edge of the precipice and sat on his head, while George climbed atop the prostrate mule to loose the lash rope -- (Mac's) hind feet were thrust out into empty air a good fifty feet above a jagged rocky talus. He was safe if one stubby cowpuncher could hold down his head and prevent his struggles while the bulky pack was removed."

Judd also noted the state of tourism to the bridge at that time:

"The trail of today is much easier than the one we built (in 1909); its more dangerous portions have been smoothed out or avoided. Three hundred individuals, including the late ex-President Roosevelt and a score of travelers from abroad, have followed in the footsteps of the discoverer and few have returned disappointed. Not all of these strangers came with reverence in their hearts. One, for instance, seems to have had as his chief object in visiting Non-ne-zo-shi the opportunity to be able to tell his grandchildren that he was the first civilized man to drive a golf ball over the stately arch."

In Earl Morris's 1922 chronicles of Bernheimer's 1921 expedition to Rainbow Bridge, he noted another travel hazard of the times:

"Johnson, the new arrival, was to have joined us at Kayenta. When he failed to put in an appearance, we became apprehensive lest in coming across country from Blanding, Utah, he had been waylaid by the Piutes, who just then were in an ugly mood owing to the leaden vengeance which had been visited upon one of their number subsequent to thefts from the herds of the Mormons."

In his 1922 report, F. J. Pack mentioned some recent and planned improvements in access to Natural Bridges National Monument:

The La Sal Mountains, through Jeep Arch, Poison Spider Mesa.

"(The natural bridges) in San Juan County are -- 160 miles from Thompsons, the nearest railroad point in the State. One hundred and twenty miles of this distance is covered by automobile road, and the remaining 40 miles must be traveled on horseback. Efforts are now under way, however, to make this section of the road passable to automobiles."

As reported in a November 1947 article in Desert Magazine about Zeke Johnson, first custodian of Natural Bridges National Monument:

"In 1929 a road was built to the monument but it was rough

> and often impassable because of mud. Zeke wore out five of his own cars and two government trucks carrying passengers over this early road. Now (1947) a good desert road is maintained from Blanding --"

The author of this book took his first trips to the monument over this "good desert road" and is willing to testify that other descriptive terms might be more appropriate. The road was realigned, paved and dedicated as a "Bicentennial Highway" in 1976. It is now an easy and highly scenic drive to Natural Bridges National Monument by paved road from any direction.

Access to the other big southern Utah natural span, Rainbow Bridge, was not to improve until some time later, however. Zane Grey took a trip to the great bridge in 1913, which he described in his 1922 book, "Tales of Lonely Trails." In the book he had this to say about access to the span:

> "There is no guide except Wetherill -- this is the only place I have ever visited which I am not sure I could find again alone. The tourist, the leisurely traveler, the comfort-loving motorist would never behold it. Only by toil, sweat, endurance and pain could any man ever look at Nonnezoshe. It seemed well to realize that the great things of life had to be earned. Nonnezoshe would always be alone, grand, silent, beautiful, unintelligible --"

Beyond a doubt, Zane Grey would have wept at the sight of the hundreds of "comfort-loving motorists" who now view his silent and alone "Nonnezoshe" each day of the long Utah travel season. Yet it cannot be wrong for many to see and appreciate the beauty and majesty of Rainbow Bridge where only a few could before. Despite Zane Grey's view that such opportunities must be "earned," it is certainly not necessary to suffer in order to appreciate outstanding beauty. Anyone who has ever stood entranced beneath Rainbow Bridge can attest to that, no matter how he got there.

ACCESS TO KNOWN SPANS

In a report published in the February 1923 issue of National Geographic, Charles Bernheimer noted that in his first trips to Rainbow Bridge he tried but failed to get there from the west. In 1922 he finally succeeded, but not without having literally to blast a trail out of solid rock at one point, to get from one sheer-walled canyon to another:

> "The major operation was the blasting of one of the three vertical adjoining colossal rock masses and filling in the forty foot hole beyond, in order to allow safe passage to our loaded animals. Six days were consumed in making a trail over a distance of one mile, four of which were taken up in blasting a way through some 400 feet."

Bernheimer's party also had trouble getting enough water for the pack animals along his western route. They got water at one point by "digging a trench eight feet deep and ten feet long" in a wash.

In an article in the October 1923 issue of The Geographical Review, authored by U.S.G.S. geologists Miser, Trimble and Paige, the idea of access to Rainbow Bridge by water was mentioned:

> "It -- can be reached only by traversing in boat the long canyons of the swift Colorado and San Juan Rivers or by traveling with pack train the few and little-used trails. (It) may be seen from the lofty summit of Navajo Mountain, from which it appears as a tiny arch --"

The report described several routes to the bridge, including Bernheimer's western approach, and mentioned that as of October 1921, only about 250 people had visited the bridge, but cautioned that "the services of a guide for pack train parties are indispensable."

The report also noted that "three or four government parties" had reached Rainbow Bridge in 1921 by descending the Colorado River, then going up Aztec and Bridge canyons to the span, and that another group had reached the bridge the same year by coming up the Colorado from Lees Ferry "in a flat-bottomed boat equipped with a motor-driven paddle wheel at the stern," a trip of 68 miles that took five days. It was also remarked that several other attempts to reach the bridge via the river had failed "on account of inability to identify the mouth of Aztec Canyon."

In his 1925 book, "Mesa, Canyon and Pueblo," Charles Lummis indicated that Rainbow Bridge was "very difficult of access," then compounded the problem by placing it "on a tributary of the San Juan River," which was incorrect. It is in a tributary canyon of the Colorado River gorge whose mouth is almost ten river-miles below the Colorado-San Juan confluence.

Robert Frothingham was more accurate and provided a report on Rainbow Bridge visitation in his 1932 book, "Trails Through the Golden West":

> "If the place were accessible by railroad, its musical Navajo name, Non-ne-zo-she Na-gee-lid, would be on the tongue of every traveler through our Southwest. But since the only way to reach it is by pack train over a rough and arid country, it has been seen only by approximately a thousand white men since its discovery a little more than twenty years ago."

After describing his trip to the bridge, Frothingham commented that:

> "This journey that I have been describing is one of the few left in the Southwest that it will never be possible to make by railway or motor-car -- but an automobile will take you within thirteen miles of Rainbow Bridge --"

He was referring, of course, to the trading post near the base of Navajo Mountain from which most pack trips to Rainbow Bridge originated from then until the mid-1960s. In his 1933 book, "Beyond the Rainbow," Clyde Kluckhohn made some interesting observations about visitation to the bridge:

> "The National Monument Register, which is kept in an iron box under the bridge, is altogether a most interesting human document. There are columns, for date, name, address, transportation, and remarks. Under 'date' one notices the infrequency of parties in early years -- perhaps one yearly. Among the names one finds many that are well known: Theodore Roosevelt, Zane Grey, Harold Bell Wright, Robert Frothingham, Jesse Lasky, Doris Kenyon, Lucien Hubbard, Dr. G. W. Crile, John D. Rockefeller Jr. Many names, beside those of John Wetherill and his assistants, appear more than once. Zane Grey, Bernheimer, Mrs. Lillian William Robertson have each come at least four times."

Kluckhohn then went on to remark about other information in the register, including several visitor comments, some of them typical of today, more than a half-century later:

> "Very cute."
> "First girl to turn cartwheel on top."
> "Lost left rear seat of trousers sliding to top of the Bridge."

In an April 1957 National Geographic article, author Ralph Gray described "Three Roads to Rainbow," one by pack train from Navajo Mountain Trading Post, and a second from Lees Ferry via a propeller-powered boat operated by Art Green. The round trip upriver and back by air boat took three days. Gray signed the monument register as visitor number 10,741. The third "road" was a long flight from Cortez, Colorado, which provided an aerial view of Rainbow Bridge. Gray offered this comment about the flight:

> "Though quickest and cheapest, the aerial road to Rainbow lacks the intimacy of the other two methods: you see the bridge, but don't actually visit it, since there is no place to land. However, the sight is so magnificent that I'm sure this method of reaching Rainbow will become more popular."

In 1966, as Lake Powell waters first began to fill Glen Canyon and its many tributaries, Neil Judd took a third trip to Rainbow Bridge, which he described in the August 1967 issue of Arizona Highways:

> "In midsummer, 1966, I saw Rainbow Bridge for the third time. My first two visits were made on horseback. On the third we motorboated down Lake Powell from the comforts of Wahweap Lodge and National Park Service hospitality to what

is now called Forbidden Canyon, and thence up a mile-long sandy trail to our destination. During this latest, amazingly easy approach, memory carried me back to that initial journey when white men first saw Rainbow Bridge, August 14, 1909. The discovery party, of which I was only a student member, reached its goal tired and footsore; the horses and pack mules, shoes mostly worn off and some hoofs bleeding, were too weary to graze the neighboring hillside.

"Today one may float in comfort to within a short walk of Rainbow Bridge. That easy approach, in my opinion, is unfortunate. With one of the world's greatest wonders to be seen almost without effort I detected, or thought I detected, a noticeable lack of appreciation. The average visitor, if I saw correctly, appears content to scramble up the path from the boat landing, take a picture to prove he was really there, and scramble back again."

Today, of course, it is possible to drive on modern paved roads to any of four marinas on Lake Powell, then take a private or commercial boat to within sight of Rainbow Bridge. The reservoir water actually goes beneath the bridge, but boats are required to dock nearby. From the dock it is only a short walk to the magnificent bridge. The only hazard left in getting to this unique and marvelous natural span is the chance of getting lost in the labyrinthine canyons of the immense reservoir. This happens all too often to people in private boats who are not careful map readers and navigators, even though the National Park Service has marked all major side canyons, including the one that leads to Rainbow Bridge, with floating signboards.

A 1967 National Geographic article on Lake Powell mentioned that within two years of the completion of Glen Canyon Dam, more visitors had seen the bridge "than in all the previous years since its discovery in 1909." The article also noted that in 1927 John Wetherill was taken around Navajo Mountain on his first plane trip, and that he could hardly believe how small Rainbow Bridge looked from that vantage point.

Many people resented the inundation of Glen Canyon and its many tributaries, and because of this some have refused to use the lake for access to the beauty that boating has made accessible for so many. This is unfortunate. Construction of the reservoir was an ecological and archaeological crime, but it is now there, and it has made a lot of beauty accessible to a lot of people, perhaps enough to make up in some degree for the beauty and prehistoric cultural knowledge that was lost forever.

Zane Grey and Neil Judd, and probably many others too, believed that easy access reduced appreciation of Rainbow Bridge's outstanding beauty. It is more probable, however, that the "lack of appreciation" noted by Judd was simply an observation of how aesthetic appreciation varies within the more generalized cross section of humanity that now reaches the bridge.

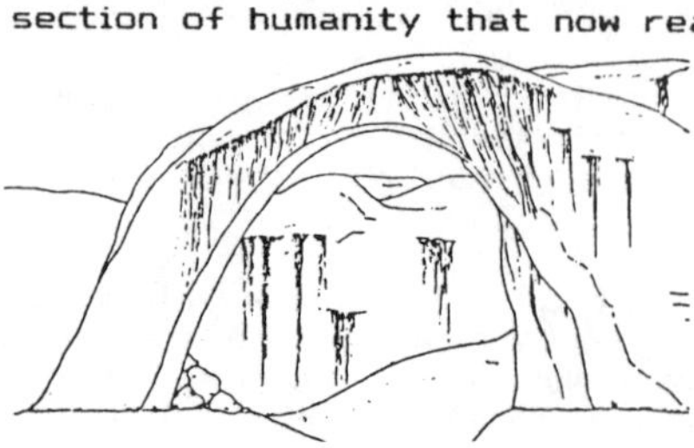

ACCESS TO KNOWN SPANS

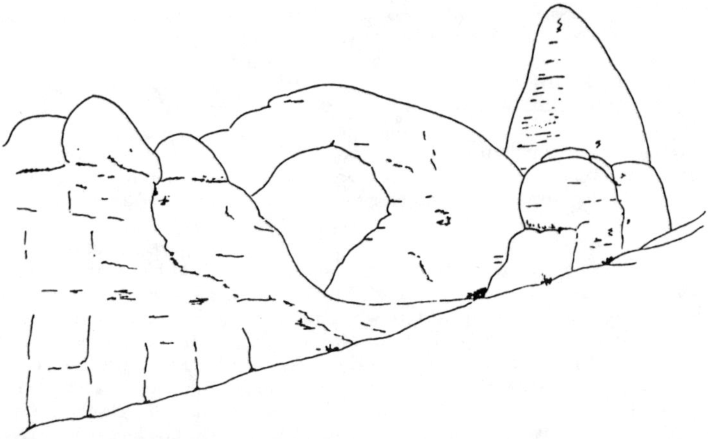

For the first five years after the April 1929 establishment of Arches National Monument the policy was evidently to study but not develop it. But in a letter accompanying his March 1934 monthly report of this continuing "scientific expedition," monument director Frank Beckwith noted a change of policy. During the month, the monument had been visited by three Park Service officials, including a Mr. George W. Norgard, who seems to have initiated the change:

> "His visit put a new line of thought into the expedition which up to that time was thinking in economical terms, with bare necessities of the first year or second year of tourist travel. Mr. Norgard saw it in terms of greater development and what was required to assist that expansion. The expedition feels much benefitted by his change of outlook on the project and now thinks in larger terms, better roads and more expanded development, to get the fullest out of the territory."

The letter ended by stating that:

> "Director Beckwith is spending his energies in a thorough study of the scenic and tourist values of the area, with a view to presentation in terms of tourist appeal."

Access to the monument was then via a 4.8 mile dirt road to Willow Springs, from the main road to the west. This rough monument access road was made impassable by even minor rain runoff. It was necessary to take horses from Willow Springs on to the various monument features. This kind of access severely limited visitation.

In an August 1947 National Geographic article, author Jack Breed told how the Willow Springs road came to be extended, almost to the Windows Section of the monument:

> "In the fall of 1936, Harry Goulding of Monument Valley, in his specially equipped car, managed to traverse the rugged sand and rock of the Arches region and thus became the first person to drive a car right into the Arches. Soon afterward a bulldozer followed Harry's tracks and made a passable trail. Little improvement was done on this rough road until recently."

ACCESS TO KNOWN SPANS

At that time, in 1947, the present main entrance road into Arches above the visitor center was a hiking route to the Courthouse Towers area, although, as Breed noted, "some work had been started during C.C.C. days to build a road by means of hairpin switchbacks up the face of the escarpment."

The 1941 WPA guide to Utah noted that the road into Arches then went to within "three miles round trip" of the Windows Section, probably from the Balanced Rock, and told of other spur roads that came within hiking distance of other monument features. It cautioned that the roads were "unimproved, but patrolled six days each week." The book also gave detailed information about access to other well-known Utah natural spans.

By 1947, another very rough dirt road penetrated Salt Valley from the north and continued down the valley to within half a mile of Turnbow Cabin, now called Wolfe Cabin, from which it was a fairly easy hike to Delicate Arch. At that time, access to the present Devils Garden Section was also via the Salt Valley road. It was necessary to hike east out of the valley to Double-O Arch, using the Dark Angel spire as a guide, then southeasterly to Landscape Arch and other features to the east of Salt Valley. Because of the bad road and difficult hiking it took, few people even tried to reach Landscape Arch, and of those who did, some became hopelessly lost. Landscape Arch can now be reached by an easy walk of one mile from the paved road that ends in the Devils Garden area of the park.

In 1947, getting to Little Rainbow Bridge (Corona Arch) required boating down the Colorado River from Moab Valley. Utah 279 now goes down the gorge, allowing easy access to this beautiful arch and the others near it, Bowtie and Gold Bar. A railroad tunnel emerges from solid rock far below Corona Arch.

In his 1947 National Geographic article, Jack Breed also described visiting Dead Horse Point, now a Utah State Park. In describing the spectacular panorama to the south of the point he remarked that:

> "Here in one magnificent vista was one of the largest areas of incompletely explored country remaining in the United States, forbidding, colorful, silent, inaccessible."

That land is still incompletely explored for its non-mineral values, even though part of it became Canyonlands National Park in 1964. At this writing, no comprehensive inventory of natural spans within the park has been made, even though there may be more sizable spans within Canyonlands than within Arches.

In his 1955 National Geographic article detailing the first successful expedition to the spans within Utah's Escalante canyon system, Robert Moore told of the difficulties encountered in getting into some of the canyons, in particular Davis Gulch. He quoted one of the guides about one place:

> "This spot is tricky. We had a horse do a two-and-a-half gainer off the cliff here, and once a pack mule landed in the cottonwood trees below. So take it easy."

On down the canyon they passed a place where a wild bull had stopped earlier explorers, then encountered so many beaver dams that travel became quite difficult. Today, one of the Escalante spans is beneath Lake Powell, but some of the others can be reached easily by boat. Spans in the upper stretches of the Escalante can still be reached only by hiking, although it is possible to view the larger ones from the air, as they were first located and reported in the mid-1950s.

Aerial view of Coyote Natural Bridge, Coyote Gulch, an Escalante Canyon tributary.

ACCESS TO KNOWN SPANS

In a May 1962 National Geographic article, Robert Moore described exploring some of the area soon to become the Needles District of Canyonlands National Park. The expedition surveyed the region from the air, explored Lavender, Salt, and Horse canyons using four wheel drive vehicles, then drove over infamous Elephant Hill to Chesler Park. They noted many of the area's natural spans, including Caterpillar, Cleft, Gothic, Castle, Fortress, Druid, Two Bobs and Ring, plus others in upper Salt Creek Canyon.

The article noted that Bates Wilson, then superintendent of Arches National Park, had first seen Druid Arch from the air, but had found a way to it on the ground only after two years of searching. That first approach was from Chesler Park, a difficult hike at that time.

Druid can still be reached from Chesler Park, via an improved foot trail, but also by way of a spectacular and fairly easy trail from the foot of Elephant Hill, starting at the end of a three mile dirt road. Access to most of the many beautiful natural spans within Canyonlands National Park still requires either hiking or the use of a four wheel drive vehicle, or both.

Despite several magazine articles about the spectacular arches of Rattlesnake Canyon near Colorado National Monument, they were largely ignored in earlier years by the BLM, the federal agency that administers the area, and received little visitation until the 1980s. Then, as required by law, a wilderness survey of the area by the BLM produced a detailed inventory of its resources, including most of the arches. From this, maps were made showing how to reach the area via four wheel drive trails. The area still receives few visitors, however, because of the nature of these trails and the strenuous hiking required to reach some of the arches.

ACCESS TO KNOWN SPANS

There are many ways to look for and reach known natural spans. Some spans are easy to find and reach. Others are more difficult. Access to a few requires considerable effort and diligence -- and an element of luck.

Almost any mode of transportation can be used to gain access to at least a sample of the many Colorado Plateau natural spans, including highway vehicle, airplane, four wheel drive vehicle, trail cycle, mountain bicycle, boat, saddle animal or feet. Combinations of these are usually required.

Quite a few natural spans are easy to reach, either by private highway vehicle or commercial tour bus. Some of these are listed below. All of the listed spans can be viewed either from a paved road, or from a viewpoint a short and easy walk from a paved road, except Rainbow Bridge. It can be reached by private boat or commercial tour boat out of Lake Powell marinas, plus an easy walk. The first four spans listed are located high in the sandstone cliffs above the roads, as these roads wind along the bottom of the spectacular Colorado River gorge in southeastern Utah.

ACCESS TO KNOWN SPANS

EASY ACCESS SPANS

NAME OF SPAN	STATE	LOCATION (highway and general location)
Little Arch	S/E UT	Utah 279, 3.5 miles from U.S.191 (1)
Gold Bar Arch	S/E UT	Utah 279, 10 miles from U.S.191
Jughandle Arch	S/E UT	Utah 279, 13.5 miles from U.S.191
Updraft Arch	S/E UT	Utah 128, 1 mile from U.S.191
Wilson Arch	S/E UT	U.S.191, 26 miles south of Moab
Looking Glass Rock	S/E UT	Dirt road, near La Sal Junction
Bridge Mountain Arch	S/W UT	Utah 9, Bridge Mountain, Zion N.P. (2)
Double Arch	S/E UT	Windows Section, Arches N.P.
North Window	S/E UT	Windows Section, Arches N.P.
South Window	S/E UT	Windows Section, Arches N.P.
Turret Arch	S/E UT	Windows Section, Arches N.P.
Broken Arch	S/E UT	Park road, Arches N.P.
Delicate Arch	S/E UT	Park road, Arches N. P. (3)
other smaller spans	S/E UT	Along park roads, Arches N.P.
Woodenshoe Arch	S/E UT	Park road, Canyonlands N.P. (4)
Mesa Arch	S/E UT	Park road, Canyonlands N. P. (5)
Washerwoman Arch	S/E UT	Park road, Canyonlands N. P. (5)
Bryce Natural Bridge	S/W UT	Rim Drive, Bryce Canyon N.P.
Wall of Windows	S/W UT	Bryce Point, Rim Drive, Bryce Canyon N.P.
Liberty Castle Window	S/W UT	Bryce Point, Rim Drive, Bryce Canyon N.P.
other unnamed spans	S/W UT	Various viewpoints, Bryce Canyon N.P.
Kachina Natural Bridge	S/E UT	Loop road, Natural Bridges N.M.
Owachomo Natural Bridge	S/E UT	Loop road, Natural Bridges N.M.
Sipapu Natural Bridge	S/E UT	Loop road, Natural Bridges N.M.
Rainbow Bridge	S/E UT	Bridge Canyon, Rainbow Bridge N.M.
Angels Window	N/W AZ	Cape Royal, North Rim, Grand Canyon N.P.
Window Rock	N/E AZ	From Navajo Route 12
Elephant Rock	S/E NV	Valley of Fire State Park
La Ventana	N/W NM	Beside NM 117, S/E of Grants
Ela Natural Bridge	S/W CO	Graded road, Glade Park

NOTES:

1. Little Arch is above Utah 279, but is best viewed from Cane Creek Road, on the other side of the river.
2. This span is visible from the area of the park visitor center, but binoculars are required for a good view.
3. Delicate Arch is only visible edgewise from the paved park road, but can be seen better from a graded dirt road that goes near the span.
4. Woodenshoe Arch is best seen from the paved loop road that goes past the park residential area, in the Needles District.
5. Mesa Arch is on the rim of the Island in the Sky. Washerwoman Arch is visible in the distance from the vicinity of Mesa Arch.

ACCESS TO KNOWN SPANS

There is a long history of natural spans being discovered from the air. Today, it is easy to see a sample of spans either by private plane or by the commercial air tours that are available from most public air fields within the Colorado Plateau. Private pilots should be aware of the special dangers in flying over the severely vertical terrain of this region, however, and should observe the special regulations that apply to flying over areas administered by the National Park Service. Commercial air tours provide a variety of scenic flights, including some oriented around natural spans.

Among the best areas for seeing natural spans from the air are Lake Powell and vicinity, which offers Rainbow Bridge and the Escalante spans, and the Arches and Canyonlands national parks vicinity. Corona Arch in the Colorado River gorge near Moab attracts many air viewers. At least one local pilot has been known to fly his plane through the span's opening, but this is a very dangerous practice even for experienced "rock pilots."

Flying through Corona Arch. Tom Kuehne photo.

Colorado Plateau backcountry dirt roads and four wheel drive trails provide access to a great many natural spans, especially if some hiking from these routes is added. Some of the literature available to aid in this form of access is noted later in this chapter. Since most of the region is either public land or Indian reservation, most of its backcountry roads and trails are accessible to privately owned vehicles, although special permits are necessary for the use of most roads and trails on Indian lands.

Commercial land tour operators also offer access to many areas of Plateau backcountry, either one-day tours or several-day campout trips. Many of these tours offer access to natural spans, such as those in Monument Valley and Arches and Canyonlands national parks. Some tour operators offer special trips for those interested in natural spans.

Arch-hunters who have their own four wheel drive vehicles or sand buggies have a special opportunity that is not available to others -- the privilege of actually driving onto or over one or more of the very few natural bridges that can actually act as bridges for four-wheeled vehicles. Of course, these and many others can be crossed by two-wheeled trail cycles. Samples of natural bridges that can be driven onto or over by four wheeled vehicles are:

Gemini Bridges -- each of these wide and massive spans can be crossed with care, and has been many times.

Uranium Arch -- small vehicles can drive out onto this span from one end, although very few have, but cannot leave by the other.

Burrito Bridge -- this strange bridge is not easy to find, but can be driven across easily.

Musselman Arch -- light-weight vehicles were driven across this natural bridge a number of times before it was included in Canyonlands National Park. It is now not permissible to drive a vehicle of any sort onto the span. It never was a safe thing to do. The span is too narrow and thin to safely support a vehicle.

Driving across Burrito Bridge

As noted in earlier chapters, many Colorado Plateau natural spans can be reached by boats of various kinds. In the vicinity of Lake Powell, Rainbow Bridge and some of the Escalante spans can be reached by either private boat or commercial tour boat, as well as a number of less well known spans adjacent to the lake. Floating through the Green River gorge, either by private raft or commercial tour, provides access to a number of natural spans, although most are high in distant canyon walls and hence can be viewed only from the boat or riverbank. They are quite difficult to reach, but can be studied through good binoculars. A few can be photographed using telephoto lenses.

Many Colorado Plateau spans can be reached by saddle trips, either on privately owned horses or on those of the commercial outfitters that operate from various locations within the Plateau. Some outfitters offer both one-day rides and several-day pack trips into the backcountry.

Walking can provide access to many natural spans, especially if added to other transport methods. The earlier chart lists spans that can be reached from good roads by casual walking. If day-hiking with a light pack is added, still more spans become accessible, and almost every known span in the Plateau is accessible to backpacking. The only exceptions are spans that are accessible only by water, such as several in the Lake Powell vicinity, and those that can be reached only by hikers with rock climbing skills and equipment. A few commercial outfitters in the region offer backpacking guide services and support.

Undine Bridges, Labyrinth Canyon

ARCH COLLECTING

Where and how should a new "arch collector" start? Perhaps the first thing to do would be to obtain and study available literature that lists natural spans and tells how to reach them. The second step would be to obtain the equipment necessary to reach a large percentage of Plateau spans. Air flights can be taken with air tour operators, boats can be rented at the various marinas at Lake Powell and river trips are best taken with experienced river-runners. Most of the region's visitor centers offer information about local commercial tours of various kinds. But for serious arch collecting, an off-road vehicle of some sort is required, plus day-hiking and backpacking equipment.

Then, since most Colorado Plateau natural spans are within Utah, and within that state some of the highest concentrations of spans are within its southeastern corner, a good place to start is from the town of Moab. This small community of several thousand is ideally located between two national parks that abound in natural spans, and is within easy driving and hiking range of hundreds more. The town also offers a variety of traveler services, including commercial tour operators of all kinds, some of whom are quite knowledgeable about the area's spans. Altogether, Moab is an excellent base from which to begin collecting known natural spans. Other Utah towns from which many spans can be collected are Blanding and Escalante.

In Arizona, the town of Page can be used as a base from which to collect spans accessible from Lake Powell, plus others in the general vicinity. Kayenta can serve as a base for gaining access to the many spans in Monument Valley, plus others in that region of the Navajo Indian Reservation. Most of the significant natural spans in Colorado can be collected using Grand Junction as a base.

It should be noted, however, that those who are prepared for lengthy camping in the Colorado Plateau's rugged backcountry can set up arch-hunting bases almost anywhere on federal or state public land, and on Indian reservations with prior permission from tribal authorities.

HELPFUL LITERATURE

Searching for known natural spans within the Colorado Plateau is not something that should be attempted casually. Knowledge of the land and its hazards is essential to safe and effective arch collecting. The region is different in many respects from other regions of the United States. It is a patchwork of land administration, for example, and each agency has its own regulations governing use of the land and access by land, water and air. These regulations can vary from time to time, but are summarized in a Wasatch Publishers guide booklet entitled "Canyon Country Exploring."

Details about access to the backcountry in some areas are provided by locally available maps and books which list and describe the roads and trails in the areas. The general Moab vicinity is well covered by the other Wasatch Publishers books and maps listed in the back of this volume.

Both the Bureau of Land Management and the National Park Service publish literature that shows the locations of some of the natural spans under their jurisdiction, and commercial publications available at various museums and visitor centers show still more. Some geology field guide books indicate natural spans on or near Colorado Plateau paved roads and backcountry tour routes, including the rivers that are commonly rafted.

The U. S. Geological Survey publishes several series of topographic maps that can be helpful in navigating Colorado Plateau backcountry, although on most such maps the backcountry road and trail information is incomplete. Maps in the U.S.G.S. "Miscellaneous Investigations Series" that show "geology-related scenic features" provide useful information about many such features in the areas covered, including many natural spans.

If they can be found, the small catalog booklets published by arch historian Vreeland can be helpful in locating some of the more prominent spans in the Colorado Plateau. The booklets are not widely available, and some are not priced for general public sale.

FOUR ARCHES CANYON

It was winter in canyon country, the winter of 1974. It was the usual mild winter, with light crusty snow lying on slopes shaded from the low-angled sun, but little elsewhere. Huge ice cascades marked seeping springs that were usually inconspicuous, and myriad smaller curtains of ice had formed from the nighttime freezing of daytime snow melt. The sun shone brightly.

As usual, when the weather permitted, we were exploring, this time in our four wheel drive Toyota Land Cruiser. We had visited our old favorites, the Gemini twin bridges. These colossal red sandstone spans were even more impressive wearing their winter-white patinas. Now we were dropping down into Crips Hole, a rather open valley within the upper Bull Canyon system. The valley was set with mushroom-shaped masses of Wingate sandstone and surrounded by high, grotesquely convoluted walls of the same colorful rock.

As we drove along the old jeep trail that crossed the meadow in the valley bottom, we could see that two deep and narrow gorges drained into its western end. The barely passable trail entered one, then soon ended in a ghost-city of silent towers and lofty, crenellated walls of ancient red-hued rock. We admired for a time this eerie scene of dark, cold shadows and brightly lit spires, then drove out and turned toward the other gorge.

This somewhat wider canyon could not have been unknown. A line of power-poles marched into it and disappeared around a curve, carrying electricity to distant Dead Horse Point State Park, but there seemed to have been few if any vehicles in the canyon since the powerline installation.

The long-unused trail was badly eroded by flash-flooding from the canyon, even though the wash was normally bone-dry. I inched the Toyota down several steep banks, over flood debris, through a couple of boulder patches, then continued up the canyon on a better stretch of trail. As I concentrated on driving, stealing glimpses as I could, my wife watched the canyon walls, exclaiming at the beauty of ice-drapery and, as usual, watching for arches. The complex Bull Canyon system is prime arch country.

Suddenly I stopped, my eye caught by a peculiarity high in the left cliff. I stepped out into the chill air and studied a huge cliff-rim alcove with a large sandstone fin in its center, edgewise to us. The fin cast a perfect shadow on the alcove wall, and that shadow had a big sunny hole in it. The fin was an arch, with an opening at least 50 feet high! There was no doubt. We could see the arching opening in the side of the fin, as well as the sunshine through the opening. The arch cast a graceful shadow on the alcove wall, which led us to name it Shadow Arch.

Excitedly, we put on jackets, gloves and hats, getting ready for an attempt at climbing to the arch. We struggled up about 150 feet of the shaded and very steep and rocky slope, with every step made hazardous by ice-hard snow. Then we reached a vertical wall that could only be ascended by an experienced climber, and that only in some other season.

Disappointed, yet elated at our discovery, we warmed up briefly in the car with cups of hot tea, then drove on up the canyon. I urged my wife to stay alert, reminding her that arches are like mice -- if you see one, there are bound to be others nearby.

We had hardly gone a hundred feet when she squealed "stop!" and pointed up to a stretch of canyon wall on the right that I couldn't see from the driver's seat. I stepped out of the Toyota for a look, and sure enough! There was another arch, a splendid one quite different in shape from the first one. The span itself resembled an elephant's trunk, or a gigantic mosquito proboscis.

That resemblance led to its name, Mosquito Arch.

A few minutes later we drove on, now with two new arches in my camera -- and perhaps with more ahead. Within half a mile we found two more -- one much like the first one and in the same canyon wall, and a smaller, tunnel-like window not in the same league as the others.

Then, as we reached the end of the trail, where the power lines climbed out of the deep gorge in graceful catenary curves, we turned around and drove back out -- only to spot a big jughandle arch in the same wall as Mosquito Arch! We had driven right by it on the way in, without seeing it.

Of course, we were not the first to see the four lovely spans in Four Arches Canyon, but we were the first to report them and publish their photographs, in a travel guide dated August 1975, making us the "discoverers of record." That spring, we drove back to the canyon. We wanted to share our four lovely arches with friends, and I wanted to climb up to lofty Mosquito Arch and photograph the scenic lower canyon through its opening.

We also tried to get to the top of Shadow Arch by driving out onto the slender peninsula between the two canyons, but failed. The peninsula was slashed by a deep cut that was impassable to either wheels or feet, leaving the arch isolated on a sheer-walled sandstone island. I hoped some day to find a way up onto that island and view Shadow Arch from above, but later explorations indicated that this will require either climbing skills that I don't yet have -- or a helicopter!

The names of the other two arches? As their "discoverers," we call the second span in the south rim of Four Arches Canyon "Crip's Arch" and the jughandle "Bullwhip Arch." The names are not, of course, "official." Four Arches Canyon is on federal land and the Board on Geographic Names has not given the names we chose its blessing. But in time, general usage will make that unimportant. At least our four arches have names. They are too beautiful in shape and setting for anonymity.

A year after we had publicized Four Arches Canyon, with photographs of all four major spans there, another arch collector placed his name in a container beneath Shadow Arch, then laid claim to its discovery. Such is the zeal developed by some who pursue this fascinating avocation.

Mosquito Arch,
Four Arches Canyon

Newberry Window

Armijo Arch

Dimmocks Jughandle

Three spans on the Macomb Expedition route. Rodney Greeno photos.

PHOTOGRAPHY

HISTORICAL PHOTOGRAPHY

When explorers first began to penetrate the Colorado Plateau, they tried to depict the natural spans they saw. At that time, shortly before the Civil War began, photography was quite primitive. It required bulky and cumbersome equipment and special skills that few explorers had. Even so, the members of some exploring parties did manage to take photographs of the natural spans they encountered, and some of these pictures were published.

The Ives Expedition of 1858 and the Macomb Expedition of 1859 were documented by John S. Newberry, who made many field sketches of the strange landforms he saw. One sketch of an arch appeared in the official Ives Expedition report. While there were no natural spans among Newberry's published sketches of the Macomb Expedition, there is a good chance that there were some among the many not published. As noted in the history chapter of this book, the Macomb Expedition went within sight of three natural spans and passed close to three others.

The 1869 Powell Expedition down the Green and Colorado rivers was also depicted by sketching, but photographers accompanied later Powell expeditions, beginning with his 1871 trip down the Green and Colorado. Jack Hillers took photographs of that trip and of other Powell expeditions in the same region. Hillers' diary of the 1871 to 1875 Powell explorations, together with a sampling of his photographs, was published under the title "Photographed All The Best Scenery" by the University of Utah Press in 1972. Photographing the Colorado Plateau was a challenge and adventure from the very first, and Hillers' diary was a compendium of all the troubles an early photographer could possibly have while trying to capture its wild and rugged scenery on photographic plates.

Frederick S. Dellenbaugh, who was also on the 1871 Powell Expedition, published a book in 1904 which contained two photographs of arches, but with no indication of when the pictures were taken. These might have been the earliest photographs taken of Colorado Plateau natural spans, but they were not the first published.

That honor probably belongs to Frederick Gardener, whose photograph of Black Rock Natural Bridge appeared in the July 24, 1885, issue of Science. Photographs of Landscape Arch and another smaller span were published in the April 22, 1898, issue of Science, although they were not taken by the article's author, Arthur Winslow, and he did not identify the photographer beyond mentioning that he lived in Moab, Utah.

The first depictions of the big White Canyon natural bridges were also sketches. These appeared in W. W. Dyar's 1904 Century Magazine account of the Long-Scorup discovery trip of 1903. Although no photographs appeared in that account, Dyar noted that the trip was made in part because Scorup, who had known of the spans since 1895, wanted "to obtain photographs of the bridges." They had problems photographing two of the three bridges, problems that modern photographers also encounter:

"Unfortunately, owing to the winding course of the canyon at this point and the consequent lack of perspective, it was impossible to obtain photographs conveying to the eye any adequate impression of the magnitude of (Kachina's) dimensions.

"Here again the curving walls of the canyon and the impossibility of bringing the whole of (Sipapu) into the narrow field of the camera, except from distant points of view, render the photographs unsatisfactory. But the lightness and grace of the arch is brought out by the partial view which Long obtained by climbing far up the canyon-wall and at some risk crawling out on an overhanging shelf."

They were able to do better with Owachomo:

"The slenderness of this aerial pathway, and the fact that the canyon here opens out into a sloping valley beyond, rendered it possible for the camera to give a proper impression of loftiness. Indeed, judging from the photographs alone, one might suppose this to be the highest of the three bridges --"

The 1905 expedition to the bridges headed by artist Culmer was all set to produce a wealth of graphics, including sketches, paintings and photographs. Culmer was to do the art work and take a few "snaps" with his personal camera, while architect S. T. Whitaker, who was with the party, was the official photographer. In addition to Culmer's camera, the expedition took with it:

"-- some of the best photographic devices ever brought to Utah, one being the telescopic or long distance camera costing $1500, a property of Mr. Whitaker, an expert in the art of handling it; also a panoramic camera with a capacity for taking three sides of a section in one great sweep."

Carrying and using this equipment became a problem almost from the beginning, as Culmer related in his diary of the trip, which was last published in the Winter 1972 edition of the Utah Historical Quarterly:

"The first adventure was -- in crossing (the storm-swollen waters of) Butler Wash. I took photos of the party coming down the trail, then stopped to renew my films. By the time I came to the wash, the others were all across, but my handsome horse, misnamed Dobbin, dashed down into the quicksand and rushing torrent and up the impossible rocks with a speed that took my breath away. I had an audience that was scarcely over the excitement of crossing and I guess they concluded that I was no tenderfoot the way Dobbin carried me through.

"The day was not without mishaps. Among other things, 2 of the 3 mules gave out right after lunch and they lay in the sand by the river as forlorn a sight as one might wish to see. But they dissimulated. As their loads were released, one of them turned loose with his business end and sent some of our food over into the Navajo Reservation. A fine shot at a can of Bents Crackers filled the air with dust and sent the larger pieces out into Monument Park 20 miles away. So they say."

Dobbin's take-charge attitude managed to make Culmer's travels and photography a continuing adventure for the entire journey, such as the episode on the return trip which he related in his diary:

> "I have not yet fallen from my horse, but it is not the fault of the rest of the crew, who love to see Dobbin prance in his high spirited way. While we were crossing (Comb Wash), one of them fired a Winchester 3 times and I thought Dobbin would jump over the Comb Reef; but we stayed together and it was George Perkins who nearly bit the sand. I had the advantage, however, of suspecting what was intended, and was watching the gun out of the corner of my eye."

An account of the first part of the expedition appeared in the May 1, 1905 edition of the Salt Lake Tribune. It told of some early troubles with all the heavy photographic gear:

> "While making the trip to (Caroline) bridge the mule carrying the cameras and other valuable property of the party got into a quick(sand) hole and sunk out of sight except for its ears and nose. One of the guides quickly cut the pack ropes and the mule, after a few frantic kicks, worked its way out and hit off for parts unknown."

An account written by Culmer for the May 14 edition of the Tribune elaborated on this episode:

> "The impatient mule that carries our most precious load of portmanteaus, containing all our films and personal effects, chooses a way that we have avoided, and after floundering in the quicksands, rolls over and sinks in a deep pool of water. The men dash in and quickly cut the pack ropes, and the mule scrambles ashore, snorting and kicking and distributing his load in every direction. With a final series of somersaults he frees himself from all entanglements and lights out down the canyon for a mile or so before he is caught.

PHOTOGRAPHY

"Meanwhile the water is quickly drained from our boxes and bags; but this does not prevent serious damage to many of our most precious possessions. We have put too many good eggs in one basket, for all our photographic plates, both exposed and unexposed, are now soaked and the sketches I have made are wringing wet. The result would have been worse had we not wrapped many packages in waterproof coverings. By great good fortune, also, we carry four of our five cameras on our backs."

Culmer's diary added more details to this:

"His was the choicest load of the lot having the big box camera and nearly all our plates, (our) two telescopes with all my films, clothing etc and finally our two grips, having all our little gimcracks. That is why our camp tonight looks like a laundry and this book is stained and generally disreputable. It was an anxious moment as we unpacked, and even now we do not know whether our films are ruined or not."

Later, after leaving the White Canyon bridges and heading down into Grand Gulch, the expedition had more problems with pack animals that affected not only photography but the whole expedition. Culmer's diary told it this way:

"This had been a day of trouble with our animals, the several days' rest having made them full of the old nick. A sorrel that had been named after a certain Senator on account of his mild and retiring disposition ran amuck among the rest and shed his pack. He caught sight of a dozing mule on the brink of the river bank and deliberately bunted him over with his pack into our swimming pool. He then proceeded to kick up the camp seriatim, and seven men for a while failed to catch him.

"The next mishap was that we heard a wild rumor that the camera mule, bringing up the rear and being led, had fallen over the precipice. An hour later he joined us, little the worse for wear, but he had a frightful slide to the brink of a cliff and would have gone over but for the two men's help.

"While we were waiting for them to come up, the sorrel again became festive. He had borne most of the food supplies and the long climbing rope 250 feet in length was coiled on his pack. With a few well arranged pitches he soon got these in a fine mixup. The air was full of flying missiles. While the coffee can was in the sky, he smashed two potatoes against it with a well directed blow; but before he was completely undressed the line became entangled in the branches of a cedar and he proceeded to do business with it. He made a ring around the tree and nearly included several horses and men in his round up. They had to step lively to prevent being bound to the tree. He himself miraculously escaped from the tangle and bounded for the same mule who was now with Whitaker, who had not dismounted, gazing into the depths of the abysses below.

"But muley was not going to be shoved again. One double slug in the breast and then a couple of rapid fire shots in the ribs turned the sorrel and probably saved Whitaker's

life. The circus continued, but finally ended happily with peace restored and we were on our way again.

"As soon as we got clear of the cedars and reached the Scorup winter range, the wind was so fierce that we looked like a party of Bedouins traversing the desert. I tried to Kodak the pack(train) but my horse was so restive that I could not even get a snap. The cook rode up to hold my bridle. He was an expert horseman, but we got badly mixed up and he was thrown, with a broken bridle and stirrup in hand. His horse careered over the plain and it took us a long time to catch him. I am not quite certain what happened to me. I managed to keep my saddle, Kodak in hand, open for use, but whether I made a snap shot or not only the developer can tell."

Culmer added a note later, saying that "by some miracle I had taken a good picture."

In an August 1967 article in Arizona Highways about his return trip to Rainbow Bridge, Neil M. Judd described the moment when the 1909 discovery expedition he accompanied first saw the magnificent span:

"I was the last to join the group -- but first to photograph Rainbow Bridge. The color photographs of today bring life and beauty to the red and brown cliffs of Bridge Canyon and to the blue sky above, but there was no color photography in 1909."

Judd used a Kodak 3-A camera to take that first photograph of Rainbow Bridge. It was reprinted in the same issue of Arizona Highways.

In his 1922 Natural History account of the 1921 Bernheimer Expedition to Rainbow Bridge, Earl Morris noted that the expedition had "taken probably the most complete (photographic) record of the Rainbow Bridge which has been made up to the present."

Delicate Arch, Arches National Park.
Negative-image prints can sometimes dramatize natural spans.

In its description of Arches National Monument, the 1941 WPA guide to Utah noted an early problem with photographing this scenic wonder:

> "Harry Reed, an eastern photographer, came to Moab in 1935 to test descriptions of the Arches scenery. He found few people who knew where the Arches were, and a smaller number who had visited them. Inquiries among cowboys disclosed that 'they wa'nt nothin' out there. Jest a lot of holes in rocks' -- this notwithstanding that one of the 'holes' was just nine feet short of a Utah city block - or a football field - in length."

As noted in earlier chapters, some of the first photographs of the big natural spans in the Escalante canyon system were taken from the air in 1955. This method continues to be the most practical for some of the large Colorado Plateau spans that are difficult to reach by other methods.

MODERN HAZARDS TO PHOTOGRAPHY

While most modern photographers will not have the kind of troubles Culmer experienced on his trip to the White Canyon bridges, the rugged nature of the Colorado Plateau still offers plenty of challenges to those who wish to photograph its many outstandingly photogenic natural spans.

Problems with access to natural spans were discussed in an earlier chapter. It should be added here that while few people now use saddle and pack animals to reach natural spans, there are hazards to photographic equipment no matter what mode of transport is used to gain access to Plateau backcountry. Cameras not properly secured can be, and have been, dropped over cliffs and out of airplane windows. Even a short fall onto the solid rock that dominates so much of the region can render most photographic gear useless. Cameras can be damaged by impact with rock as they swing by their straps on a hike or climb, and by the bouncing around they can get while being carried in off-road vehicles or packs. They can also get wet from sudden showers, sand-blasted by gusty winds, or lost overboard from boats.

Climate and season are variables that photographers encounter everywhere, but within the Plateau these variables are more extreme. Both summer and winter can produce temperatures that are hazardous to the photographer, his equipment and his film. While the climate is generally dry, sudden thunderstorms can produce violent rain, wind and even hail.

Even without rain, wind is a hazard to photography, especially during the spring and summer months, not so much in itself but because of the dust and grit it puts into the air and into cameras. Modern precision cameras are especially sensitive to this, as are cameras that use roll film that does not have paper backing. Such film is very prone to scratching from even minor particles of grit that might get into the camera during film changes. Unfortunately, this includes 35mm film, the most popular size used, and the 220 size film that many medium format cameras use. The cut film used in larger cameras is not seriously affected by this hazard.

The dark colors of most cameras can also be a problem in desert country. The bright sunlight that prevails there can be absorbed by a dark camera, making it hot enough to damage the film it contains, and perhaps even electronic components if it has any.

This brief summary of the special hazards to photography within the Colorado Plateau is not intended to discourage the pleasurable avocation of collecting the region's natural spans with a camera, but rather to inform photographers in advance so that suitable equipment can be used and adequate precautions taken. This high desert region exacts a price from all who are ignorant of, or ignore, its unique characteristics. This is especially true of those who choose to record its mysteries using the complex artform of photography.

SOME SUGGESTIONS

Following are some practical suggestions for photographing the region's natural spans, with minimum trouble and maximum results. The information outlined is gleaned from more than two decades of picture-taking in desert terrain, in particular the high-desert Colorado Plateau. They do not represent the only system for taking good pictures under such conditions, just one that has produced consistently good results over many years and under the wide variety of conditions that can occur in Plateau country.

As all photographers know, photographic gear can be infinite in variety, yet the serious arch hunter who wants to do his collecting with a camera needs certain basic equipment. This includes, besides a camera and film, a selection of lenses, accessories such as a sunshade, filters, and perhaps a tripod, and a pack in which to carry everything. Two extras that are also very useful, even though not indispensable, are binoculars and a 50 meter length of good climbing rope, plus the skills to use it.

The camera can be whatever kind the photographer likes and is accustomed to using, but only a camera with a wide range of shutter speeds and lens apertures, interchangeable lenses, and optional manual control of all variables, can handle all the conditions that will be encountered in photographing natural spans. For most photographers, a good quality 35mm camera is suitable. Professional photographers may prefer larger format cameras, but will find that the heavier weight of their equipment can be very burdensome when trying to reach spans not accessible by vehicle.

In addition to a standard lens, wide angle and telephoto lenses can be very helpful. Neither should be extreme in focal length. For a 35mm camera, a wide angle lens with a focal length of less than 28mm will produce excessively distorted images. A 135mm telephoto lens is suitable for most purposes. Larger cameras should use lenses with equivalent focal lengths.

Lens filters are also a matter of choice, and will vary with the kind of film used, but in bright desert country where brilliant sunlight and abnormal reflectivity are problems, some kind of filter is recommended. A polarizing filter can be used with both color and black and white film, with appropriate filter factors, and if skillfully used can minimize reflectivity and reveal the subtle colors that dominate desert terrain.

To protect lenses and filters from the undesirable effects produced by direct sunlight, suitable sunshades are necessary on all lenses. Filter-sunshade combinations should be tested before serious use to make certain they do not cause vignetting, or the darkening of picture corners.

Most photographers select their film on the basis of personal preference, but not all films respond the same to the intense light, subdued colors and brilliant skies found in desert country. Some films do not produce good color when used with the polarizing filters that are essential to the best results in such areas. One color positive film that has been found to be excellent under desert conditions with a polarizing filter is Ektachrome. Most black and white films will give good results, but those using 35mm cameras should use a fine grain film such as Panatomic-X in order to allow for cropping during enlargement without significant loss of detail.

Another personal matter is the use of a tripod. For practical purposes, it is better to do without one in Colorado Plateau backcountry, because tripods are heavy and awkward to carry around while hiking in rough terrain. For the most part, natural spans are going to be photographed under good lighting conditions, so normal exposures can be made with hand-held cameras. Even in the few cases where extreme depth of focus is desirable, no tripod is necessary with either 35mm or medium format cameras. A tripod is necessary for larger format cameras, and can be useful with any camera if the photographer wants to use a delayed shutter and get into the picture for scale.

For arch hunters, binoculars are almost essential. They make the search for known spans much easier, and can lead to the discovery of others. Climbing rope is needed at times for safely reaching natural spans and the best photographic angles, although agile photographers who are skilled at free-climbing may find little use for the aid of a rope. For example, the several arches in Rattlesnake Canyon, Colorado, are all within a cliff that is below the access trail. To photograph them, it is necessary to climb down the steep slickrock slope beneath one arch. More conservative photographers will find it desirable to use a safety rope for this descent, and later return. With a few spans, such as Aqueduct Arch, climbing rope and skills are essential to reaching the best camera angles.

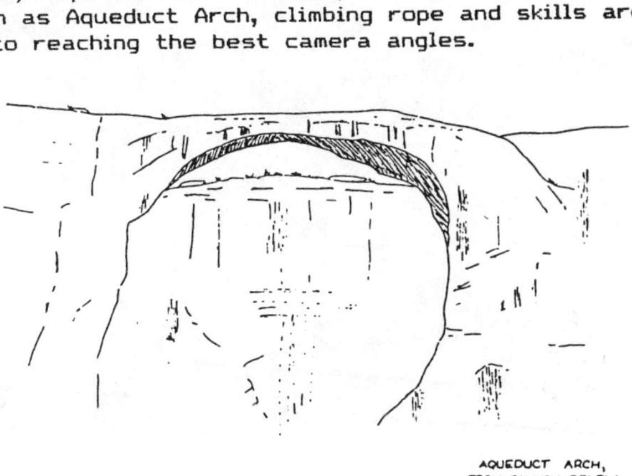

AQUEDUCT ARCH, FROM CANYON BELOW

PHOTOGRAPHY

Access to most Colorado Plateau natural spans requires special transport equipment. While some spans can be reached using highway vehicles and moderate day-hiking, most require the use of a vehicle with off-road capabilities. While some kinds of motorcycles can be used for this, they are simply not as safe to use in rugged back country, either for the driver or his photographic equipment. Enclosed high-clearance four wheel drive vehicles offer the most reliable transport and provide protection from sun, wind and dirt for cameras, if they are carried properly.

Cameras should be protected during transport over backcountry roads and trails. They are susceptible to being thrown around as a vehicle bounces over rough terrain. They are also exposed to dirt and to heat from direct sunlight. A camera in a case or pack lying in the sun in the back of an enclosed vehicle can get hot enough to ruin both film and camera, even during cooler weather.

For ideal transport in a vehicle during the warm season, a reflective aluminum camera case with padded interior is best, and even that should be carried where it will get minimal direct sunlight. For hiking, a water and dust-proof padded camera-pack

Covert Arch, Lost Spring Canyon

is essential, preferably one that can be carried so that it will not impact with rock or swing around during climbs. Arch country can be hard on both photographers and photographic equipment. Both should be given every kind of practical protection that is available.

All monochrome photographs in this book, with the exception of those contributed by other photographers, were taken with a medium format single-lens reflex camera using either standard, wide angle or telephoto lenses, Plus-X panchromatic black and white film and a polarizing filter. The color photographs on the covers of this book were taken with Ektachrome 64 film using a polarizing filter. No tripod was used to take any of the pictures. The equipment was carried either in a padded aluminum case in a vehicle, or in a padded camera-pack while hiking.

The monochrome film was developed in D-76 developer. Each negative was then proofed on 4 by 5 paper. After each selected proof was marked for its best composition, it was printed for publication on 8 by 10 paper. Both proofs and final prints were made on Polycontrast-F paper and processed in Dektol developer at standard strength. The films, printing paper and developers noted are all reliable, time-tested products of Eastman Kodak Company.

As noted earlier in this chapter, while this system of photography is only one of many, it has proved to be effective for professional use over several decades, and should produce equally satisfactory results for serious arch collectors.

PHOTOGRAPHING NATURAL SPANS

Once a photographer has reached a natural span with suitable photographic equipment, there are several variables that must be considered in taking its picture, such as distance, angles, framing, lighting and personal creativity.

Some spans can be photographed from almost any distance and angle. Others are so positioned that picture angles are restricted. A general rule to follow is "work the span," that is, take a variety of pictures. Some spans are easy to reach, so a poor shot can be retaken later, but many are so difficult to reach, or the photographer has traveled so far to reach them, that only by taking several pictures can the photographer be certain of getting the best a span has to offer at that time.

One way to do this is to take one or more pictures from distances that will show the entire setting of the span, then others from closer and even from behind the opening if this is possible. For best composition, setting shots should include the ground line below the span. Many spans are so small or so located that from behind it is not possible to depict the entire span, even with a wide angle lens. In such cases, excellent and dramatic pictures can sometimes be taken by framing the background through the span's opening, with only the underside of the span showing in silhouette.

Some natural spans may be too high in a cliff for this ideal treatment, but can still be captured with a telephoto lens. In any event, it is well to study a span before taking any pictures, so as to choose angles that clearly show its opening. Open sky visible through the opening should be used if possible, providing this does not require shooting upward at an extreme angle. Pictures taken up or down at greater angles than, say, thirty degrees will be confusing when later viewed in print or slide form.

If sky cannot be shown through a span's opening, an angle

should be chosen that contrasts the span against something behind it, such as an alcove shadow. In some cases it may be necessary to wait for a different time of day, so that changed lighting can provide contrast. If the span does not contrast with something, it may not be readily apparent in the final picture. In a few cases, it may be necessary to use the shadow of the span on brighter rock beneath it to make it visible in the final picture.

If there is a choice, a span should be photographed from various angles, not just broadside. Unusual angles can produce dramatic effects, as long as the basic elements of composition are followed, and as long as high angles are avoided. Pictures of slender, unattached arcs of stone across expanses of sky may seem to be dramatic, but usually leave viewers unsatisfied and are almost always poor composition.

Photographing a span from above is generally a last resort, because it usually entails steeply downward angles and seldom provides good contrast for the span. These shortcomings also make aerial photography of arches and bridges tricky. With most large spans, however, such problems can be overcome if the photographer is knowledgeable, communicative and quick, and the pilot is skillful, cooperative and patient.

Since there are few objects in arch country that are familiar enough to the average viewer to provide scale, and even fewer near most natural spans, whenever it is possible a photographer should include something in the picture to give scale. If access is safe, a human figure beneath or on the span can do this. If not, including a nearby tree may be better than nothing, even though the generally unfamiliar pinions and junipers common to the Colorado Plateau are not ideal. Framing the scene with a tree limb or other nearby object can give scale and a sense of depth to some pictures, but may diminish the apparent size of the span.

The lighting on a span can affect its contrast and hence visibility in the final picture, but early or late lighting can also be used to create dramatic and aesthetically pleasing pictures of natural spans. A backlit span silhouetted against a sunset or a sky full of bright clouds can make a moody picture. A side-lighted span, with its sunlit outline contrasted against a shadowed cliff, can be doubly impressive. The low-angled rays of the rising or setting sun can spotlight and lend drama to a span that may be barely visible in the brilliance of midday.

Wow Arch, Fish Creek Canyon. Harvey Halpern photo.

Even overcast days can be used creatively when photographing natural spans. Bright days produce high contrast. Cloudy days soften the lighting and permit the taking of low-key pictures that show the subtle colors of desert sandstone to better advantage, although not all color films respond well to subdued lighting. With black and white film, contrast in the final picture can be controlled during the printing.

Whenever a camera angle is used that might allow direct sunlight to hit the lens or filter, either the photographer or an assistant should shade the lens. Direct sunlight striking a lens or filter can cause hazing, spots or streaks in the picture. Although some photographers occasionally use sun streaks or spots for dramatic effect, most pictures are better without such distractions.

If the right equipment is used, protected from damage in the field, and used skillfully, then collecting natural spans with a camera can be a very rewarding avocation. There are uncounted arch hunters already, and most of them use cameras to bring home their trophies. It is hoped that the foregoing ideas will help encourage still others to pursue this creative and non-destructive use of Colorado Plateau public land.

Escalante Natural Bridge, Escalante River gorge

A LACY ARCH

We had gone looking for "five arches in one place." A cattleman friend who had known about them for years had told us where to find them, along a route he took while driving his herd to winter pasture.

So we had taken that route, one which was probably safe for vehicles only in the autumn when both water flow and the chance of flash-flooding were low. We had dropped into a wild, narrow and appallingly deep canyon that had its own flowing creek, and that miles later had joined a shallow river in a broader and more colorful canyon. We had driven down that gorge for many miles more, looking for the giant alcove with its "five arches."

We had found the arches, only to be reminded that to some Utahns the word "arch" can mean either an arch-roofed alcove or, more conventionally, an arch of rock over an opening. The site we had been seeking had, indeed, had five arches. One was a large and quite photogenic arched opening of red-hued sandstone, and another was an oddly shaped opening in a fin, with part of the arch itself resembling a hiker carrying a backpack. The other three "arches" were splendid arch-roofed alcoves that did not have openings.

Still, the search route had been rewarding in many ways. The entire length of both canyons had been spectacularly beautiful, and we had sampled a bit of history in the form of some historic structures and an old movie set of a western town near where we had finally emerged from the canyon.

We had spent the night in a motel some miles away, and were returning north via one stretch of "The Cockscomb" that stretches across the border between Utah and Arizona. As we drove along that colorful monocline, we admired the stark beauty of the weird geologic distortion with its ancient, almost vertical rock strata. The dry stream bed that paralleled the dusty road was fed by countless equally dry tributary gorges that cut sharply into the enormous sandstone reef. We remarked time and again about how we wished we had time to hike up each one of the mysterious gorges that sliced into The Cockscomb like the traces left by some gigantic knife blade.

Shortly before noon, I braked our four wheel drive vehicle to a stop and backed up a few yards for a better look at something I had spotted from the corner of my eye. Then, after yanking the vehicle into its low range, I eased off of the road down into a sandy gully, carefully following an old and very eroded trail. I then turned up a sandy wash that entered one of those slash-narrow gorges that cut into the enormous reef. I was able to drive for only a short distance before being stopped by huge boulders that blocked further wheeled access up the narrow, sheer-walled canyon.

We got out and walked ahead for a few more yards, gazing upward at the lofty sandstone walls that closed in on us and grew higher with each step. In voiceless accord, we retraced our steps, gathered up lunch makings, locked the vehicle and headed into the canyon, looking for a sunny spot. It was early November and there were still some colorful leaves on the cottonwoods, but the air was decidedly cool in the dark shadows of the narrow, echoing gorge we were penetrating.

Within a few hundred feet, the gorge angled north between the near-vertical strata. This allowed the low autumn sun to penetrate its depths, and offered us a warm slab of slickrock for a lunch table. We had just settled down to enjoy the setting when I let out a yelp and pointed toward the opposite cliff, which was densely wooded and steeply terraced. There, some hundred feet above us and partly obscured by pinyon trees, was an enormous flat-topped opening! Call it arch or bridge, we had found another span to add to our trip's collection! While my wife spread out the lunch, I excitedly hurried back to get my camera. As I rounded the bend and entered the cooler, darker depths of the gorge I glanced upward, now alert for more openings -- and immediately saw several more! There, high in a slanting, knife-thin ridge of eroded rock was a row of small windows.

In a few more yards I saw an even thinner wall of white sandstone even higher on the opposite side of the gorge, and in it was one of the most delicately lovely natural rock openings I have ever seen! The rock it was in was so white and porous it seemed like fragile heirloom lace. The sun was angled so that it caught the thin ridge edgewise, making it appear almost luminescent against the deep blue autumn sky. Lacy Arch was enchanting, a fairy span, an Indian Summer vision of beauty!

I hurried back with the camera and both wide angle and telephoto lenses. Taking pictures of this batch of spans was going to be a challenge! Before eating, I photographed the big one we had first seen, then took more shots of it a few minutes later as the lighting improved. After a hurried sandwich and gulp of tea, I started to climb the bluff behind us, seeking a still better vantage point for the big span. As I did this, my wife explored a deep and narrow cut that penetrated the cliff farther upcanyon.

Shortly, she urged me to follow her back into that cleft to its very end, to admire the eerie, echoing beauty of being a tiny entity at the bottom of a slender gorge barely wider than our reach and hundreds of feet deep, a gorge that angled ever upward and ended in a round-walled amphitheater that could have been formed by a mythical giant plunging an enormous pole vertically into hardening rock, then making a knife-cut exit from that deep hole. The setting and the mood it created were strange, primordial.

On our return trip to the vehicle, I showed my row of windows and fairy arch to my wife and, as she stood admiring them, I took their pictures, first with the wide angle lens to show their height above the canyon floor, then with my telephoto lens to capture their delicate beauty in detail. I took the setting with my wife near the base of the towering wall for scale -- she looked minute -- and then searched for angles that would lend good composition to telephoto views of the relatively small spans, without shooting upward at too steep an angle.

Lacy Arch, so high, white and bright in the sun, was indeed a challenge! It would probably take some careful darkroom work to enhance its contrast with the sky and make its fragile beauty apparent in a monochrome print, but surely the results would be worth the effort. How else could I share this gem of an arch with others? It was too far from any paved road for much public visitation, and it was too lovely to stay unknown.

Could I capture on film all that I felt that day? Probably not. Too much of what people see and feel about such an enchanting place of beauty cannot be seen by a camera or captured on film. Yet photographers must try, and the degree of their success as artisans can be measured in how nearly their pictures invoke in viewers the feelings originally experienced by the photographer. I keep trying.

Lacy Arch, top-center -- too high for photography

Dragoneye Arch, The Cockscomb

SPANS IN SPECIAL AREAS

Within the Colorado Plateau there are a number of special areas which either are devoted to natural spans or have unusual concentrations of spans, or are areas set aside for their natural values, among which are examples of natural rock openings. In the following pages, each of the areas listed below is described briefly and examples of the natural spans within it are shown. These spans depicted are merely representative. Because of space limitations in a book of this nature, no attempt has been made to be comprehensive in any area. In most of the areas listed there are many spans in addition to those depicted. There are also a great many spans throughout other areas of the Plateau, but they are generally more widely dispersed.

SPECIAL AREA	STATE
Arches National Park	Utah
Bryce Canyon National Park	Utah
Canyonlands National Park	Utah
Capitol Reef National Park	Utah
Desolation Canyon	Utah
Escalante canyons	Utah
Glen Canyon National Recreation Area	Utah/Arizona
Moab area	Utah
Monument Valley-White Mesa area	Utah/Arizona
Natural Bridges National Monument	Utah
Paria River area	Utah/Arizona
Petrified Forest National Park	Arizona
Rainbow Bridge National Monument	Utah
Rattlesnake Canyon area	Colorado
San Rafael Swell	Utah
Zion National Park	Utah

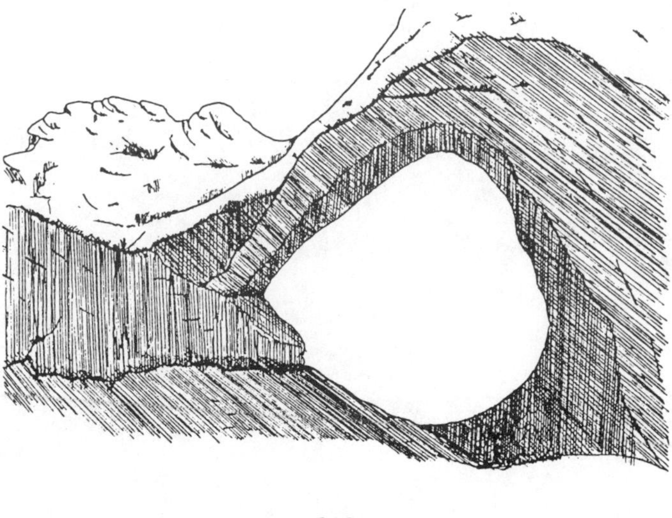

ARCHES NATIONAL PARK
Utah

Arches National Park was the third area within the Colorado Plateau to be dedicated to natural spans. It was set aside as a National Monument in 1929. After several boundary changes, it became a National Park in 1969. Despite the name of the park, some of its natural rock openings are better defined as bridges, windows and tunnels. This is partly because most names were assigned before any scientific study of rock openings had been made.

Many of the hundreds of rock openings within the park are readily accessible by highway vehicle. Others can be reached only by hiking various distances on established trails. Many are in backcountry and can be found and reached only by diligent searching and trail-less hiking. Established four wheel drive vehicle trails within the park can be used for easier access to a few spans, such as Eye-of-the-Whale and Tower Arch.

Maps of the park that show many of its natural spans are available at its entrance station and visitor center. Commercial publications available at the park and in the nearby town of Moab describe how to reach many others. Park records describe the locations of all spans cataloged within its boundaries to date.

The unusual concentration of natural rock openings found within this park is due to several factors. One is the combination of rock strata, a grouping found exposed on the surface in very few other places. These strata are Navajo Sandstone plus three distinct forms of Entrada Sandstone -- the Dewey Bridge, Slickrock and Moab members. Another major factor is that millions of years ago these strata were thrust upward by a rising ridge of ancient mineral deposits, a "salt anticline." This upwarping caused massive jointing within the rock strata above it. When exposed to surface erosion and weathering, this tectonic jointing contributed to the formation of systems of sandstone fins, such as those in the park's Fiery Furnace, Devils Garden, Eagle Park and Klondike Bluffs areas. These same processes also formed and shaped the park's many natural openings. Within the park, Salt Valley is the collapsed and eroded center of the salt anticline that played such an important part in the the creation of the strange and beautiful natural features unique to Arches National Park.

Landscape Arch

Delicate Arch setting

Klondike Arch

Dewey Window

Turnip Tunnel

Tunnel Arch

Black Arch, as seen from the hiking trail, and from behind, below left.

Black Arch

Tower Arch

Clover Canyon Bridge

Eye-of-the-Whale Arch

Wall Arch

Sand Dune Arch

Double-O Arch

Ribbon Arch

Dark Angel through
lower Double-O

Broken Arch

Tapestry Arch

Navajo Arch

Partition Arch

Ring Arch

BRYCE CANYON NATIONAL PARK
Utah

Bryce Canyon National Monument was established in 1923. Its name was changed to Utah National Park the following year, then changed again to Bryce Canyon National Park in 1928. There were several boundary changes over the next few years.

The most conspicuous features of the park are a number of immense amphitheaters below wooded rimlands. The colorful rock strata in the park, collectively called the Wasatch Formation, consist of mineral-laden sediments deposited within an ancient lake bed. These layers vary in color and hardness. When exposed to erosion and weathering they form bright-hued mazes of spires and ridges. Holes have eroded through some of the ridges, forming arches and windows. Most of these holes are not very big, but a few are quite large. None will be very long-lived, compared to natural spans in more durable rock. All are colorful and highly photogenic, especially when viewed from below.

Within the park, a paved road that roughly parallels the bluff rim offers access to a number of spectacular overlooks from which some of the park's arches and windows are visible. The miles of hiking trails that penetrate the gigantic maze of colorful towers provide closer views of the same spans and many others, as well as excellent picture angles.

Nearby Cedar Breaks National Monument is similar to Bryce except that it has a single large amphitheater. It also has similar natural openings, and a large natural bridge has formed in a tributary of Ashdown Gorge, through which the immense basin drains.

Wall of Windows from rim viewpoint

Tower Bridge, N.P.S. photo

Wall of Windows from below, N.P.S. photo.

Bryce Natural Bridge

Mossy Cave Windows, N.P.S. photo

Typical rim erosion that forms arches, with one complete.

Chinese Wall, N.P.S. photo

Liberty Castle

One of Two Bridges

CANYONLANDS NATIONAL PARK
Utah

Canyonlands National Park was established in 1964. It was enlarged to approximately one-third of a million acres in 1971. From the viewpoint of natural span formation, the park can be divided into three distinct geographic districts, the Island in the Sky, the Maze and the Needles.

The Island in the Sky is the immense, sheer-walled, multi-armed peninsula that dominates the northern part of the park, plus its White Rim Sandstone base. The Island's few natural spans are within either the Navajo, Kayenta or Wingate sandstones that form the peninsula. Two of these are Mesa Arch, which is an easy walk from a park road, and gigantic Washerwoman Arch, which is visible below the Island rim from the vicinity of Mesa Arch. The White Rim Sandstone benchland that is visible below the Island in the Sky also has very few spans. The largest is Musselman Arch, which can be reached by four wheel drive vehicle or hiking.

The dominant features within the park's Maze and Needles districts eroded from Cedar Mesa Sandstone, a red-banded beach-sand formation in which flowing and seeping water readily forms natural openings of unusual beauty. There are many natural spans within these two districts. Those in the Maze can be reached only by four wheel drive vehicle plus hiking, or by extensive backpacking. One span in the Needles district, Woodenshoe Arch, can be seen from the paved park road. Others are accessible by four wheel drive vehicle, but most require either four wheel drive vehicle plus hiking, or extensive backpacking.

Within the Needles district, there are large natural spans within Elephant, Salt Creek, Horse, Davis and Lavender canyons. Four wheel drive vehicle routes provide limited access to all of these but Elephant Canyon, which is accessible only by hiking.

Upper Salt Creek Canyon is closed to vehicles from just beyond the spur trail to Angel Arch, but several of the spans there can be viewed from Big Pocket Overlook, a lofty peninsula that juts into the southern end of the park. Access to this viewpoint requires either a four wheel drive vehicle or a high-clearance, low-geared highway vehicle plus a two-mile hike.

Park literature and several commercial publications available locally provide access details and other information about the park and its various features, including many of its natural spans.

Wedding Ring Arch

Castle Arch

Aztec Butte Arches

Washerwoman Arch, aerial view

Harvey Halpern photo

N.P.S. photo

Typical pothole arches found in the Cedar Mesa Sandstone in the Needles and Maze districts of Canyonlands National Park.

Muffin Arch, N.P.S. photo

Pachyderm Arch, Linda Kuehne photo

Woodenshoe Arch, on the skyline

Fisheye Arch

Caterpillar Arch. For scale, note figures on top of span and on the ledge, bottom-right.

Teapot Arch

Druid Arch

Tibbet Arch
Steve Manning photo

Davis Bridge

Mesa Arch

Taj Arch, Harvey Halpern photo

Mastodon Arch

Cleft Arch

Beehive Arch, N.P.S. photo

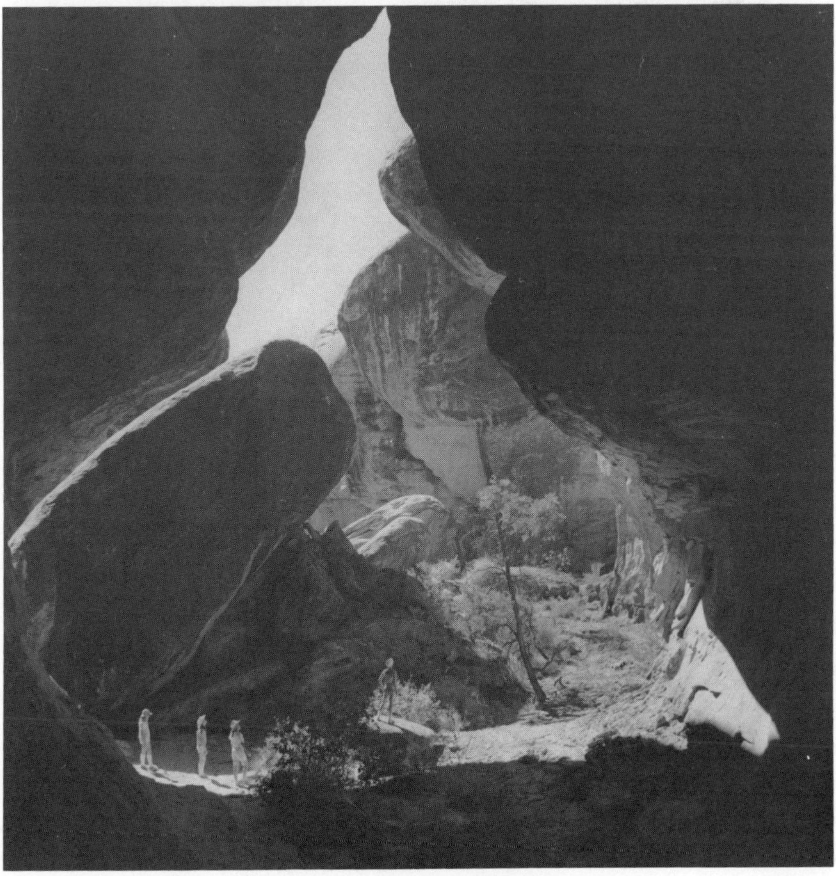

Joint Trail Tunnel

CAPITOL REEF NATIONAL PARK
Utah

Capitol Reef National Monument was established in 1937. After three boundary adjustments, the monument was given National Park status in 1971. Most of the park consists of the uptilted geologic strata of a gigantic monocline and the colorful, eroded rock on either side of this spectacular "reef." There are many natural spans within the park, most of them within the harder reef strata. Hickman Natural Bridge and a small double span can be reached by hiking from the paved road that crosses the park.

Access to the several large spans in Upper Muley Twist Canyon is by a graded dirt road, then either a four wheel drive trip up the canyon plus a hike, or a much longer hike from the road.

Park brochures and maps and commercial literature available at the park visitor center provide detailed guidance to park highlights, including most of its major natural spans.

Brimhall Double Arch
Photo courtesy Pearl Baker

Double Arch. For scale,
note figures, bottom-center.

Hickman Natural Bridge

Cassidy Arch
N.P.S. photo by Bob Reynolds

Muley Arch

Rimhandle Arch

Peek-a-Boo Arch

Nels Johnson Bridge, a double span near the trail to Hickman Natural Bridge.

DESOLATION CANYON
Utah

Between the Uinta Valley and the town of Green River, Utah, the Green River flows for many miles through an immense, wild gorge. Major John Wesley Powell, who first explored this gorge in 1869, named its upper part Desolation Canyon, and the lower part Gray Canyon. This spectacular river gorge is almost as wild now as it was then, but many people each year travel Powell's route in private or commercial rubber raft trips of three or more days' duration. Although there are other ways to penetrate the gorge in a few places, the only practical way to see its many natural spans is by river. Even then, close viewing of the arches requires extensive hiking.

Special permits are required well in advance for private trips through Desolation Canyon. Rigid craft such as canoes are not suitable for running this stretch of the Green River, but skillfully handled kayaks and other types of special boats can be used with suitable support by rubber raft. Commercial river outfitters must be licensed. Trips with them usually require making reservations several months in advance.

A graded dirt road penetrates the lower end of the gorge, but may not always be passable to highway vehicles, and does not provide access to any major natural spans. It is possible to backpack still farther up the gorge, but not all the way through it.

The Bureau of Land Management has issued a recreational map of Desolation Canyon, and several commercial books and maps available in nearby towns provide still more details, some of them on a mile-by-mile basis.

Broken Finger Arch, on the distant skyline

ESCALANTE CANYONS
Utah

As noted earlier, there are numerous outstanding natural spans within the canyon system of the Escalante River and its many tributaries. Some of these spans are within Glen Canyon National Recreation Area. Others are on adjacent public land administered by the Bureau of Land Management. Most of the canyon system's spans are within the Glen Canyon Group of rock strata, the Wingate, Kayenta and Navajo sandstones, a combination that produces natural openings almost everywhere it is exposed to significant erosion.

Access to the spans in the lower part of the canyon is by boat only. Spans above the level of Lake Powell can be reached by hiking or backpacking various distances from the lake, or from paved or graded dirt roads in the area, some of which may not always be passable to highway vehicles. Several arches, such as Escalante Natural Bridge and Phipps and Bowington arches, can be reached by fairly short hikes from a paved road, Utah 12, between the towns of Boulder and Escalante. Others such as Mano, Metate and Cedar Wash arches can be reached by hiking from the graded Hole-in-the-Rock road, which heads south from Utah near the town of Escalante.

The Bureau of Land Management has issued a hiking map of the canyon system. Regional maps published by the Utah Travel Council and distributed throughout the state show most backcountry access roads. A few commercial publications, including some referenced in this book, offer more information concerning the locations of Escalante natural spans. A U.S.G.S. map titled "Geology-Related Features in the Kaiparowits Region" shows the locations of many of the spans in the upper tributaries of the Escalante.

Broken Bow Arch, Bell Brothers photo

Metate Arch, Devils Garden Outstanding Natural Area

Gregory Natural Bridge after inundation. For scale, note boat.

Scorpion Arch, Rudi Lambrechtse photo

Cedar Wash Arch, from canyon rim and below

Aerial view of Jacob Hamblin Arch

Mano Arch, Devils Garden Outstanding Natural Area

Boyington Arch, Rudi Lambrechtse photo

Stevens Arch -- aerial and
canyon views. Inset photo
by Rudi Lambrechtse.

Double pothole arch Lamanite Bridge
Rudi Lambrechtse photos

Aerial view of Cliff Arch

Jughandle Arch
Rudi Lambrechtse photo

Coyote Natural Bridge
Steve Manning photo

Mamie Natural Bridge -- views from both directions.
Rudi Lambrechtse photos

Little Death Hollow Arch, Rudi Lambrechtse photo

Maverick Natural Bridge, Rudi Lambrechtse photo

Escalante Natural Bridge

GLEN CANYON NATIONAL RECREATION AREA
Utah and Arizona

Glen Canyon National Recreation Area was established in 1965. It encompasses Lake Powell and a considerable amount of surrounding land, although much of the lake's south shore above a specific elevation lies within the Navajo Indian Reservation. Many of the area's spans have been inundated by the lake water. Access to most of those that remain above water is by boat. Some require hiking beyond where a boat can go.

As with those in the Escalante canyon system, most of the area's spans have formed in the Glen Canyon Group of sandstones -- Wingate, Kayenta and Navajo. This combination of rock layers was named after Glen Canyon, which they dominate. Navajo Sandstone is especially prone to form natural openings, although some can also be found in the layered Kayenta and in the top of the Wingate.

There are natural spans in the vicinity of Lake Powell which can be reached by hiking, such as those in the strangely eroded highlands to the west of the Wahweap Arm of the lake, and a few that can be reached by strenuous backpacking, such as those in the broken terrain around the northern base of Navajo Mountain. Rainbow Bridge can be reached by private boat or by one-day commercial boat tours from most Lake Powell marinas.

The National Park Service, which administers Glen Canyon National Recreation Area, issues a map of Lake Powell and other interpretive literature. Various series of U.S. Geological Survey topographic maps also show the locations of some of the natural spans in the area, especially the series that depicts "geology-related scenic features." Commercial guidebooks and maps available at all marinas, and at most nearby communities, indicate where some of the area's natural spans are located. Individuals looking for these spans will need boating, navigating and map-reading skills and considerable time, because the spans still above water are widely scattered, and most require at least some hiking.

Miser Bridge, now inundated.

Mystery Bridges as they were, now inundated.

MOAB AREA
Utah

The little town of Moab, Utah, is within an area of the Colorado Plateau that has produced a great many natural spans. Many of these are outside of the area's national parks and monuments, on public land administered either by the Bureau of Land Management, the U. S. Forest Service or the State of Utah. These organizations have not made any special effort to locate and protect the many exceptional spans under their administration, although a few are within areas being considered for federal wilderness status. Several other local areas that contain outstanding spans have been dropped from wilderness consideration.

The geology of the Moab area is quite varied, and natural spans can be found in many of the formations exposed on the surface. Although the majority are found in such strata as the Cedar Mesa, Wingate, Kayenta, Navajo and Entrada, examples can also be found in the Elephant Canyon, Cutler, Moenkopi, Chinle and Cedar Mountain formations.

While there are a few paved roads in the Moab area, access to most of its backcountry is by graded dirt roads that may not always be passable to highway vehicles, or by off road vehicle trails. Many of the area's natural spans can be reached by vehicle alone. Others require additional hiking or backpacking.

Maps of the Moab area issued by the Utah Travel Council show the area's paved roads and some of its backcountry roads and trails, but are not reliable for off-pavement travel. Although the entire area is covered by U. S. Geological Survey topographic maps made to various scales, these maps are not current for backcountry roads and trails. Commercial maps based on topographic maps but with accurate backcountry details are available at various outlets in Moab, as are a variety of backcountry guide books. These can be helpful in finding and reaching many of the area's natural spans.

There are several commercial tour guides and outfitters based in Moab who offer backcountry trips of one day or longer. Some of these provide access to a sampling of natural spans, both within and outside of federal parks. A few guides are willing to arrange trips specializing in natural spans. Arch collectors who want to visit any appreciable percentage of the natural spans in the Moab area should plan to spend considerable time there.

Tukuhnikivista Arch, from both sides

Cache Arch, Ber Knight photo

Picture Frame Arch

Pritchett Arch, with historic photo showing horses on the span.

Aerial views of Corona Arch and Rainbow Bridge (inset), showing why Corona is also called "Little Rainbow Bridge."

Mosquito Arch

Crips Arch

Shadow Arch

Bullwhip Arch

The four arches in Four Arches Canyon

Otho Natural Bridge
Rodney Greeno photo

Triggerguard Arch
Ber Knight photo

Hellroaring Window

La Boca Arch

M & M Window, showing its setting and the view of Monitor and Merrimac Buttes through the window.

Updraft Arch

Fraggle Rock

Sallys Arch

Arrowhead Arch

Wilson Arch. For scale, note figure in opening.

Lin-Lynn Arch

Elevator Arch

Aviary Arch

Keen Bridge

Jughandle Arch

Squaw Window

Courthouse Arch

Lopez Arch or Baby Arch

Silhouette Window

Cummings Arch

Steer Arch

Buttress Arch

Gold Bar Arch

Bowtie Arch

Slitwindow Arch

Hellroaring Bridge

Troll Bridge

Cavern Arch

MONUMENT VALLEY - WHITE MESA AREA
Arizona and Utah

The Monument Valley-White Mesa area lies within the Navajo Indian Reservation. It is under tribal administration, and visitors there are subject to tribal law. The Navajos permit limited visitor access to Monument Valley and the White Mesa area. For details, contact tribal offices at Kayenta, the Tribal Park within the valley, or commercial tour operators in the area who are licensed to operate within the reservation.

Most of the natural spans in Monument Valley are in the De Chelly Sandstone that dominates the region in the form of soaring cliffs, spires, towers and domes. This member of the Cutler Formation is similar to the Cedar Mesa, Wingate, Navajo and Entrada sandstones in its tendency to form gracefully contoured natural openings.

Public access to Monument Valley is by way of one paved highway, plus paved spurs and several miles of graded dirt road in the vicinity of the Tribal Park. There are also many miles of four wheel drive trails in the Valley, but these are generally closed to public vehicles. Monument visitors planning off-pavement travel must first get tribal permission.

The best way to "collect" the many beautiful natural spans in the area is by taking four wheel drive tours with licensed guides. Most Monument Valley tour routes offer a variety of natural spans. Paved highways Arizona 98 and U. S. 160 go to the north and south of White Mesa, respectively, and a number of dirt roads provide closer access, but this area of the Navajo reservation should be entered only when accompanied by a licensed guide.

Literature describing Monument Valley can be found at the Tribal Park visitor center and nearby tourist facilities. An informative book on the geology, geography and history of the Valley was published by the New Mexico Geological Society following its 1973 field conference. This volume of scientific reports can still be obtained through some book stores.

White Mesa Natural Bridge, photo courtesy Bill Crawley

Margaret Arch, photo courtesy Bill Crawley

Ear-of-the-Wind Arch

Stout Arch

Big Hogan Arch

Pine Tree Arch

Eggshell Arch, photo courtesy Bill Crawley

Hope Arch, photo courtesy Bill Crawley

Honeymoon Arch

Three Level Arch

Moccasin Arch

Standing Cow Arch
Photo courtesy Bill Crawley

The Spectacles

Full Moon Arch

Swinerton Arch
Photo courtesy Bill Crawley

Needle-Eye Arch

Window Rock
Bell Brothers photo

Cly Arch

Royal Arch, photo courtesy Bill Crawley

Eye-of-the-Sun Arch

NATURAL BRIDGES NATIONAL MONUMENT
Utah

Natural Bridges National Monument was established in 1908. Its boundaries were adjusted in 1909, 1916 and 1962. It was the first federal park area in the Colorado Plateau dedicated primarily to natural spans, in this case its three magnificent natural bridges. Some of its colorful early history is described in previous chapters of this book.

While the monument's highlights are its three massive natural bridges, there are other smaller spans within its borders. All formed in the Cedar Mesa sandstone that dominates the entire region surrounding southeastern Utah's Abajo Mountains. One of the natural bridges, Sipapu, formed when the stream in White Canyon broke through one of its own meanders. Another, Kachina, formed when the stream in White Canyon broke through a wall just upstream of its original junction with Armstrong Canyon. The third bridge, Owachomo, formed when a tributary of Armstrong Canyon twice broke through into Armstrong just above their original junction.

These three major spans can be viewed from canyon-rim turnouts on a paved loop road in the monument. An easy walking trail leads to Owachomo. More difficult hiking trails descend into the canyon to Kachina and Sipapu. A one-day loop hike from the paved road down into White Canyon at Sipapu, on downcanyon to Kachina, then up Armstrong Canyon to Owachomo and out again to the paved loop road provides a variety of views of all three bridges, plus a look at other smaller spans.

Park Service and commercial maps and other literature about the monument can be obtained from visitor centers in the monument and nearby communities.

Kachina Natural Bridge. For scale, note figure under span.

Owachomo Natural Bridge

Sipapu Natural Bridge.
For scale, note figure, bottom-right.

PARIA RIVER AREA
Arizona and Utah

The Paria River area includes the river gorge itself plus a nearby monocline called The Cockscomb. The two cross a few miles north of U.S. 89, forming a gigantic letter X in the Utah-Arizona borderlands to the northwest of Page, Arizona. Most of the Paria River gorge south of U.S. 89 is a BLM Wilderness Area, and some of the river's northern stretches have been proposed for wilderness status. The Cockscomb is not protected. The unique beauty of this outstanding geologic feature on public land has already been marred by a power line corridor, and other industrial developments that will affect it are in various stages of planning.

Access to the lower Paria is only by backpacking, for which a BLM permit is required. For most of the season a contact station is manned at the head of the main canyon, just south of U.S.89. Information about entering via a hazardous tributary canyon farther west can be obtained from the same station, or from BLM offices in nearby towns. The upper stretches of Paria Canyon can presently be backpacked or driven by four wheel drive vehicles, but may be closed to vehicles if the area is given wilderness status.

Backcountry dirt roads that may not always be passable to highway vehicles closely parallel The Cockscomb for many miles. The northern leg of this road travels between U.S. 89 and the small Utah town of Cannonville. The Arizona leg of the road may not show on state or regional maps but can be found by local inquiry.

Most of the natural spans within the Paria River area are in the Navajo, Kayenta and Wingate sandstones of the Glen Canyon Group. As in the monocline reefs in Capitol Reef National Park and the San Rafael Swell, the drastically tilted rock strata in The Cockscomb have eroded in novel ways, creating some unusually interesting natural openings.

The Bureau of Land Management provides a hiker's map of the Paria Canyon Wilderness Area. U.S. Geological Survey 1:24,000 scale maps of the entire Paria-Cockscomb area are fairly current and are the only maps that provide the details needed in searching for the area's natural spans. There is no commercial literature on the area at this time, although a few magazine articles about lower Paria Canyon have been published.

Dunham Arch

Paria Arch, above the Paria River

Backpacker Arch

PETRIFIED FOREST NATIONAL PARK
Arizona

Petrified Forest National Monument was established in 1906. After five boundary adjustments, it was given National Park status in 1962.

Access to the park is by paved road. Established hiking trails and routes provide access to the park's arid backcountry. Park and commercial maps and other literature available at park visitor centers and in nearby towns provide guidance to its features.

The few natural spans within the park are unique in that they formed when petrified logs were undercut by erosion, leaving the logs bridging normally dry water courses. One of these, Agate Bridge, can be reached by an easy walk from a paved turnout. Others are accessible only by desert hiking. Arch collectors planning such backcountry hikes should register with park authorities before starting, and take note that collecting petrified wood and other specimens within the park is forbidden. This regulation is strictly enforced.

Petrified Forest National Park, typical view

Agate Bridge

Onyx Bridge -- N.P.S. photo

RAINBOW BRIDGE NATIONAL MONUMENT
Utah

Rainbow Bridge National Monument was established in 1910. It was the second federal park area within the Colorado Plateau to be dedicated to natural spans, and has remained 160 acres in size since being created.

The monument is shown on maps of Glen Canyon National Recreation Area, and described in many other federal and commercial publications. Public access is by boat. U.S. Geological Survey geologist Herbert E. Gregory wrote what was probably the first scientific description of Rainbow Bridge and how it formed in a 1917 paper titled "Geology of the Navajo Country:"

> "In the landscape of Rainbow Plateau the bridge is not a conspicuous feature, being lost in a labyrinth of profound canyons and lofty towers. From the rim of Bridge Canyon it appears as a hoop bent across an inner gorge. In near view its grandeur is realized and its structural plan revealed. The arch rests on a bench of bedded sandstone below which the inner gorge of Bridge Creek is sunk to a depth of 80 feet. Between the chord uniting the springers at the level of this supporting bench and the lower side of the keystone point the vertical distance is 187 feet, and the arch at its summit is 42 feet thick.
>
> "The Rainbow Bridge is carved from a single massive cross-bedded stratum of Navajo Sandstone; the ends of the arch rest on bedded rock. One limb of the arch rises free from a broad platform and is encircled by an abandoned meander curve; the other limb is firmly buttressed against the canyon wall. In the light of these facts, the steps in the formation of the bridge may be traced.
>
> "In its first expression the block from which the arch is carved extended from the canyon wall as a narrow meander cusp. The attack of the erosive forces was two-fold: disks and shells outlined by curved laminae of the porous Navajo sandstone were removed from both faces of the spur and at the same time ground water was making its way through joint cracks and along the surface of the underlying relatively impervious beds. The spur was thus continuously narrowed and undermined.
>
> "Whether the scaling reached a stage when the spur wall was perforated is not known; but it is believed probable that the wall was pierced by a window that stood above the newly developed subterranean channel. As time went on Bridge Creek adopted the shorter, steeper course beneath the spur, abandoning the ancient meanders. Undermining and scaling could then proceed at an increasing rate; the bridge was narrowed and increased in height and in length of span until the present magnificent arch was formed. The bridge thus owes its position to stream adjustment; its form is due to erosion of homogeneous sandstone guided by tangential cross-bedding."

Rainbow Bridge, as it was.

Aerial view of Rainbow Bridge as it is

Ground view of Rainbow Bridge as it is

RATTLESNAKE CANYON AREA
Colorado

The Rattlesnake Canyon area includes that spectacular gorge plus nearby Mee, Knowles and Sieber canyons. All of them are in the broken country to the west of Colorado National Monument, near Grand Junction, Colorado. The first three canyons are in public land administered by the Bureau of Land Management. Sieber Canyon is largely private land. Most but not all of the natural spans in the area have been inventoried by the BLM.

Access to Sieber Canyon is by an all-weather ranch road, from which two arches on private land can be viewed, but the other canyons can be reached only by high-clearance four wheel drive vehicles. Getting to and below the many natural spans then requires moderate to strenuous hiking.

Most of the spans in the Rattlesnake Canyon area formed in two of the same rock strata that are exposed in Arches National Park, the Moab and Slickrock members of Entrada Sandstone. The tectonic jointing that was critical to span formation in Arches also occurred in the Rattlesnake Canyon area, which is on the northern downwarp of the ancient Uncompahgre Uplift. The openings formed in the more erodable Slickrock member, leaving spans of hard, white Moab member sandstone. The same tectonic jointing that helped form the canyons also helped create the arches in the canyon rims.

The BLM has issued a recreation map showing how to reach Rattlesnake Canyon from the Grand Junction area. A U.S.G.S. 1:50,000 scale topographic map of Mesa County provides the best details of the area, although it does not show all existing backcountry trails. Copies of this map can be obtained from U.S. Geological Survey map outlets and the Colorado National Monument visitor center.

Cedar Tree Arch

Hole-in-the-Bridge Arch

The "rattlesnake" of
Rattlesnake Canyon

Eye Arch

Trap Arch

East Rim Arch

SAN RAFAEL SWELL
Utah

Most areas within the San Rafael Swell that contain natural spans are public land administered by the Bureau of Land Management. The Swell is a gigantic oval anticline, or uplift, some fifty miles long that has been highly eroded during the last several million years. The rock strata in the western edge of the uplift are gently inclined. Those in the eastern edge are steeply tilted, forming an immense ridge or "reef" of harder rock that lies to the west of Utah 24 for many miles south of Interstate 70.

There are natural spans in the walls of canyons carved in the center of the Swell, and others in its eastern reef. Most of these spans have formed in the same rock strata that are prone to opening formation elsewhere, such as the Navajo and Wingate sandstones. Tectonic jointing, created ages ago as the Swell uplifted, has contributed to the arch-forming process.

Interstate 70 crosses the San Rafael Swell, and a number of public gravel and graded dirt roads and four wheel drive trails give access to much of its interior, including a few of its natural spans such as Hondoo Arch. Access to other spans requires hiking or backpacking.

U.S. Geological Survey maps of the San Rafael Swell area can be obtained wherever such maps are sold. The Utah State Travel Council maps of the area can be used for travel on the main roads in the Swell, but are not reliable in its backcountry. Other than a very few magazine articles, little has been published about the San Rafael Swell or its natural spans.

Entering the San Rafael Swell from the east via Interstate 70

Solomons Arch, Ward Roylance photo

Hurst Bridge, photo courtesy Pearl Baker

Dutchmans Arch, Dana Ertel photo

Camelhead Arch, photo courtesy Pearl Baker

Eagle Canyon Arch   Wild Horse Window
Steve Manning photos

Spirit Arch
Jim Chandler photo

Hondoo Arch
Bell Brothers photo

ZION NATIONAL PARK
Utah

Mukuntuweep National Monument was established in 1909. It was renamed Zion National Monument in 1918 and given National Park status a year later. In 1937, an adjacent area was designated Zion National Monument. That area was incorporated into Zion National Park in 1956.

Paved roads enter Zion Canyon from the south and east. A paved spur road penetrates the northwest corner of the park a short distance. Access into its hinterlands is by a network of established hiking trails and routes. Park and commercial literature available at the park and nearby communities provides guidance to its features, including its natural spans.

Most of the natural spans within the park formed in Navajo Sandstone and are accessible only to hikers, but magnificent Bridge Mountain Arch can be seen in the distance from the vicinity of the park visitor center, high on the slopes of nearby Bridge Mountain. This arch is an outstanding example of a span formed largely by stress-relief jointing. Kolob Arch is the largest span in the park and, by some standards, in the entire Colorado Plateau. This magnificent arch can be reached only by hiking.

Bridge Mountain Arch, Victor Jackson photo courtesy Zion Natural History Association

Kolob Arch, Allen Hagood photo
courtesy Zion Natural History Association

Lee Pass Arch -- N.P.S. photo

FAREWELL, FAREWELL

It was an October Tuesday, a season and day when the hustle-bustle of all the usual hyperkinetic activities on Lake Powell were but a dim memory. The water had stilled from the summer's churning by restless desert winds and the hurried passage of countless rushing boats. The lake lay calm and reflective, mirroring the vertical shores of the Escalante arm of the immense reservoir, with walls darkly shadowed on one side and sun-bright on the other. A scene of high contrast, created by a desert sun lower in the sky and less fierce by far than a quarter-year earlier.

I had knifed our boat up the Escalante, slicing the aqueous mirror that doubled the scene ahead, leaving behind a bubbly path and two angled corduroy rows of smooth waves, each reflecting its own funhouse image of the scene.

Perhaps five miles up the twisting canyon I angled the boat toward the mouth of a side canyon then, as it neared, dropped the boat off plane and eased it into the canyon at an idle. It was afternoon, and the low autumn sun penetrated the canyon only when it curved in just the right direction. Otherwise the narrow gorge was dark, lit only by the light reflected from the top half of one sunlit wall, creating an autumn mood even without a trace of frost-tinged vegetation in sight.

As I eased our quiet boat up this somber, sheer-walled gorge, turning first this way, then that, the walls closed in still further. But we idled on, continually lured by the promise of another sun-bright stretch just ahead. The tops of the cliffs undulated, like the canyon had been cut from a time-frozen mass of gigantic, rolling desert dunes -- as indeed it had.

We watched ahead, studying each sheer wall as it came into view, looking for a hole, a very special hole. The year was 1968. A friend who had visited this canyon during the summer past had told of barely being able to swim under the gigantic arc of Gregory Natural Bridge, with the rising reservoir water just a foot or so from the top of the huge span's opening. We wanted to see this famous sandstone bridge just once more before it disappeared forever beneath the dark waters.

Forever? Effectively, yes, for anyone now living. Perhaps some day the immense concrete block of Glen Canyon Dam will either be removed, or the relentless Colorado River will have been allowed to eat its way around the dam's ends and slowly carry away the countless megatons of sediments that will by then have filled the reservoir to its brim.

But that will be hundreds of years from now, and by then will Herbert Gregory's magnificent natural bridge still be intact -- or will it have long since collapsed from its centuries of submersion? Standing water is not kind to sandstone. It dissolves its cementing, very slowly but very surely, and weakens its structure, quite rapidly. So we wanted to say a last goodbye to Gregory Natural Bridge.

We rounded another bend, and there it was -- a slender arc of light at the base of a high wall of dark sandstone. As we approached this arc of brightness we could see that the reservoir's water level had fallen since its summer peak. We could see sunlit slopes of slickrock beyond the opening.

But how much had the water level fallen? There was a bathtub ring of minerals just above the water all around us, but that white band could have been four feet wide, or twenty. There was nothing nearby to help gauge. The reflective water didn't help. It doubled everything we saw, causing still more optical confusion. The narrow bridge opening became a perfectly symmetrical image, a bright, narrow elipse within a dark cliff that soared above us and also plunged an equal distance into the inky depths. Would our last passage beneath Gregory also have to be made by swimming?

I stood up in our boat as it idled slowly, slowly toward the eliptical opening that I knew to be only half its apparent height. As the distance narrowed and narrowed, I still could not judge the distance between water and rock. Would I have to duck at the last instant? Would the boat get wedged in under the opening -- metaphorically caught between the rock and a wet place?

The boat drew closer still. I quickly yanked the idling motor into neutral, letting the boat's own momentum carry us on across the dark water. I was entranced by the great squinted eye we were apparently entering, by the bright world beyond the dark one which enfolded us. The last trace of sky disappeared above us and we were under the bridge, under thousands of tons of ancient rock held up only by water-soaked abutments.

I watched the opening's roof cautiously. The boat drifted slowly onward. I stood there, barely breathing, studying the age-stained underside of the immense natural bridge as it passed just inches above my head. My wife was equally awed and silent.

We reached about the midpoint of the wide span -- and I felt the hair on top of my head brush the top of Gregory's opening. I wonder if anyone else bade such an intimate, ultimate farewell to this wondrous span?

FINIS

BIBLIOGRAPHY

Baars, Donald L., "Red Rock Country," Doubleday, 1972, reprinted under title "The Colorado Plateau," University of New Mexico Press, 1983.

Bailey, Alfred M., "Desert River Through Navajo Land," National Geographic Magazine, August 1947.

Baker, Arthur A., "Geology of the Monument Valley-Navajo Mountain Region, San Juan County, Utah," U.S. Geological Survey Bulletin 865, USGPO, 1936.

Baker, Pearl, "Rim Flying Canyonlands With Jim Hurst," 1973.

Balch, F. H., "The Bridge of the Gods," McClurg & Co., 1890.

Bales, Nancy S., "Utahn Discovers New Arch in Bryce Canyon," Salt Lake Tribune, 17 January 1984.

Barnes, F. A., "Utah's Gemini Twins," Outdoor World Magazine, November/December 1970.

---- "-- A Brand New Look at Much-Photographed Angel Arch," Moab Times-Independent, 28 January 1971.

---- "-- Gemini Bridges Have a Short But Interesting History Following Discovery," Moab Times-Independent, 11 February 1971.

---- "-- Arches By Air," Moab Times-Independent, 25 February 1971.

---- "Lovely Twin Bridges Will Soon Be Lost to Sight Under Lake Powell Waters," Moab Times-Independent, 25 March 1971.

---- "-- Another View of Popular Delicate Arch --," Moab Times-Independent, 17 June 1971.

---- "Arch, Tunnel, Bridge -- Or A Hole In The Rock," Moab Times-Independent, 1 July 1971.

---- "Arches: America's Newest National Park," Moab Times-Independent, 14 October 1971.

---- "Canyon Country's Flat Topped Spans," Moab Times-Independent, 28 October 1971.

---- "Colorado Jeep Club -- Explores Back Country Near Gemini Twins," Moab Times-Independent, 2 December 1971.

---- "Beautiful Arch Near Moab Resembles Rainbow Bridge," Moab Times-Independent, 9 December 1971.

---- "Funnel Arch -- Close at Hand, But Hard to Find," Moab Times-Independent, 16 December 1971.

---- "Rainbow Bridge -- A Tough Question," Moab Times-Independent, 20 January 1972.

---- "Arch Collector Finds Bonanza in Utah," Deseret News, 26 February 1972.

---- "A Pair of Desert Castles --," Moab Times-Independent, 12 April 1972.

---- "Rainbow Bridge, An Objective Summary," High Country News, 12 May 1972.

---- "Two Arches -- Behind The Rocks," Moab Times-Independent, 8 June 1972.

---- "Desert Safari -- to Gemini Twin Bridges," Four Wheeler Magazine, July 1972.

---- "Hike To Gold Bar Arch --," Moab Times-Independent, 28 December 1972.

---- "Hunter Canyon --," Moab Times-Independent, 19 October 1972.

---- "Lavender Canyon Rich In Natural Arches," Moab Times-Independent, 7 December 1972.

---- "Some Arches Seldom Seen," Golden Circle Outdoors, February 1973.

---- "Moab's Annual Jeep Safari, Pritchett Canyon," Golden Circle Outdoors, March 1973.

---- "Under The Horse's Belly," Golden Circle Outdoors, April 1973.

---- "Springtime Opens Horse Canyon --," Moab Times-Independent, 5 April, 1973.

---- "Salt Creek Hosts More Visitors --," Moab Times-Independent, 17 May 1973.

---- "Ships of the Desert," Golden Circle Outdoors, June 1973.

---- "-- Visit Natural Bridges National Monument," Moab Times-Independent, 2 August 1973.

---- "How Many Arches --," Moab Times-Independent, 18 October 1973.

---- "Arches, Bridges and Canyonlands," Golden Circle Outdoors, October 1973.

---- "Water Under the Bridge," Landslide Magazine, Fall 1973.

---- "Behind The Rocks --," Moab Times-Independent, 15 November 1973.

---- "Magnificent Tower Hidden In Klondike Country," Moab Times-Independent, 25 April 1974.

---- "-- Arches From The Air," Moab Times-Independent, 2 May 1974.

---- "Canyonlands by Air," Colorado West Magazine, Grand Junction Daily Sentinel, 27 May 1974.

---- "Corona-Bow Tie Arches --," Moab Times-Independent, 19 September 1974.

---- "Davis Canyon's Castle in a Cave," Moab Times-Independent, 10 October 1974.

---- "Area Arches -- Biggest in the World," Moab Times-Independent, 31 October 1974.

---- "Trails Into West Arches," Off Road Magazine, November 1974.

---- "Canyon Country -- A Pictorial Guide," Canyon Country Publications, November 1974.

---- "A Natural Arch Has to Begin Somewhere," Moab Times-Independent, 12 December 1974.

---- "Hiking to Druid Arch," Canyon Country Trails & Highlights #1, Canyon Country Publications, April 1975.

---- "Arches and Bridges, a Sampler of Wild Ones," Desert Magazine, April 1975.

---- "Lavender Canyon," Colorado West Magazine, Grand Junction Daily Sentinel, 4 May 1975.

---- "Up Salt Creek Without A Paddle," Off Road Magazine, May 1975.

---- "Salt Creek Canyon," Canyon Country Trails & Highlights #2, Canyon Country Publications, June 1975.

---- "Three Bridges With Hopi Names," Colorado West Magazine, Grand Junction Daily Sentinel, 8 June 1975.

---- "Three Ways To See Utah's Natural Bridges National Monument," Desert Magazine, August 1975.

---- "Four Arches," Canyon Country Trails & Highlights #3, Canyon Country Publications, August 1975.

---- "To See An Angel," Desert Magazine, January 1976.

---- "Druid Arch," Desert Magazine, February 1976.

---- "The Walking Rocks," Desert Magazine, May 1976.

---- "Lake Powell By Air," Desert Magazine, July 1976.

---- "Utah Is Land Of Bridges," Colorado West Magazine, Grand Junction Daily Sentinel, 25 July 1976.

---- "Giant Rainbow Bridge Has Its Own Look-Alike Near Moab," Colorado West Magazine, Grand Junction Daily Sentinel, 3 October 1976.

---- "Can Rainbow Bridge Survive?", Utah Outdoors Magazine, Autumn 1976.
---- "Canyon Country Exploring," Wasatch Publishers, 1977.
---- "Canyon Country Scenic Roads," Wasatch Publishers, 1977.
---- "Canyon Country Hiking and Natural History," Wasatch Publishers, 1977.
---- "Canyon Country Off-Road Vehicle Trails, Island Area," Wasatch Publishers, 1978.
---- "Canyon Country Off-Road Vehicle Trails, Canyon Rims and Needles Areas," Wasatch Publishers, 1978.
---- "Canyon Country Off-Road Vehicle Trails, Arches and La Sals Areas," Wasatch Publishers, 1978.
---- "Canyon Country Geology for the Layman and Rockhound," Wasatch Publishers, 1978.
---- "San Rafael Swell," Off Road Magazine, October 1978.
---- "Four Wheeling Wyoming's Wind River Range," Off Road Magazine, November 1979.
---- "The Angel No One Sees," National Parks Magazine, September/October 1982.
---- "Utah Canyon Country," Utah Geographic Series, 1986.
Beck & Haase, "Historical Atlas of New Mexico," University of Oklahoma Press, 1969.
Beckwith, Frank, "Report of Arches National Monument Scientific Expedition," Arches National Monument Monthly Report, March 1934.
Bernheimer, Charles L., "From Kayenta to Rainbow Bridge," Natural History Magazine, November/December 1920.
---- "Encircling Navajo Mountain With a Pack Train," National Geographic Magazine, February 1923.
---- "Rainbow Bridge," Doubleday, Doran & Co., 1929.
---- Field Notes, Bernheimer Expeditions of 1922, 1923, 1924, 1926, 1927, 1929 and 1930, unpublished.
Bickers, Jack, "Guide To Arth's Rim's Canyons and Arches," November 1983.
Blair, et al, "Origin and Classification of Natural Arches in Southern Utah," Four Corners Geological Society Eighth Field Conference, 1975.
Blake, Reed, "Utah's Best Kept Secret," Mountain West Magazine, May 1977.
---- "Taking the Measure," Deseret News Magazine, 2 October 1983.
---- "Measuring the Span of the Great Arch at Zion National Park, Kolob Section," Brigham Young University, 1984.
Bradley, William C., "Large-Scale Exfoliation in Massive Sandstones of the Colorado Plateau," Bulletin of the Geological Society of America, May 1963.
Breed, Jack, "Utah's Arches of Stone," National Geographic Magazine, August 1947.
---- "First Motor Sortie Into Escalante Land," National Geographic Magazine, September 1949.
---- "Roaming the West's Fantastic Four Corners," National Geographic Magazine, June 1952.
Breed, William J., Curator of Geology, Museum of Northern Arizona, personal letter, 9 July 1973, subject: danger to Rainbow Bridge.
Bryan, J. Y., "Rainbow Bridge: 1973," National Parks & Conservation Magazine, November 1973.
Carter & Sargent, "Geology-Related Scenic Features in the Kaiparowits Plateau Area, Utah," U. S. Geological Survey, 1983.
Cazier, Lola, "Surveys and Surveyors of the Public Domain, 1785-1975," undated.
Cleland, Herdman F., "North American Natural Bridges, With a Discussion of Their Origin," Bulletin of the Geological Society of America, 2 July 1910.
Crampton, C. Gregory, "Standing Up Country," Knopf, 1965, reprinted Peregrine Smith, 1983.
---- "Land of Living Rock," Knopf, 1972.
Crispin, Douglas, "Notes and Comments Regarding Natural Bridges National Monument," personal communication, 1 November 1983.
Culmer, H. L. A., "Country of Natural Bridges," The Technical World Magazine, September 1908.
Cummings, Byron, "The Great Natural Bridges of Utah," National Geographic Magazine, February 1910.
---- "The Great Natural Bridges of Utah," Bulletin of the University of Utah, November 1910.
Dansie, L., "Arches and Bridges of the Maze District, Canyonlands National Park," unpublished, 1983.
Dellenbaugh, F. S., "The Causes of Natural Arches," Science Magazine, 20 May 1898.
---- "The Romance of the Colorado River," Putnam's Sons, 1902, reprinted by Rio Grande Press, 1965.
---- "A Canyon Voyage," Yale University Press, 1926.
Doelling, Hellmut H., "Geology of Arches National Park," Utah Geological and Mineral Survey, 1985.
Doolittle, Jerome, "Canyons and Mesas," Time-Life Books, 1974.
Douglass, William B., "Field Notes of Survey of Reservation Embracing Natural Bridges National Monument," Department of the Interior, General Land Office, September 12 - October 3, 1908.
---- "The Discovery of Rainbow Natural Bridge," Our Public Lands, U.S. Department of the Interior quarterly, April/May/June 1955.
Dyar, W. W., "The Colossal Bridges of Utah," Century Magazine, Volume 68, 1904.
Edwards, Walter M., "Lake Powell, Waterway to Desert Wonders," National Geographic Magazine, July 1967.
Findley & Edwards, "Realm of Rock and the Far Horizon -- Canyonlands," National Geographic Magazine, July 1971.
Frazier, Deborah, "Arch Discoverer Roams Utah Park," Rocky Mountain News, 27 November 1983.
Frothingham, Robert, "Trails Through The Golden West," Robert M. McBride & Co., 1932.
Gardiner, Frederick, "An Arizona Natural Bridge," Science Magazine, 24 July 1885.
Gardner, Thomas W., "The History of Part of the Colorado River and its Tributaries: An Experimental Study," Four Corners Geological Society Eighth Field Conference, 1975.
Gillmor, F. and Wetherill, Louisa W., "Traders to the Navajos," author's copyright 1934, reprint, University of New Mexico Press, 1952.
Gilluly, Waters & Woodford, "Principles of Geology," Freeman & Co., 1975.
Gregory, Herbert E., "Geology of the Navajo Country," U. S. Geological Survey Professional Paper 93, USGPO, 1917.
Gregory, Herbert E. and Moore, Raymond C., "The Kaiparowits Region," U. S. Geological Survey Professional Paper 164, USGPO, 1931.

BIBLIOGRAPHY

Gregory, Herbert E., "The San Juan County: A Geographic and Geologic Reconnaisance," U. S. Geological Survey Professional Paper 188, USGPO, 1938.

Gregory, Herbert E. and Anderson, J. C., "Geographic and Geologic Sketch of the Capitol Reef Region, Utah," Bulletin of the Geological Society of America, 1 December 1939.

Gregory, Herbert E., "Geology and Geography of the Zion Park Region, Utah and Arizona," U. S. Geological Survey Paper 220, USGPO, 1950.

---- "The Geology and Geography of the Paunsaugunt Region, Utah," U.S. Geological Survey Professional Paper 226, USGPO, 1951.

Gray, Ralph, "Three Roads to Rainbow," National Geographic Magazine, April 1957.

Grey, Zane, "Tales of Lonely Trails," Harper & Brothers, 1922.

Halliday, William R., & Woodbury, Angus M., "Protection of Rainbow Bridge National Monument," Science Magazine, 19 May 1961.

Hansen, James, "The River Vanishes," Science 84 Magazine, January/February 1984.

Hansen, Wallace R., "A Geologic Examination of Rainbow Bridge National Monument," U. S. Geological Survey Administration Report, 1959.

---- "Discussion: Water Under the Bridge," response to Barnes, 1973, "Water Under the Bridge," Landslide Magazine, unpublished.

Harris, Stephen L., "American Indian Legends: The Bridge of the Gods," American West, July/August 1985.

Heald, Weldon, "Who Discovered Rainbow Bridge?", Sierra Club Bulletin, October 1955.

Henderson, Junius, "Caverns, Ice Caves, Sinkholes and Natural Bridges," University of Colorado Studies, March 1933.

Hillers, Jack, "Photographed All The Best Scenery," University of Utah Press, 1972.

Hintze, Lehi F., "Geologic History of Utah," Brigham Young University Geologic Studies, Volume 20, Part 3, July 1973.

Hoffman, John F., "Arches National Park," Western Recreation Publications, 1981.

Holbrook, Stewart H., "The Columbia," Rinehart & Co., 1956.

Holmes, E. F., "The Great Natural Bridges of Utah," National Geographic Magazine, March 1907.

Hunt, Charles B., "Geology and Geography of the Henry Mountains Region, Utah," U. S. Geological Survey Professional Paper 228, USGPO, 1953.

---- Johns Hopkins University, personal letter, 19 June 1973, subject: danger to Rainbow Bridge.

Huser, Verne, "Rainbow in the Rock," American Forests Magazine, February 1974.

Ives, Joseph C., "Report Upon the Colorado River of the West," USGPO, 1861, reprinted by Da Capo Press, 1969.

Janssen, Raymond E., "Nature's Bridges," The Scientific Monthly, September 1943.

Jett, Stephen C., "Testimony of the Sacredness of Rainbow Natural Bridge to Puebloans, Navajos and Paiutes," Plateau Magazine, Spring 1973.

---- University of California at Davis, personal letter, 8 June 1973, subject: danger to Rainbow Bridge.

Judd, Neil M., "Discovery of Rainbow Bridge," National Park Service Report, 30 July 1919.

---- "Beyond The Clay Hills," National Geographic Magazine, March 1924.

---- "Rainbow Trail to Nonnezoshe," National Parks & Conservation Magazine, November 1927, reprinted November 1973.

---- "Return to Rainbow Bridge," Arizona Highways Magazine, August 1967.

Keele, Marsha, "Paiute Indian Honored for Guiding White Men to the Rainbow Bridge," Deseret News, 5 July 1984.

Keller, Edward A., "Environmental Geology," Merrill Publishing, 1979.

Kelly, Charles, and Martin, Charlotte, "Zeke Johnson's Natural Bridges," Desert Magazine, November 1947.

Kelsey, Michael R., "Canyon Hiking Guide to the Colorado Plateau," Kelsey Publishing Company, 1986.

---- "Hiking Utah's San Rafael Swell," Kelsey Publishing Company, 1986.

Klinck, Richard E., "Land of Room Enough and Time Enough," University of New Mexico Press, 1953.

Kluckhohn, Clyde M., "To The Foot of the Rainbow," Century Co., 1927, reprinted Rio Grande Press, 1980.

---- "Beyond the Rainbow," Christopher Publishing, 1933.

Lakes, Arthur, "The Natural Bridges of Utah," The Mining World Magazine, 11 May 1907.

Lambrechtse, Rudi, "Hiking the Escalante," Wasatch Publishers, 1985.

Larson, E. E., et al, "Development of the Colorado River System in Northwestern Colorado During the Late Cenozoic," Four Corners Geological Society Eighth Field Conference, 1975.

Lohman, Stanley W., "The Geologic Story of Canyonlands National Park," U. S. Geological Survey Bulletin 1327, USGPO, 1974.

---- "The Geologic Story of Arches National Park," U. S. Geological Survey Bulletin 1393, USGPO, 1975.

Luce, Willard, "Arches and Bridges of Stone," Natural History Magazine, July 1964.

Luckert, Karl W., "Navajo Mountain and Rainbow Bridge Religion," Museum of Northern Arizona, 1977.

Lummis, Charles F., "Mesa, Canyon and Pueblo," The Century Co., 1925.

Lyman, William Denison, "The Columbia River," Putnam's Sons, 1909.

Mather, Stephen T., (Arches National Monument), New York Times Magazine, 9 May 1926.

McCarrick, Edward, "Status of Inventorying and Documenting Arches as of 12/1/83," National Park Service Memorandum, Arches National Park, 21 March 1984.

Miller, Joseph, "Tonto Natural Bridge," Arizona Highways Magazine, September 1946.

Minor, Will, "Rattlesnake Canyon," Colorado West Magazine, 27 February 1972.

Miser, Trimble & Paige, "The Rainbow Bridge, Utah," The Geographical Review Magazine, October 1923.

Miser, Hugh D., "The San Juan Canyon," U. S. Geological Survey Water-Supply Paper 538, USGPO, 1924.

Moore, Raymond C., "Origin of Enclosed Meanders On Streams of the Colorado Plateau," The Journal of Geology, January/February 1926.

---- "Significance of Enclosed Meanders in the Physiographic History of the Colorado Plateau Country," The Journal of Geology, February/March 1926.

Moore, W. Robert, "Escalante: Utah's River of Arches," National Geographic Magazine, September 1955.

---- "Cities of Stone in Utah's Canyonlands," National Geographic Magazine, May 1962.

Morris, Earl H., "An Unexplored Area of the Southwest," Natural History Magazine, November/December 1922.

Mutschler, Felix E., "Desolation and Gray Canyons," River Runners' Guide to the Canyons of the Green and Colorado Rivers, Volume IV, Powell Society, undated.

Newell, Maxine, "The Old Spanish Trail," National Park Service, February 1984.

Pack, Frederick J., "Natural Bridging in High Plateaus of Utah," Geological Publishing Co., 1922.

Patton, Alexander & Kramer, "Physical Geography," Wadsworth Publishing, 1974.

Pearl, Richard M., "Stone Rainbows," Part I, Earth Science Magazine, July/August 1976.

---- "Stone Rainbows," Part II, Earth Science Magazine, January/February 1977.

---- "Stone Rainbows," Part III, Earth Science Magazine, May/June 1977.

Perkins, Nielson & Jones, "Saga of San Juan," San Juan County Daughters of Utah Pioneers, 1957.

Pogue, Joseph E., "The Great Rainbow Natural Bridge of Southern Utah," National Geographic Magazine, November 1911.

Powell, John W., "The Exploration of the Colorado River and its Canyons," Flood & Vincent, 1895, reprinted by Dover Publishing, 1961.

Price, Dona H., "Escalante Man Finds Natural Arch," Salt Lake Tribune, 7 May 1977.

Rahm, David A., "Reading The Rocks," Sierra Club, 1974.

Raymo, Chet, "The Crust of Our Earth," Prentice-Hall, 1983.

Redfern, Ron, "Corridors of Time," Orbis, 1980.

---- "The Making of a Continent," Times Books, 1983.

Reiger, George, "The Trammeling of Rainbow Bridge," Audubon Magazine, November 1977.

Rigby, Keith J., "Northern Colorado Plateau," K/H Geology Field Guide Series, Kendall/Hunt Publishing, 1976.

---- "Southern Colorado Plateau," K/H Geology Field Guide Series, Kendall/Hunt Publishing, 1977.

Robinson, Edwin S., "Mechanical Disintegration of the Navajo Sandstone in Zion Canyon, Utah " Bulletin of the Geological Society of America, September 1970.

Rogers, J. David, "The Development of Natural Rock Arches at Arches National Park, Utah," Second Conference on Scientific Research in the National Parks, November 1979.

Roosevelt, Theodore, "Across The Navajo Desert," The Outlook Magazine, 11 October 1913.

Roylance, Ward J., "Seeing Capitol Reef National Park," Wasatch Publishers, 1979.

---- "Utah, A Guide to the State," Utah Arts Council, 1982.

Ruth, Kent, "Colorado Vacations," Knopf, 1959.

Scher, Zeke, "The Man Who Discovered Rainbow Bridge," Empire Magazine, Denver Post, 9 December 1973.

Schmidt, Henry G., "Monthly Report," Arches National Monument, March 1941.

---- "The Arches of Rattlesnake Canyon," Our Public Lands, U. S. Department of the Interior quarterly, Winter 1981.

Schneeberger, Jon, "Escalante Canyon -- Wilderness at the Crossroads," National Geographic Magazine, August 1972.

See, Dorothy, "Kolob Arch Confirmed As World's Largest," Deseret News, 10 August 1983.

Shimer, John A., "Field Guide to Landforms in the United States," Macmillan, 1972.

Shoemaker, Eugene M., California Institute of Technology, personal letter, 14 July 1973, subject: danger to Rainbow Bridge.

Steen, Charlie R., "The Natural Bridges of White Canyon: A Diary of H. L. A. Culmer, 1905," Utah Historical Quarterly, Winter 1972.

Stevens, Dale J., "A Classification of Natural Sandstone Rock Openings in Southeastern Utah," Great Plains-Rocky Mountain Geographical Journal, Volume III, 1974.

---- "The Dimensions of Kolob Arch in Zion National Park, Utah," Geography Department, Brigham Young University, July 1984.

---- "The Dimensions of Landscape Arch in Arches National Park, Utah," Geography Department, Brigham Young University, January 1985.

Stokes, W. Lee, "Scenes of the Plateau Lands," Publishers Press, 1969.

Stone, Julius S., "Canyon Country," Putnam's Sons, 1932.

Sumner, David, "Lost: 86,000,000 Acres of Wilderness," Mariah Magazine, Fall 1976.

Thaden, Trites & Finnell, "Geology and Ore Deposits of the White Canyon Area, San Juan and Garfield Counties, Utah," U. S. Geological Survey Bulletin 1125, USGPO, 1964.

Threet, Richard L., "The Wind At Work," Natural History Magazine, May 1959.

Treadway, Douglas, "Glade Park Area Arches," personal communication and map, BLM District Office, Grand Junction, Colorado, March 1980.

Vokes, H. E., "Rainbows of Rock," Natural History Magazine, October 1942.

Vreeland, Robert H., "Nature's Bridges and Arches," self-published, Volume 1, 1976; Volume 2, 1977; Volume 5, 1976; Volume 6, 1977; Volume 7, 1976; Volume 8, 1986, Volume 9, 1982; Volume 10, 1979; Volume 11, 1979; Volume 12, 1985, Volume 18, 1982.

Walker & Bufkin, "Historical Atlas of Arizona," University of Oklahoma Press, 1979.

Weisbecker, Leo W., "Snowpack, Cloud-Seeding and the Colorado River," University of Oklahoma Press, 1974.

Wetherill, Mrs. John and Gillmor, Francis, "Traders to the Navajos," Houghton Mifflin Co., 1934.

Wilson, Richard, "The Arch Catalog," compiled 1968-1971, unpublished.

---- "Still-Unknown Arches in River Canyon Described by Powell Party in 1869," Moab Times-Independent, 10 July 1969.

---- "Hall's Bridge," letter to the editor, Moab Times-Independent, 9 September 1971.

Winchell, Newton H., "The Colossal Bridges of Utah, Editorial Comment," The American Geologist Magazine, September 1904.

Winslow, Arthur, "A Natural Bridge in Utah," Science Magazine, 22 April 1898.

BIBLIOGRAPHY

Winstead, Manya, "The Tonto Natural Bridge," Arizona Highways Magazine, August 1980.

Winther, Oscar O., "The Great Northwest," Knopf, revised edition, 1950.

Woodbury, Angus M., "Protecting Rainbow Bridge," Science Magazine, 26 August 1960.

Yost, Billie Williams, "Bread Upon the Sands," Caxton Printers, 1958.

NO AUTHOR GIVEN, LISTED BY DATE

"Colossal Natural Bridges of Utah," National Geographic Magazine, September 1904.

"A Notable Expedition," Deseret News, 1 April 1905.

"Latest News From Utah Explorers," Deseret News, 22 April 1905.

"No Such Thing As Cliff Dwellers," Deseret News, 1 May 1905.

"In Land of the Cliff Dweller," Salt Lake Tribune, 1 May 1905.

"In Wilds of Utah -- H. L. A. Culmer's Story of His Trip to Unexplored Region," Salt Lake Tribune, 14 May 1905.

"Hall Bridge," photograph in Moab Times-Independent, 17 November 1911.

"Wonderlands of Utah," Cedar City Commercial Club, ca. 1920.

"Administrative History, Establishment of Arches National Park," File H1417, letters and reports authored by Mather, Wadleigh, McKinley, Ringhoffer, Oestler, Ryan and the Department of the Interior General Land Office, dated 1923 to 1929.

"Wonderful Scenic Attractions at 'The Windows' to be Set Aside as Monument," Moab Times-Independent, 17 July 1924.

"A Proclamation Establishing Arches National Monument, Utah," United States Statutes at Large, Volume 46, Part 2, 71st Congress, 1929-1931.

"Utah, A Guide to the State," a WPA project, Hastings House, 1941, completely rewritten by Roylance, Ward J., and republished by Utah Arts Council, 1982.

"Arizona, The Grand Canyon State," Hastings House, 1940, 1956.

"The Discovery of Rainbow Bridge," Cummings Publication Council, August 1959.

"Natural Bridges," Encyclopedia of Geomorphology, Reinhold, 1968.

"National Parks and Landmarks, Areas Administered by the National Park Service and Related Properties as of January 1, 1972," USGPO, 1972.

"Sonic Booms Here a Thing of the Past," Moab Times-Independent, 20 April 1972.

"Rainbow Bridge Decision Hailed," High Country News, 30 March 1973.

"New Prediction of Danger to Rainbow Bridge," Not Man Apart, July 1973.

"Fallen Arch Examined --," Moab Times-Independent, 10 July 1975.

"Canyonlands Country," Fassett, James E., editor, A Guidebook of the Four Corners Geological Society Eighth Field Conference, September 1975.

"Dictionary of Geological Terms," Anchor Press/Doubleday, 1976.

"Natural Bridge," Illustrated Encyclopedia of the Earth Sciences, Abrams, Inc., 1976.

"Encyclopedia of the Geological Sciences," McGraw-Hill, 1977.

"Hiking the Escalante River," Bureau of Land Management map, January 1979.

"New Data 'Shrink' Rainbow Bridge," Salt Lake Tribune, 24 February 1979.

"Arch Is All But Unknown In Area," Salt Lake Tribune, 20 May 1979.

"River Management Plan, Desolation and Gray Canyons of the Green River," Bureau of Land Management, Moab District, June 1979.

"Maps to Rattlesnake Canyon (Colorado)," Bureau of Land Management District Office, Grand Junction, Colorado, ca. 1979.

"Decisions on Geographic Names in the United States, January through March 1983," U. S. Board on Geographic Names, U. S. Department of the Interior, 1983.

"Natural Bridge Discovered," Moab Times-Independent, 19 January 1984.

"Morning Report," National Park Service, Canyonlands Complex, 10 February 1984.

"Arches National Park," National Park Service brochure, USGPO, 1984.

"How To Accession A Natural Arch or Natural Bridge," National Park Service, Arches National Park, undated, unpublished.

"Legislative History of Natural Bridges National Monument," National Park Service, undated.

"Legislative History of Arches National Park," National Park Service, undated.

INDEX OF SPANS IN THIS BOOK

Agate Bridge, Petrified Forest National Park ---------- 389
Angel Arch, Canyonlands National Park -------- 10, 178, 197
Angels Window, Grand Canyon National Park ------------- 148
Aqueduct Arch, Harts Draw, Moab area ------------------ 172
Arch 117, New Mexico ---------------------------------- 126
Armijo Arch, Harts Draw, Moab area -------------------- 296
Arrowhead Arch, Squaw Park, Moab area ----------------- 366
Aviary Arch, Sevenmile Canyon, Moab area -------------- 368
Aztec Butte Arches, Canyonlands National Park---------- 332

Backpacker Arch, Paria River gorge -------------------- 387
Balcony Arch, Behind the Rocks, Moab area ------------- 273
Bartlett Bridge, Bartlett Wash, Moab area ------------- 195
Beehive Arch, Canyonlands National Park---------------- 341
Big Hogan Arch, Monument Valley area ------------------ 376
Black Arch, Arches National Park ---------------------- 317
Boulder Bridge, Bull Canyon, Moab area ---------------- 91
Bowtie Arch, Colorado River gorge, Moab area ---------- 372
Boyington Arch, Escalante canyons --------------------- 352
Bridge Mountain Arch, Zion National Park -------------- 402
Brimhall Double Arch, Capitol Reef National Park ------ 342
Broken Arch, Arches National Park ---------------- 118, 320
Broken Bow Arch, Escalante canyons ---------------- 58, 348
Broken Finger Arch, Desolation Canyon ----------------- 347
Bryce Natural Bridge, Bryce Canyon National Park ------ 326
Bullwhip Arch, Bull Canyon, Moab area ------------ 192, 363
Burrito Bridge, Blue Hills, Moab area ----------------- 291
Buttress Arch, Behind the Rocks ----------------------- 371

Cache Arch, Cache Valley, Moab area ------------------- 361
Camelhead Arch, San Rafael Swell ---------------------- 400
Cassidy Arch, Capitol Reef National Park -------------- 345
Castle Arch, Canyonlands National Park ---------------- 332
Caterpillar Arch, Canyonlands National Park ----------- 336
Cathedral Arch, Arch Canyon --------------------------- 223
Cavern Arch, Lost Spring Canyon, Moab area ------------ 373
Cedar Tree Arch, Rattlesnake Canyon -------------- 149, 394
Cedar Wash Arch, Escalante canyons -------------------- 351
Chinese Wall, Bryce Canyon National Park -------------- 327
Clara S. Bernheimer N. B., Monument Valley area ------- 45
Cleft Arch, Canyonlands National Park ------------ 148, 340
Cliff Arch, Escalante canyons ------------------------- 354
Cliffhanger Bridge, Labyrinth Canyon, Moab area ------- 79
Clover Canyon Bridge, Arches National Park ------------ 318
Cly Arch, Monument Valley area ------------------------ 380
Corona Arch, Colorado River gorge, Moab area 227, 290, 362
Courthouse Arch, Courthouse Pastures, Moab area ------- 369
Covert Arch, Lost Spring Canyon, Moab area -- 191, 247, 306
Coyote Natural Bridge, Escalante canyons --------- 287, 355
Crips Arch, Four Arches Canyon, Moab area ------------- 363
Cummings Arch, Pritchett Canyon, Moab area ------------ 371

INDEX OF SPANS IN THIS BOOK

Davis Bridge, Canyonlands National Park --------------- 337
Delicate Arch, Arches National Park ----------- 1, 302, 315
Dellenbaugh Tunnel, Spring Canyon, Moab area ---------- 150
Dewey Window, Arches National Park -------------------- 316
Dimmocks Jughandle, Harts Draw, Moab area ------------- 296
Dolores River Arch, Dolores River gorge --------------- 126
Double Arch, Arches National Park --------------------- 240
Double Arch, Capitol Reef National Park ---------- 152, 343
Double-O Arch, Arches National Park ------------------- 320
Dragoneye Arch, The Cockscomb ------------------------- 312
Druid Arch, Canyonlands National Park ------------ 178, 337
Dunham Arch, Paria River gorge ------------------------ 385
Dutchmans Arch, San Rafael Swell ---------------------- 400

Eagle Arch, near Kanab -------------------------------- 117
Eagle Canyon Arch, San Rafael Swell ------------------- 401
Ear-of-the-Wind Arch, Monument Valley area ------------ 376
East Rim Arch, Rattlesnake Canyon ----------------- 60, 397
Eggshell Arch, Monument Valley area ------------------- 377
El Morro Arch, El Morro National Monument ------------- 126
Ela Natural Bridge, Glade Park, Colorado -------------- 265
Elephant Canyon Window, Canyonlands National Park ----- 93
Elephant Rock, Valley of Fire State Park, Nevada ------ 161
Elevator Arch, Short Canyon, Moab area ---------------- 368
Escalante Natural Bridge, Escalante canyons ------ 309, 357
Eye Arch, Rattlesnake Canyon -------------------------- 395
Eye-of-the-Sun Arch, Monument Valley area ------------- 381
Eye-of-the-Whale Arch, Arches National Park ----------- 318

Fable Arch, Fable Valley, Needles area ---------------- 235
Fisheye Arch, Canyonlands National Park --------------- 335
Fortress Arch, Canyonlands National Park -------------- 250
Fraggle Rock, Sevenmile Canyon, Moab area ------------- 366
Full Moon Arch, Monument Valley area ------------------ 379
Funnel Arch, Cane Creek Canyon, Moab area ------------- 189

Gemini Bridges, Bull Canyon, Moab area ------ 149, 176, 207
Gold Bar Arch, Colorado River gorge, Moab area -------- 371
Gothic Arch, Canyonlands National Park ---------------- 163
Granary Arch, Canyonlands National Park --------------- 234
Grand Central What?, Taylor Canyon, Moab area --------- 88
Gregory Arch, Arch Canyon ----------------------------- 185
Gregory Natural Bridge, Escalante canyons -------- 350, 405
Grosvenor Arch, Kodachrome Basin area ----------------- 54

Halls Bridge, Pritchett Canyon, Moab area ------------- 52
Hand Holt Arch, Canyonlands National Park ------------- 262
Harts Draw Bridge, Harts Draw ------------------------- 104
Hellroaring Bridge, Hellroaring Canyon, Moab area ----- 372
Hellroaring Window, Hellroaring Canyon, Moab area ----- 364
Hickman Natural Bridge, Capitol Reef N. P. -- 151, 249, 344
Hole-in-the-Bridge Arch, Rattlesnake Canyon ------ 188, 395
Hondoo Arch, San Rafael Swell ------------------------- 401
Honeymoon Arch, Monument Valley area ------------------ 378
Hope Arch, Monument Valley area ----------------------- 377
Hunter Arch, Hunter Canyon, Moab area ------------ 131, 224
Hurst Bridge, San Rafael Swell ------------------------ 399

Inchworm Arch, Lost Spring Canyon, Moab area ---------- 63
Ives Arch, northern Arizona --------------------------- 24

Jacob Hamblin Arch, Escalante canyons ----------------- 351
Jeep Arch, Poison Spider Mesa, Moab area -------------- 280
Joint Trail Tunnel, Canyonlands National Park --------- 341
Jughandle Arch, Colorado River gorge, Moab area ------- 368
Jughandle Arch, Escalante canyons --------------------- 355

Kachina Natural Bridge, Natural Bridges N. M. --------- 382
Keen Bridge, Short Canyon, Moab area ------------------ 368
Keystone Arch, Arch Canyon ---------------------------- 47
Kirk Arch, Canyonlands National Park ------------------ 243
Klondike Arch, Arches National Park ------------------ 316
Kolob Arch, Zion National Park -------------- 46, 202, 403

La Boca Arch, Yellowcat, Moab area -------------------- 364
La Ventana, New Mexico -------------------------------- 138
Lacy Arch, The Cockscomb ------------------------------ 312
Lamanite Bridge, Escalante canyons -------------------- 354
Landscape Arch, Arches National Park --------- 26, 212, 315
Lee Pass Arch, Zion National Park --------------------- 403
Liberty Castle, Bryce Canyon National Park ------------ 328
Lin-Lynn Arch, Arths Rim, Moab area ------------------- 367
Little Death Hollow Arch ------------------------------ 356
Longbow Arch, Poison Spider Mesa, Moab area ----------- 149
Looking Glass Rock, Moab area ------------------------- 13
Lopez Arch, Hook and Ladder Gulch, Moab area ---------- 370

M & M Window, Sevenmile Canyon, Moab area ------------- 365
Macomb Arch, East Canyon Wash ------------------------- 271
Mamie Natural Bridge, Escalante canyons --------------- 355
Mano Arch, Escalante canyons -------------------------- 352
Margaret Arch, Monument Valley area ------------------- 375
Mastodon Arch, Canyonlands National Park -------------- 339
Maverick Natural Bridge, Escalante canyons ------------ 356
Mesa Arch, Canyonlands Natural Park ------------------- 338
Metate Arch, Escalante canyons ------------------ 208, 349
Mill Creek Arch, Mill Creek Canyon, Moab area --------- 91
Mineral Canyon Arch, Mineral Canyon, Moab area --- 275, 276
Miser Bridge, Glen Canyon N. R. A. -------------------- 358
Moab Rim Arch, Behind the Rocks, Moab area ------------ 114
Moccasin Arch, Monument Valley area ------------------- 378
Morning Glory Arch, Negro Bill Canyon, Moab area ------ 205
Mosquito Arch, Four Arches Canyon, Moab area ----- 295, 363
Mossy Cave Windows, Bryce Canyon National Park -------- 326
Muffin Arch, Canyonlands National Park ---------------- 334
Muley Arch, Capitol Reef National Park ---------------- 345
Musselman Arch, Canyonlands National Park ------------- 83
Mystery Bridges, Glen Canyon N. R. A. ------- 260, 261, 359

Navajo Arch, Arches National Park --------------------- 321
Needle-Eye (Rabbit Eye) Arch, Monument Valley area ---- 380
Nels Johnson Bridge, Capitol Reef National Park ------- 346
Nevills Arch, Owl Creek Canyon ------------------------ 25
Newberry Window, Indian Creek Canyon ------------------ 296
North & South Windows, Arches National Park ----------- 97
North Window, Arches National Park ----------- 97, 252, 314

INDEX OF SPANS IN THIS BOOK

Ojo Arch, Little Canyon, Moab area -------------------- 195
Onyx Bridge, Petrified Forest National Park ------ 136, 389
Otho Natural Bridge, Mill Creek Canyon, Moab area ----- 364
Owachomo Natural Bridge, Natural Bridges N. M. 29, 85, 383

Pachyderm Arch, Canyonlands National Park ------------- 334
Paria Arch, Paria River gorge ------------------------ 386
Partition Arch, Arches National Park ------------------ 322
Paul Bunyans Potty, Canyonlands National Park --------- 237
Peek-a-Boo Arch, Capitol Reef National Park ----------- 346
Phipps Arch, Escalante canyons ------------------------ 56
Picture Frame Arch, Behind the Rocks, Moab area ------- 361
Pine Tree Arch, Arches National Park ------------------ 80
Pine Tree Arch, Monument Valley area ------------------ 376
Pocket Arch, Sand Flats, Moab area -------------------- 239
Pritchett Arch, Pritchett Canyon, Moab area ------- 38, 362

Rainbow Bridge, Rainbow Bridge N. M. ----- 36, 77, 110, 167
200, 217, 220, 281, 362, 391, 392, 393
Ribbon Arch, Arches National Park --------------------- 320
Rimhandle Arch, Capitol Reef National Park ------------ 345
Ring Arch, Arches National Park ----------------- 268, 323
Royal Arch, Monument Valley area ---------------------- 381

Saddle Arch, Capitol Reef National Park --------------- 181
Sallys Arch, Shafer Basin, Moab area ------------------ 366
Sand Dune Arch, Arches National Park ------------------ 319
Scorpion Arch, Escalante canyons ---------------------- 350
Shadow Arch, Four Arches Canyon, Moab area ------------ 363
Shakespear Arch, Kodachrome Basin area ---------------- 62
Silhouette Window, Lost Spring Canyon, Moab area ------ 370
Sipapu Natural Bridge, Natural Bridges N. M. ----------- 27
148, 245, 384
Skylight Arch, Wahweap Basin -------------------------- 128
Skywalk Arch, Sevenmile Canyon, Moab area ------------- 192
Slitwindow Arch, Sevenmile Canyon, Moab area ---------- 372
Solomons Arch, San Rafael Swell ----------------------- 399
South Window, Arches National Park -------------------- 97
Spirit Arch, San Rafael Swell ------------------------- 401
Squaw Window, Squaw Park, Moab area ------------------- 369
Standing Cow Arch, Monument Valley area --------------- 378
Steer Arch, Bull Canyon, Moab area -------------------- 371
Stevens Arch, Escalante canyons ----------------------- 353
Stimper Window, Grand Gulch --------------------------- 126
Stout Arch (Low Level Arch), Monument Valley area ----- 376
Swinerton Arch, Monument Valley area ------------------ 380

Taj Arch, Canyonlands National Park ------------------- 338
Tapestry Arch, Arches National Park ------------------- 321
Teapot Arch, Canyonlands National Park ---------------- 337
The Spectacles, Monument Valley area ------------------ 379
Three Level Arch, Monument Valley area ---------------- 378
Tibbet Arch, Canyonlands National Park ---------------- 337
Tower Arch, Arches National Park ---------------------- 317
Tower Bridge, Bryce Canyon National Park -------------- 325
Trap Arch, Rattlesnake Canyon ------------------------- 396
Triggerguard Arch, Mill Creek Canyon, Moab area ------- 364
Troll Bridge, Pritchett Canyon, Moab area ------------- 372

Tukuhnikivista Arch, Behind the Rocks, Moab area ------ 360
Tunnel Arch, Arches National Park ---------------------- 316
Turnip Tunnel, Arches National Park -------------------- 316
Turret Arch, Arches National Park ---------------- 145, 252
Tusher Tunnel, Tusher Canyon, Moab area --------------- 151
Twin Arch, Arches National Park ------------------------ 255
Two Bridges, Bryce Canyon National Park --------------- 329

Undine Bridges, Labyrinth Canyon, Moab area ----------- 292
Updraft Arch, Colorado River gorge, Moab area --------- 366

Wall Arch, Arches National Park ------------------------ 319
Wall of Windows, Bryce Canyon National Park - 135, 324, 325
Washerwoman Arch, Canyonlands National Park ----------- 333
Wedding Ring Arch, Canyonlands National Park ---------- 331
White Craig Arch, Monument Valley area ---------------- 228
White Mesa Natural Bridge, Monument Valley area ------- 375
Wigwam Arch, Pritchett Canyon, Moab area -------------- 38
Wild Horse Window, San Rafael Swell ------------------- 401
Wilson Arch, Wilson Canyon, Moab area ----------------- 367
Window Arch, Pritchett Canyon, Moab area -------------- 38
Window Rock, Monument Valley area --------------------- 380
Woodenshoe Arch, Canyonlands National Park ------------ 335
Wow Arch, Fish Creek Canyon --------------------------- 308

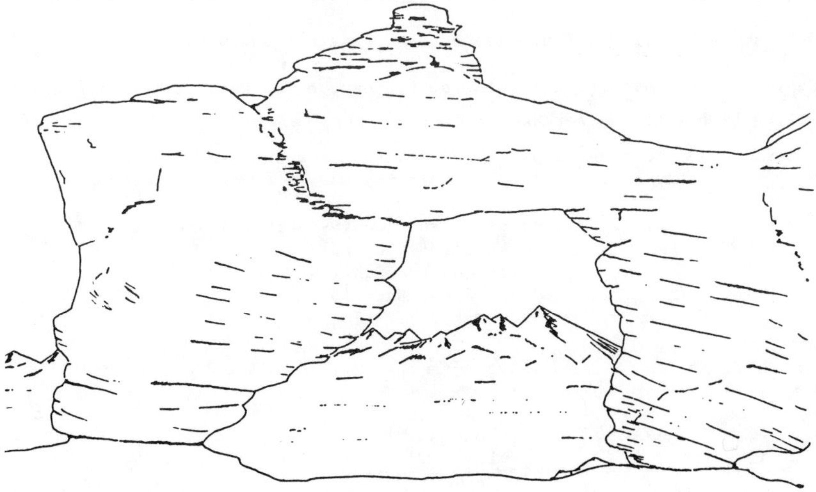

OTHER CANYON COUNTRY PUBLICATIONS

Utah Canyon Country, F. A. Barnes, Utah Geographic Series

Canyon Country Exploring, F. A. Barnes, Wasatch Publishers

Canyon Country Hiking, F. A. Barnes, Wasatch Publishers

Canyon Country ORV Trails, Island Area, F. A. Barnes, Wasatch Publishers

Canyon Country ORV Map, Island Area, F. A. Barnes, Wasatch Publishers

Canyon Country ORV Trails, Arches Area, F. A. Barnes, Wasatch Publishers

Canyon Country ORV Map, Arches Area, F. A. Barnes, Wasatch Publishers

Canyon Country ORV Trails, Needles Area, F. A. Barnes, Wasatch Publishers

Canyon Country ORV Map, Needles Area, F. A. Barnes, Wasatch Publishers

Canyon Country Camping, F. A. Barnes, Wasatch Publishers

Canyon Country Geology, F. A. Barnes, Wasatch Publishers

Canyon Country Paddles, Verne Huser, Wasatch Publishers

Canyon Country Prehistoric Indians, F. A. Barnes, Wasatch Publishers

Canyon Country Prehistoric Rock Art, F. A. Barnes, Wasatch Publishers

Hiking The Escalante, Rudi Lambrechtse, Wasatch Publishers

Canyon Hiking Guide to the Colorado Plateau, M. R. Kelsey, Kelsey Publishing

Hiking Utah's San Rafael Swell, M. R. Kelsey, Kelsey Publishing

These guide books and maps can be ordered from bookstores or purchased at many retail outlets within the Canyon Country region.

Single copies of Canyon Country Arches & Bridges
can be ordered by mail from
Canyon Country Publications
P. O. Box 963
Moab, UT 84532

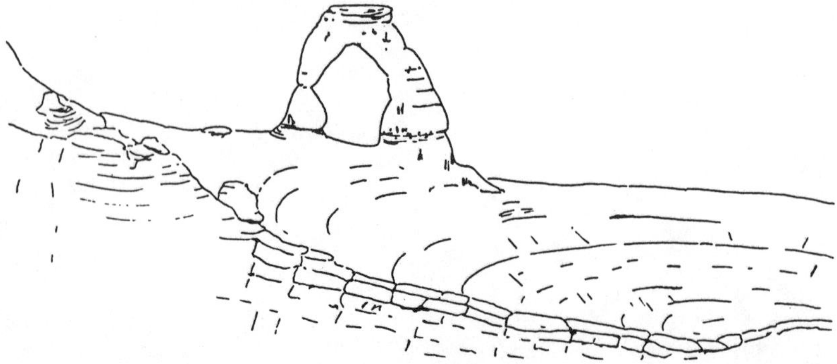